The African Mother Tongue and Mathematical Ideas

A Diopian Pluridisciplinary Approach

Abdul Karim Bangura
American University Center for Global Peace
Washington DC, USA

Series in Language and Linguistics

VERNON PRESS

www.vernonpress.com

In the Americas:
Vernon Press
1000 N West Street,
Suite 1200, Wilmington,
Delaware 19801
United States

In the rest of the world:
Vernon Press
C/Sancti Espiritu 17,
Malaga, 29006
Spain

Series in Language and Linguistics

Library of Congress Control Number: 2019957921

ISBN: 978-1-62273-818-2

Cover design by Vernon Press. Cover image by Isatu Ramatu Bangura.
Note about the Image: The Mancala is a game board that allows two players to test each other's linguistic and mathematical prowess.
Background image designed by rawpixel.com / Freepik.

Table of contents

Dedication

Afrikans everywhere, who must keep the struggle alive and triumph!

My Humble Maxim for Afrikans: Seeing is natural, looking spontaneous, and noticing can be incidental to both. Paying attention must be voluntary, examining requires attention, and analysis demands sustained attention. While earlier Afrikan thinkers set standards, we must extend insights to problems they did not discuss. The extent to which we can agree, disagree or be puzzled by any contemporary work on Afrika and its people should be a measure of its achievement.

Acknowledgments

I, and hopefully many readers, owe gratitude to the following:

Allah (SWT), for providing me *Taqwa*—i.e. being aware of Allah (SWT), of truth, and of the analytical and rational reality: i.e. piety.

Afrikan ancestors in the Motherland and the Diaspora, for providing me guidance, protection, and the willingness to seek the truth.

Mwalimu Emmanuel Babatunde and Mwalimu Kelebogile Setiloane, for providing me intellectual and spiritual support for this and other projects.

Mwalimu Rev. Alfred Munda SamForay, Mwalimu Esther Nkhukhu-Orlando and Mwalimu Ndi Chick Loveline Ayoh, for their research and intellectual contributions to this book.

My wonderful daughter, Isatu Ramatu Bangura, for designing the effulgent cover image on the book.

The anonymous reviewers, for their substantive evaluations. Asking difficult questions leads to better answers and clarifications.

My immediate and global families, for their prayers and support.

Preface

Using Cheikh Anta Diop's pluridiciplinary approach (i.e. the analysis of an African phenomenon by utilizing the tools of two or more disciplines, with the result in turn contributing to the epistemology of each discipline), this book, which is overwhelmingly sociolinguistics using mathematical ideas as evidence that even a non-mathematics student can understand, is an expansion of my award-winning article titled "Domesticating Mathematics in the African Mother Tongue" that appears in *The Journal of Pan African Studies* (vol. 6, no. 1, 2014:12-58). The article deals with why and how African-centered mathematical ideas can be a driving force in Africa's development efforts. I first argue in the article that Africa being the center of mathematics history for tens of thousands of years is hardly a matter of dispute. From the civilizations across the continent emerged contributions which would enrich both ancient and modern understanding of nature through mathematics. Yet, today, I add, scholars and other professionals working in the field of mathematics education in Africa have identified a plethora of problematic issues in the endeavor. I also argue that a major reason for these problems is that the African mother tongue has been greatly neglected in the teaching of mathematics in the continent. I then call for the situation to be changed if Africa is to benefit from the tremendous opportunities mathematics offers. I further show that while a great deal of works exists on the connections between linguistics and mathematics in general, few can be found on the nexus between African languages and mathematics in particular, and these latter works are not linguistics-theoretically grounded. Thus, I begin by identifying the objects of study of linguistics and mathematics and delineate which ones they study in common in the article. Next, since the object of study of linguistics is language, the nine design features of language are employed to examine each of the objects as it pertains to African languages. After that, mathematical ideas of sustainability and those of tipping points are suggested as means to help Africa's development efforts.

The book, therefore, aims to provide ample empirical evidence to show that African languages exhibit all nine design features that can facilitate the domestication of mathematics for effective learning. But since European languages have been privileged over African languages in mathematics education in the continent, the book must do at least two very important things, if we are to remedy the problems so that Africans can benefit from the tremendous opportunities mathematics offers. First, the book must offer material for schools to encourage the use of African languages in order to nurture and promote those languages vis-à-vis mathematics education. Second, the book must also provide evidence for why more research on the connection between language and the learning and teaching of mathematics from a sociopolitical point of view is necessary.

This book is therefore suitable as a main or supplementary text for undergraduate and graduate students taking courses in African Studies, Education, Mathematics, and Linguistics. It also is useful for professors teaching about and scholars conducting research in these areas. In addition, policy makers in these areas would be interested in the valuable information that the book provides. Furthermore, relevant political activists and advocacy groups would be interested in the book to pursue their objectives.

Abdul Karim Bangura

Washington DC, USA
Summer 2019

Foreword: Africanity and the Language of Mathematics

Toyin Falola

The Jacob and Frances Mossiker Chair in the Humanities

The University of Texas at Austin

There is no gain reiterating the normative privileging of western languages over languages from Africa as well as the marginalization of Africa in many areas of global and human development. As noted by Anchimbe, through the colonial and missionary education projects, these languages took root in the education systems in many parts of Africa.[1] In various academic disciplines around the continent, the languages of instruction since colonialism have remained acquired western languages. Fafunwa is of the opinion that the languages imposed on Africa as a result of the continent's contact with colonialism constitute a major obstacle to the effective dissemination of knowledge and skills which, in turn, impact the socio-cultural and economic development of the nation.[2] Bangura's multidisciplinary book deftly marshals evidence from linguistics, mathematics, history, archaeology, computing, engineering, development studies, sociology, education and more to link the continent's past, present and future in a manner that will intrigue and excite debate among scholars of Africa for a long time to come. Therefore, with the above background, this essay reviews Bangura's *The African Mother Tongue and Mathematical Ideas.*

Language provides contextual understanding of ideas, subjects, and society; therefore, its role in broadening horizons and accelerating assimilation in

---

[1] Eric A. Anchimbe, "Functional Seclusion and the Future of Indigenous Languages in Africa: The Case of Cameroon," in *Selected Proceedings of the 35th Annual Conference on African Linguistics*, ed. John Mugane (Somerville, MA: Cascadilla Proceedings Project, 2006), 94-103.

[2] Aliu Babs Fafunwa, *Education in Mother Tongue: The Ife primary education research project (1970-1978)* (Ibadan: University of Ibadan Press 1989).

various academic disciplines cannot be overemphasized. The use of the mother tongue will no doubt be more effective in teaching, especially given the relevance of context and the plethora of familiar examples at the teacher's disposal to instruct and achieve effective learning. For the learner, Fafunwa observes that it also affords him/her a wide imaginative space for expression and thought and therefore gives room to meaningful education.[3]

In the field of mathematics, western languages have been adopted in the impartation of mathematical knowledge and ideas. This can, however, be linked to Bangura's observation in the text that mathematics scholars like Papakonstantinou and Tapia opine that Africans had no prior knowledge of Algebra. It can, therefore, be inferred that African languages are considered unsuitable for the teaching of mathematics. This implied unsuitability has its roots in the perceived inferiority of African languages and knowledge systems which were further subdued by the incursion and dominance of colonial and missionary pedagogy. Bangura's book, however, counters this notion and explores the nexus between mathematics and African languages. It identifies distinct objects of study for the areas of linguistics and mathematics while also outlining the objects of commonality. The work equally identifies nine design features of language and uses them to examine African languages, to reveal that African languages are fully equipped with these design features and are therefore effective for the domestication and study of mathematics to suit the African context.

Judging from the above findings, the book also advocates for the centrality of African ideals and perspectives, as well as the use of the mother tongue in the teaching and exploration of mathematical ideas so as to access the benefits that the discipline has to offer. This, however, stems from Bangura's notion that mathematical ideas centered on African epistemes can be an effective force towards efforts geared at Africa's development.

In the first chapter, the book explores the mother tongue and the study of mathematics from the perspective of African centricity. This is primarily as a result of the pertinence of central African ideals in the teaching and understanding of mathematics within the continent. According to Chawane, African centricity could be defined in terms of its (or as) methodology, theory, and ideology. However, while defining it as a methodology, it can be viewed as a response to the intellectual colonialism that tends to justify and validate socio-political and economic colonialism. As a theory, the term can be defined by placing the African at the core or center of any inquiry with reference to African behavior and action. Finally, as an ideology, the concept

[3] Aliu Babs Fafunwa, *Education in Mother Tongue.*

refers to the African desire for an ideological bond against external interference.[4] In other words, from any perspective it is viewed, the concept calls for the centralization of the consciousness of an African identity in the processes of African development.

The chapter also goes further to provide a brief history of African-centricity while also exploring various research questions and theories emanating from the concept to educate the African people on the relevance of a central African perspective in their understanding of self and their application of such understanding in various areas of their development. With reference to the terminology, Mashengle is of the opinion that Africans need to answer the question of who defines their history and knowledge as well as whose history they know.[5] In other words, a consciousness of self; that is, the interpretation of knowledge from an African perspective should be the heart of African development and education.[6]

However, prior to above exploration, with a detailed and relevant background of mathematical publications, this book implies that there are some limitations to the research done and knowledge represented by some of the works analyzed therein that alienate Africa from the history of Algebra. Also, with the same degree of motivation, the work provides evidence buried within famous mathematical publications from several writers like Cheikh Anta Diop, which clearly reveal that Ancient Kemet/Egypt is the origin of Algebra that Bangura broadly sees as "Algebraic mathematical series, simple equations, quadratic equations, balance of quantities." Therefore, by providing evidence that Africa is the history of Algebra and countering contradictory opinions, Bangura's work contributes to African renaissance and tries to expel the African mathematical history from the margins of global mathematical inventions.

Also, by advocating a centrally African perspective in the study of mathematics, the book, as earlier stated, identifies nine design features of language and extensively discuses and examines each of these nine features with regards to its correlation to African languages and mathematics. The nine design features include the following; (1) mode of communication, (2) semanticity, (3) pragmatic function, (4) interchangeability, (5) cultural transmission, (6) arbitrariness, (7) discreteness, (8) displacement, and (9) productivity. To express their relevance in identifying the adequacy of African

---

4 Midas Chawane, "The development of Afrocentricity: A historical survey", *Yesterday and Today* 16, 2016: 78-99.

5 Danny Mashengele, "Africentricity: New context, new challenges, new futures," (paper presented at the 27th Annual SCUTREA Conference, University of London, 1-3 July, 1997), http://www.leeds.ac.uk/bei/Education-line/browse/all_items/42755.html

6 Midas Chawane, *The development of Afrocentricity*, 78-99.

languages in facilitating the teaching of mathematics, the work compartmentalizes their examination into nine chapters starting from the third chapter.

Having identified the nine design features of language, the second chapter engages in detailed, albeit separate definitions of language for the fields of linguistics and mathematics. It also goes further to provide a brief explanation of how the connection between both fields of study emerged. To provide the data relevant for the exploration, the study extensively examines the content of several publications to proffer necessary definitions and historical materials that provide a description of the connection between both distinct fields of study. With the relevant information provided in this chapter, the book sets a tone for the succeeding chapters which provides in-depth knowledge on the nine design features of linguistics with reference to African languages and the teaching of mathematics.

In the third chapter, using three sections, the work examines language as a mode of communication. It explores the linguistic expositions on the mode of communication and within this linguistic expositions the work identifies ten themes, which include: (1) system of communication, (2) communication theory, (3) communicative grammar, (4) types of communication, (5) communicative effect taxonomy, (6) indices of communication and communicativity, (7) communicative act and strategies, (8) communicative function, (9) communicative power, and (10) language development. The second section examines the linguistic evidence that shows that African languages have successfully devised several methods in which language can be used for effective communication just like other languages across the globe. In the final section, the work provides evidence that shows that African languages possess mathematically communicative acts in the vocalization and gestures of counting that serve communication purposes within the continent.

However, by elaborately analyzing these communication elements and approaches in African languages, the work reaffirms the suitability and adequacy of African languages in effectively passing across information irrespective of the complexity or abstract nature of the information. Also, by providing evidence that shows that African languages possess mathematical communicative acts, the work gives credence to its initial claims that Africa is the origin of Algebra and her languages possess the necessary tools for the teaching and understanding of mathematics. This, therefore, implies that the continued use of European languages in the teaching of mathematical ideas is hardly as a result of an absence of choice amongst African indigenous languages but rather the colonial conditioning of western language superiority and the lasting effect of language imperialism. It also enforces the opinion that for more effective teaching and learning of the subject, scholars of mathematics should employ the use of indigenous African languages.

In the fourth chapter, the work examines the place of meaning (semantics) in the communication of mathematical ideas. This part of the text implies that the primary need for the adoption of the mother tongue as the language of instruction for mathematics is the benefit of efficient mathematical communication between the speaker and hearer that share common languages and therefore more likely to understand the ideologies being discussed. This must be in line with Trudell and Shroeder's opinion that "learning to read and write is a psycholinguistic and social process."[7] They also go ahead to affirm that African language speakers tend to assimilate and learn better when taught in the language they speak. For Love, "languages are systems of correspondences between two levels of structure, one of which is called form and the other which is called meaning. The native speaker of a language is said to know the set of forms comprised by the language and the set of meanings that correspond to them".[8] Therefore, communication between teacher and student is possible and even effective because both teacher and student understand the correspondence of form and meaning resident in the language in use. To expand on the semantics in language this chapter also addresses three subtopics, including linguistic expositions on the subject; African linguistic evidence; and African mathematical evidence.

The third of the nine design features of language, which constitutes the fifth chapter of the text deals with the pragmatic function of languages. It explores the idea of pragmatic function and then goes further to establish and analyze the nexus between African languages and mathematics. Language as a communication system tends to serve a useful purpose, and this purpose characterizes the pragmatic function of the language. However, to extensively discuss the pragmatic function of language, the chapter is sectionalized into three parts like the chapters before it. This part includes the linguistic expositions on the subject, the African linguistic evidence, and the mathematical evidence. In the first part, the work explores extensively, a variety of pragmatic function in language that Professor Bangura labeled as (a) deictic functions; (b) conversational implicatural functions; (c) presuppositional functions; (d) speech act functions; (e) conversational

---

[7] Barbara Trudell and Leila Schroeder, "Reading Methodologies for African Languages: Avoiding Linguistic and Pedagogical Imperialism," *Language, Culture and Curriculum* 20, 3 (2007): 165-18. https://www.tandfonline.com/doi/abs/10.2167/lcc333.0.

[8] Nigel Love, "Translational semantics: A discussion of the second edition of Geoffery Leech's Semantics: the study of meaning," *Stellenboch Papers in Linguistics* 11 (1983): 115-136.

structural functions; (f) ideational, interpersonal, and textual functions; (g) tact maxim functions; (h) communicative grammatical functions; and (i) anaphora controlled functions.

These functions are also carefully identified in African languages, while the second segment provides linguistic evidence of some of these pragmatic functions. While exploring the pragmatic functions of African languages based on several scholarly opinions and definitions, among others, the text notes that the Association for the Development of Education in Africa (ADEA) considers that one of the basic pragmatic functions of African languages is to serve as a tool for education and sustainable development. It is also identified that African languages serve the pragmatic function of consolidating African identity and expelling the continent from the global margins. It also acknowledges and decries the death of African languages as a fast track to the loss of identity and culture of the African people and a force to successful external imperialism. In "A Motivational Treatise on Parenting," Bola Dauda noted that,

> Language is critical not only for the survival of any culture but also for its revival. Language is the software for the storage, retrieval, processing, and use of culture. Any threat or danger to language spells doom to a culture and consequently to the essence and identity of a people. … This trend of bringing up Africans without a mother tongue and without competence in any language is not only a danger to African culture but it is of a worrying concern. After all, without being proud to be African, we are paradoxically condemned to be second class citizens at home and abroad. And how can we develop our economy when we are not proud of our skin, language, food, clothes, and any homemade goods? How can we promote decency in relations, be it in politics or in business, that's, rule of law and fair trade, without a good mastery of language? How can we eradicate air and water-borne diseases, superstition, ignorance, and poverty without a language to promote public enlightenment? Language is critical to the future of African culture. And without being sexist or attempting to be pretentious, mothering, and if you want me to be politically correct, parenting, has a critical role in the child's language development and consequently in imbibing the African culture and identity."[9]

---

[9] Bola Dauda, "A Motivational Treatise on Parenting." in *Gender, Sexuality, and Mothering in Africa,* eds. Toyin Falola and Bessie House-Soremekun. (Trenton: Africa World Press, 2011) 80-81.

However, Pratt *et al.* give credence to this deduction by stating that by undermining traditional languages, the roots of traditional education are gradually being uprooted.[10] Also, as African society is characterized by multilingualism, the work contemplates this linguistic feature as either a bane or a blessing. The learning of multiple languages can be seen in different urban areas of Africa; this process assists daily interactions and is also mutually beneficial for the development of the continent. It will equally assist the exchange of goods and services as well as promote efficient collaboration between cultures.

To provide the mathematical evidence of the pragmatic function of African languages, this book indicates the invention of the Lebombo bone which is considered the most ancient mathematical tool; it is equally identified as a mathematical system that enabled humans in history to master time. To establish its connection to Africa, the work denotes that "the Lebombo bone is an exemplar of early Africans' endeavor to develop a system on communication with a Pragmatic Function."[11] Also to substantiate this claim, cited in the text, the *Encyclopedia of Science* states that, "It has been argued that since the Pragmatic Function of the Lebombo Bone was for lunar phase counting, then women must have been the first mathematicians, since keeping track of a menstrual cycle calls for a lunar calendar".[12] The work also examines various languages from different parts of Africa so as to identify their different numerical systems that can serve as evidence to the mathematical efficiency or potential in those languages. The text examines the Mende language spoken by people in Sierra Leone and Liberia to reveal that the language exhibits "a numerical system based on the decimal structure as the Western system."[13] It also explores the variations in both systems to prove that, unlike the Western numerical system, the Mende people only count up to the number 19 and the language does not possess a word signifier for zero quantity. Also, the text further portrays that Africa has also had a huge impact on modern computing. It reveals that the ancient Kemetians/Egyptians invented the number zero (0) which has so far been the basis of the binary number system used for modern computing.

---

[10] Yvonne Poitras Pratt, Dustin W. Louie, Aubrey Jean Hanson, and Jacqueline Ottmann, "Indigenous education and decolonization", *Oxford Research Encyclopedia of Education* (2018): 1-32. https://www.researchgate.net/publication/322695370_Indigenous_education_and_decolonization

[11] Abdul Karim Bangura, *The African Mother Tongue and Mathematical Ideas.*

[12] *The Internet Encyclopedia of Science*, 2007:1.

[13] Ibid.

As the work progresses, the sixth chapter examines the subject of interchangeability, which, as explained in the text, refers to the conveyance and reception of information through language by persons or groups of people. This is, however, due to the human ability to convey messages and information using gestures, spoken and written works and are also able to decode messages from others by mere looking, reading, or listening. The subject of interchangeability makes it possible to understand the reason for the human ability to receive information through language with no restrictions on the nature of the information they can give or hear and even read. However, just like the subsequent chapters, this chapter carries out this examination by exploring the linguistic and African linguistic expositions on the subject, as well as the African mathematical evidence with regard to the subject. To fully understand the linguistic expositions of the subject, the text analyzes the following themes; (1) synonymy, (2) complete feedback, (3) zoösemiotics or design features of animal communication, and (4) design feature of communication systems. These themes provide information on the various interactions necessary for interchangeability to occur linguistically.

In the exploration of the subject of interchangeability in African languages, the work implores the input of several scholars; one of whom is Cheikh Anta Diop. He discovered several writing systems of the indigenous African people, as well as how foreign writing systems have been able to develop orthographies for some of them, and how they can also be employed to create orthographies for others. For the text, this presents proof that interchangeability is a significant process of languages emanating from Africa. It also opines that interchangeability is most likely to occur with languages that are more closely related. It also goes ahead to analyze several African languages for the interactions of complex verbs as well as the discovery of ancient syllabic scripts, past tense verbs in the black English in South Africa, relative tenses of the isiZulu language in parts of Southern Africa, and so on. Deductions from these explorations provide incontestable proof that African languages are interchangeable and intersecting.

The work also reveals that for the African environment, the concept of interchangeability is also represented in the multilingual nature of the environment. Most African children acquire more than one language and tend to use these acquired languages interchangeably in society. It, therefore, advocates for the use of interchangeability in the teaching of African and European languages in education since it has been mastered so skillfully in other areas of life. This also, therefore, counters all claims that multilingualism cannot be achieved if African languages are opted to be used first in an education system. However, to provide mathematical evidence in connection to interchangeability, the work refers to the mathematical evidence inherent in the combination of food prices like, say, 1 shilling and a wave of a hand to represent twenty cents from refugees from Rwanda. In other words, hand gestures are used in place of words, and they have mathematical representations. The work goes further to mention the concept across different discoveries and examples that prove the exchange of mathematical ideas between Africa and different parts of the world.

In chapter seven, the work examines the idea of cultural transmission as language design feature that proves that some aspects of language can only be learned by interactions with other users of the same language system. The linguistic discussion of the subject matter is subsumed in the following themes, which are elaborately and distinctly addressed for readers' clarity. They include the following; (1) evolution of human behavior, (2) cultural vocabulary, (3) cultural level of usage, (4) language transfer, (5) cultural variation, (6) cultural context, (7) cultural identity, and (8) cultural language styles.

Using various research reports from scholarly materials, the work proffers evidence that African languages have so far been a source of cultural transmission. This notion is however validated by de Voogt, who states that Africa has played a major role in the creation of writing systems, the writer reveals Egypt as the prime location for one of the earliest occurrences of writing.[14] Drawing from a scholarly opinion, the text opines that by representing language in written form using scripts, an activity that is attributed to speakers of Afroasiatic languages, cultural beliefs, and practices were transmitted to different parts of the world. Assenting to this notion of cultural transmission, Emeka Nwobia is of the opinion that "language is the conduit for transmission of people's culture, norms, ideas, and belief."[15] In other words, language is the storehouse of a people's identity and their ways of life and can, therefore, be an effective tool for the transmission of culture, ideas, and epistemologies of a given people. The work goes further to identify proofs of African linguistic-cultural transmission in different parts of Africa and beyond, while also identifying factors like the transatlantic slave trade as one of the most effective tools in the dispersal of African linguistic culture around the world.

To proffer evidence of cultural transmission of mathematical ideas, the book provides an array of examples that serve as evidence of cross-cultural transmission of mathematical ideas. The work makes reference to the transmission between the South of the African Sahara and North Africa, using

---

[14] Alex de Voogt, "The cultural transmission of script in Africa: the presence of Syllabaries," *Scripta* 6 (2014): 121-137.

[15] Ngozi U. Emeka-Nwobia, "The place of indigenous Nigerian languages in national development", *Research in Humanities and Social Sciences* 5, 12 (2015): 112-116.

the exchange of mathematical ideas. Central Sudan introduced mathematics in Kanem-Borno, which at the time was part of Islamic sciences. It was taught in secular and Muslim schools and also served purposes in the courts. Also, among many other pieces of evidence, scholarly opinions explored in the text reveals that the "geometric and hand-printed rock art in the Central Limpopo Basin located in Southern Africa" was not created in that location but is rather traced to "geometric hunter-gatherer rock art" of Central Africa.

In the eighth chapter, the subject of the arbitrariness of language is explored, and the text tries to underline the relationship between arbitrariness and African linguistic and mathematical ideas. Arbitrariness in this context is described as "the nonexistence of any tangible connection between gestures, actions, or sounds that are used to convey information and the distinct and independent things in the universe to which they allude." The concept is considered a norm in human languages and asides from some onomatopoeic words; it is very predominant in human languages. However, like the previous chapters, this chapter is equally divided into three major sections that try to show the connection between arbitrariness and African linguistic and mathematic ideas.

Also, in chapter nine, the text analyzes the concept of discreteness in language, and like the other chapters, it proffers a connection between Discreteness and African linguistic and mathematical ideas. However, to better explain the concept, David Minger, cited in the text, describes the subject of discreteness "as a linguistic rendering which can be divided into small, distinct elements that can then be reconnected with other small, distinct elements to form novel linguistic renderings." In other words, the notion of discreteness has to do with the human ability to fuse separate elements of language to expand the units of communication. To further buttress this notion, the book provides this example: "Add a plural morpheme to something like dog to get dogs [dagz]." It also goes further to explain that these units are not just restricted to morphemes but rather extends to phonetic elements as well as other elements of language. However, while trying to draw an African linguistic connection, the work provides evidence from parts of Africa that attest to the presence of discreteness in African languages. From Nigeria in west Africa, the text notes that the tones of Yoruba language exhibit features of discreteness. Also, by using factual illustrations from various parts of Africa, the present work provides evidence of discreteness in African mathematics.

Furthermore, the tenth chapter discusses displacement as one of the nine design features of language. As a concept, displacement is a language characteristic that explains the ability for people to discuss situation or persons that are spatially absent from the speaker. As an example, the text identifies the discussions or talks of a supreme God like Allah, who the speaker has never met and who is not present at the moment of discussion. It also refers to times past as well as the distant future. However for the purpose of finding the linguistic expositions and the African linguistic and mathematical evidence of this language feature, the study goes further to explore the linguistic features of displacement extensively and they include the following identified features: (1) displaced modifiers and predication, (2) displaced speech, (3) displaced entities, (4) non-verbal systems, (5) phenomena included by Charles Francis Hockett, (6) evolution of language and cognition, and (7) communication. For lucidity, the perspectives of various scholars on these items are presented in the order they are listed here. The work also goes further to provide a synopsis of the research on displacement as it pertains to African languages and mathematics.

In the eleventh chapter, productivity is explored. In succinct terms, productivity refers to creativity with language; it shares a vague similarity with the earlier explored discreteness. However, as a distinct concept, it deals with language creativity, an example cited is the 'pluralization' of words in the English language. The mere addition of 's' to a word creates a new word. For clearer understanding, the view of Richard Nordquist is most appropriate for this explanation. He describes the concept as a "language feature that is employed to utter novel entities." For him, "Productivity is also recognized as "open-endedness" or "creativity"; it can also be viewed as being used to characterize certain "forms or constructions (such as affixes) that can be used to produce new instances of the same type." However, by employing several scholarly opinions and views the work presents evidence of the concept of Productivity in African languages while making a valid connection to African mathematical ideas.

The twelfth chapter, in many ways, underlines the basic ideology and background to this text, as it examines the role of mathematicians in the African renaissance. The work is of the opinion that mathematicians have played a crucial role in the development of Africa in ancient times. However, having recognized their contribution to the development of ancient Africa, the work makes relevant suggestions for a proposed collaboration between mathematicians and experts from other disciplines for an African renaissance. So far, the work has provided the necessary evidence to prove that African languages possess all the nine design features that will make the domestication of mathematics an easy task. The work also goes further to encourage the use of African languages in teaching so as to displace them from their perceived substandard status. It also advocates for further researches on the connection between language and the teaching of mathematics in Africa.

The thirteenth chapter, which is the final chapter of this exhaustive text, provides a general conclusion to the previous chapters explored. With the detailed information and opinions sampled, the work concludes that several factors influence the African mother tongue and mathematical ideas. It identifies the nine design features of language explored as core factors. It also establishes the need for African mathematicians to get involved in the struggle for African renaissance since it is obvious that they play a role in the entire process.

In all, this book has provided strong evidence that African languages possess nine design features that will make the domestication of mathematics an easy task. The book also goes further to encourage the use of African languages in teaching so as to recognize them for what they are and debunk the myth of perceiving them as substandard tools for educating African children. The book strongly advocates further research on the connection between language and the teaching of mathematics in Africa.

Works Cited

Anchimbe, E. A. "Functional Seclusion and the Future of Indigenous Languages in Africa: The Case of Cameroon." In *Selected Proceedings of the 35th Annual Conference on African Linguistics*, edited by John Mugane, 94-103. Somerville, MA: Cascadilla Proceedings Project, 2006.

Chawane, M. 2016, "The development of Afrocentricity: A historical survey", *Yesterday and Today* 16.

Dauda, Bola, "A Motivational Treatise on Parenting." In *Gender, Sexuality, and Mothering in Africa*, 79-100, edited by Toyin Falola and Bessie House-Soremekun. Trenton: Africa World Press, 2011.

Emeka-Nwobia, N. U. "The place of indigenous Nigerian languages in national development", *Research in Humanities and Social Sciences* 5, 12 (2015): 112-116.

Fafunwa, A. B. *Education in Mother Tongue: The Ife primary education research project (1970-1978)*. Ibadan: University of Ibadan Press, 1989.

Love, N. "Translational semantics: A discussion of the second edition of Geoffery Leech's Semantics: the study of meaning." *Stellenboch Papers in Linguistics* 11 (1983): 115-136.

Mashengele, D. 1997, "Afri-centricity: new context, new challenges, new futures." Paper presented at the *27th Annual SCUTREA Conference, University of London, 1-3 July, 1997.* http://www.leeds.ac.uk/bei/Education-line/browse/all_items/42755.html.

Pratt, Y., P, Louie, D., W, Hanson, A. J., and Ottmann, J. "Indigenous education and decolonization", *Oxford Research Encyclopedia of Education* (2018): 1-32. https://www.researchgate.net/publication/322695370_Indigenous_education_and_decolonization

The Internet Encyclopedia of Science, 2007.

Trudell, Barbara., and Schroeder, Leila. "Reading Methodologies for African Languages: Avoiding Linguistic and Pedagogical Imperialism," *Language, Culture and Curriculum* 20, 3 (2007): 165-18. https://www.tandfonline.com/doi/abs/10.2167/lcc333.0.

de Voogt, A, "The cultural transmission of script in Africa: the presence of Syllabaries," *Scripta*, 6 (2014): 121-137.

List of Acronyms

ADEA	Association for the Development of Education in Africa
ADiMA	Association of African Didacticians of Mathematics
APMEP	French Association of Mathematics Teachers (Association Française des Professeurs de Mathématiques)
ATDM	Tunisian Association of Didactics of Mathematics (Association Tunisienne de Didactique des Mathématiques)
ATS	Auric Time Scale
ATSM	Tunisian Association of Mathematical Sciences (Association Tunisienne des Sciences Mathématiques)
AU	African Union
CFGs	Context-free Grammars
COCA	Corpus of Contemporary American English
COHA	Corpus of Historical American English
CSGs	Context-sensitive Grammars
CV	Consonant/ Vowel
CVC	Consonant / Vowel / Consonant
CVCV	Consonant / Vowel / Consonant / Vowel
CVN	Consonant / Vowel / Nasal
CVV	Consonant / Vowel / Vowel
DDEs	Delay Differential Equations
ECSS	Elsewhere Condition Serial Search
EMF	Francophone Mathematical Space (Espace Mathématique Francophone)
ENS	Ecole Normale Supérieure
ESL	English as a Second Language
FSAs	Finite State Automata
FSTs	Finite State Tranducers
GESDI	Gross Environmental Sustainable Development Index
HHMs	Hidden Markov Models

ICT	Information and Communication Technology
LCA	Life Cycle Analysis
LNREs	Large Number of Rare Events
LTR	Low Tone Raising
MAA	Mathematical Association of America
MDC	Movement for Democratic Change
NCSU	North Carolina State University
NLP	Neuro-Linguistic Programming
ODEs	Ordinary Differential Equations
OEIS	On-line Encyclopedia of Integer Sequences
PBUH	Peace Be Upon Him
PM	Plural Marking
PR	Phonological Representation
P-rules	Phonological Rules
SIs	Semantic Interpretations
SMS	Short Message Service
SOV	Subject-Object-Verb
S-rules	Semantic Rules
SVO	Subject-Verb-Object
STEM	Science, Technology, Engineering, and Mathematics
SWT	*Subuhana Wa Ta'ala* (meaning in Arabic 'All Glory belongs to God; God is The Most Exalted and The Most High')
T-rules	Transformational Rules
UNIA	United Negro Improvement Association
URSs	Unrestricted Rewriting Systems
USEPA	United States Environmental Protection Agency
ZANU-PF	Zimbabwe National Union-Patriotic Front
ZAPU	Zimbabwe African People's Union

Chapter 1

General Introduction

As I retell in my article titled "Domesticating Mathematics in the African Mother Tongue" (2014), the impetus for this work was birthed by the following uniformed claim:

> As we describe the Rules of False Position, it is critically important to realize that the Egyptians and the Babylonians did not know algebra, indeed it did not exist at that time, nor did they have the notion of an equation; hence, they could not make obvious simplifications and they did not work with a general rule (Papakonstantinou and Tapia, 2013:504).

Given the preceding quotation, it is disheartening that in the year 2013, such an ignoramus statement could be made by Joanna M. Papakonstantinou and Richard A. Tapia, two well-placed mathematicians. And if I may quote their brief biographies in *The American Mathematical Monthly* (vol. 120, no. 6, June-July 2013) in which the statement is made, "Joanna M. Papakonstantinou received her B.A., M.A., and M.A.T., as well as her Ph.D., from Rice University, under the direction of the second author. After completing her postdoctoral work at Rice University in the Computational and Applied Mathematics Department, she joined PROS Revenue Management as a Senior Associate where she served as a science expert in the Center of Excellence. Currently, she works as the Senior Science Consultant at Advanous. She remains involved in research in the field of optimization, writes curriculum and teaches, and actively participates in outreach activities" (Papakonstantinou and Tapia, 2013:517). "Richard A. Tapia, 2011 recipient of the National Medal of Science [the top award the United States government offers its researchers given to him by President Barack Obama—author's note], holds the rank of University Professor in the Rice Department of Computational and Applied Mathematics. He is also the Director of the Rice University Center for Excellence and Equity in Education. Due to Tapia's efforts, Rice has received national recognition for its educational outreach programs and has become a national leader in producing women and underrepresented minority Ph.D. recipients in the mathematical sciences" (Papakonstantinou and Tapia, 2013:517). Even more disheartening is that the editor of the journal, Scott T. Chapman, another well-placed mathematician at Sam Houston State University, the three anonymous referees, and the authors' colleague, Michael Trosset, who the authors credited "for suggestions that greatly improved the paper" (Papakonstantinou and Tapia, 2013:516), ignored the pernicious statement.

It is quite obvious that all of these mathematicians have not read Cheikh Anta Diop's famous work, *Civilization or Barbarism* (1981/1991), and serious works on Africain Mathématiques in which it is clearly demonstrated that Ancient Kemet/Egypt, in fact, is the origin of Algebra. Algebraic mathematical series, simple equations, quadratic equations, balance of quantities (*pesou*) were all invented in Kemet. Even the Rules of False Position, which is the main focus of Papakonstantinou and Tapia's paper, were invented by the Ancient Kemet. Add to these the many arithmetic and geometric techniques that were invented in Ancient Kemet and other parts of Africa.

Many years earlier, I was so excited about the announcement of what turned out to be one of Morris Kline's most popular books titled *Mathematics and the Search for Knowledge* (1985) that I rushed to order a copy. I was quite disappointed after reading the book and found not a single African mathematician or mathematical idea in it. In fact, the word "Africa," much less "Kemet/Egypt," is in the index. These tendencies of marginalizing African Mathematics impelled my interest in this area of mathematics.

Indeed, that Africa was the center of mathematics history for tens of thousands of years is hardly a matter of dispute. From the civilizations across the continent emerged contributions which would enrich both ancient and modern understanding of nature through mathematics. Yet, today, scholars and other professionals working in the field of mathematics education in Africa have identified a plethora of problematic issues in the endeavor. These issues include attitudes, curriculum development, educational change, instruction, academic achievement, standardized and other tests, performance factors, native speakers, etc. (for more on this, see Bangura, 2011a). In my aforementioned article, I argue that a major reason for these problems is that the mother tongue has been greatly neglected in the teaching of mathematics in Africa. Indeed, as Mamokgethi Setati (1998, 2002, 2003, 2005a, 2005b, 2008) has demonstrated numerously in the case of South Africa, even though teachers and learners who employ African languages in mathematics education position themselves in relation to mathematics and, concomitantly, epistemological access, and use those languages as instruments of solidarity, English is the dominant language in the classroom, its ascendancy privileged by procedural mathematics discourse. This situation must be changed if Africa is to benefit from the tremendous opportunities mathematics offers.

Linguists have for a long time been concerned with the context of teaching any subject/text, be it qualitative or quantitative. Students of linguistics can readily find examples where the meaning of an utterance changes with its contextual framework. The following are examples: (a) General Motors had no success selling the Chevy Nova in Mexico because Nova (*no va*) means "no go" in Spanish; (b) the "I'm a Pepper" advertisement that was successful in the United States had to be changed in England when Dr. Pepper's management realized that *pepper* is a slang word for prostitution to the English; (c) the "Pepsi, the choice of a new generation" advertisement translated into Taiwanese got many Taiwanese angry after they flocked to buy Pepsi only to realize that their dead ancestors were not coming back to life. The extent of the problem became more widely known when computer programmers were faced with the vexing task of translating texts. A famous example was an attempt to translate "The spirit is strong but the body is weak" into Russian. The resulting phrase was "The Vodka is stale and the meat is rotten." Yet still, linguists also know that any subject, even mathematics that has become extremely abstract, can be taught effectively in any language once the appropriate tools are made available.

While a great deal of works exist on the nexus between linguistics and mathematics in general, only a small number can be found on the nexus between African languages and mathematics in particular, and these latter works are not linguistics-theoretically grounded. Thus, this book begins by identifying the objects of study of linguistics and mathematics and delineates which ones they study in common. Next, because the object of study of linguistics is language, the nine design features of language are employed to examine each of the objects as it pertains to African languages and mathematics. The nine design features are (1) mode of communication, (2) semanticity, (3) pragmatic function, (4) interchangeability, (5) cultural transmission, (6) arbitrariness, (7) discreteness, (8) displacement, and (9) productivity. Since to offer examples for each of these features from each of the more than 2,100 African languages would require at least a book-length manuscript for each feature, a small number of examples from a small number of languages across the continent are presented for each feature for the sake of brevity. After that, the discussion turns to the role for mathematicians in fulfilling the African Renaissance—defined by Dani Wadada Nabudere as the initiative to recapture the basic elements of African humanism (*ubuntu, eternal life*, and *immanent moral justice*) as the path to a new humanistic universalism. He quotes Chancellor Williams as stating that this initiative "is the spiritual and moral element, actualized in good will among men (and women), which Africa itself has preserved and can give to the world" (Nabudere, 2003:4). The proposition in this chapter is that just as mathematicians had played a major role in the development of African societies during antiquity (see Bangura, 2011a), they are sorely needed now to work with experts in other disciplines to help fulfill the African Renaissance.

Thus, the observations made in this book are not directed at the discovery of any method or pedagogical panacea. They are presented in complete modesty in the belief that what matters most is not the method but the teacher. May the observations serve then, at best, as a starting point for that self-examination. Since teaching about the nexus between African languages and mathematics is one to which many of us are deeply committed, it is the hope that this book will not only inspire colleagues to give serious consideration to its suggestions and perplexities, but also strive to suggest better solutions than those proposed. The immediate question that emerges here is the following: What research methodology undergirds this book to make the analysis systematic? The following section provides an answer.

Africancentric Research Methodology

In order to systematically delineate findings that are germane to the African phenomenon being examined in this book, the Africancentric Methodology is employed. As I narrate in more depth in my books titled *African-Centered Research Methodologies: From Ancient Times to the Present* (Bangura, 2011b), *Toyin Falola and African Epistemologies* (Bangura, 2015a) and *Falolaism: The Epistemologies and Methodologies of Africana Knowledge* (Bangura, 2019), and my essays titled "The Colors of the Flag of Sierra Leone: An Africancentric Analysis" (Bangura, 2015b) and "Bo School Symbols: A Deep African-centered Analysis" (Bangura, 2015c), from the late 1980s to the mid-1990s, many, and consistent, definitions of Africancentricity were proffered by Africanists. The first definition was by Molefi Kete Asante who defined "Afrocentricity [Africancentricity/African-centered] as the placing of African ideals at the center of any analysis that involves African culture and behavior" (1987:6). The second definition was by C. Tsehloane Keto who defined the "African-centered perspective [as an approach that] rests on the premise that it is valid to position Africa as a geographical and cultural starting base in the study of peoples of African Descent" (1989:1). The third definition was by Wade Nobles who defined "Afrocentric, Africentric, or African-Centered [as being] interchangeable terms representing the concept which categorizes a quality of thought and practice which is rooted in the cultural image and interest of African people and which represents and reflects the life experiences, history and traditions of African people as the center of analyses. It is therein that the intellectual and philosophical foundation [with] which African people should create their own scientific criterion for authenticating human reality" exists (1990:47). The fourth definition was by Maulana Karenga who defined "Afrocentricity...as a quality of thought and practice rooted in the cultural image and human interest of African people [and their descendants]. To be rooted in the cultural image of African people is to be anchored in the views and values of African people as well as in the practice which emanates from and gives rise to these views and values" (1993:36). Finally, Lathardus Goggins II defined "African-centered [as being able] to construct and use frames of reference, cultural filters and behaviors that are consistent with the philosophies and heritage of African cultures in order to advance the interest of people of African descent" (1996:18).

From the preceding definitions, it is evident that Africancentricity presupposes knowledge of a commonality of cultural traits among the diverse peoples of Africa which characterize and constitute a worldly view that is somehow distinct from that of the foreign world views that have influenced African peoples. Africancentricity simply means that the universe is a collection of relationships, and an individual or a group being in that universe is defined by and dependent upon these relationships. Africans, prior to European and Asian dominance, and still to some degree now, considered the Cause or God as being a part of His creation while Europeans, on the other hand, considered God separate from His creation.

Academic Areas that Have Employed Africancentricity

Asante suggests that in the analysis of what he calls the "three fundamental Afrocentric themes of transcendent discourse—(1) human relations, (2) humans' relationship to the supernatural, and (3) humans' relationships to their own being" (1987:168—if done with an awareness of the interrelatedness of these themes, a greater understanding of the African being will be acquired. These themes are embedded in the following academic disciplines:

a) *Psychology* is the study of the way in which the mind works and the manifestation of those thoughts into actions or behaviors while recognizing the distinct character of African thought processes and behavior. In particular, this suggests the predominance or greater use of the right hemisphere, the intuitive side, of the brain when compared with other peoples and its subsequent effects on thoughts and behaviors.

b) *Anthropology* is the study of the physical, social, and cultural adaptation of African peoples to their ever-changing environment.

c) *Theology* is the study of the way in which Africans define the Supreme Being that is responsible for the creation and sustaining of the universe and all in it and their relationship to the Supreme Being, God.

d) *History* is the recording and studying of the relationships between events. From an Africancentric perspective, the African conception of time, cyclic as opposed to the European linear, must be used in application of this discipline in order for it to truly be considered Africancentric.

e) *Linguistics* is the study of the way in which language is structured and its nature, as well as the particular way in which Africans conceive and represent the universe in their languages.

There are other areas of study in which an Africancentric perspective can and should be applied in order to grasp the impact that African cultures have had in those disciplines: Egyptology, literature, music, political science, psychology, anthropology, philosophy, sociology, mathematics, the natural sciences, etc. As in the African conception of the universe, all things are defined by relationships and, therefore, Africancentricity in its application can only be successful when the interrelated natures of these disciplines are reconciled.

A Brief History of Africancentricity

The sun that once illuminated the socio-cultural landscape of Africa is once again approaching its zenith after having set for hundreds of years. Seeking understanding of the universe from an African-centered perspective was in existence long before the term "Afrocentricity" was coined by Asante. The separation of the African mind from its body was a result of the Holocaust which African peoples endured at the hands of both Asians and Europeans. This thousand-year tragedy, having its birth in Kemet and its culmination along the western coast of Africa, would take its toll on the minds, bodies and souls of Africa's children, resulting in the separation of Africans from their cultures and history. The actual focus of this subsection is on the rebirth of the African consciousness of which Africancentricity is the totalization.

Even before its conceptualization, there was a period prior which helped to determine the character and establish the context in which the two following stages could be born. There were individuals whose scholarship was pivotal in the establishment of Africancentricity. These intellectual giants helped not only to raise the consciousness of their people, they also were instrumental in the establishment of a mode of analysis that had an African character. In total, the rebirth of the African mind can be seen through three stages: (1) the pre-conceptualization period, (2) the conceptualization period, and (3) the concretion period.

During the pre-conceptualization period, African scholars, particularly in the United States and the Caribbean, became increasingly aware of their physical and cultural origins in Africa. These scholars/activists would focus their energy on understanding the conditions of their people, albeit from a Eurocentric perspective, initially, in the United States, and then the rest of the world to aid them in their fight against the oppression and subjugation of African peoples. One of these great pioneers was William Edward Burghardt (W. E. B.) Du Bois, a sociologist, poet, historian, and activist who sought to raise his people's awareness of their potential to transform their reality. Du Bois was a prolific researcher and writer. As a sociologist, he asked very pertinent questions about the validity of his White contemporaries' conclusions on the mental state and capabilities of African peoples. In his attempts at showing the fallacies inherent in these conclusions due to the influence of racism, Du Bois would also shed light on the ancient past of Africa, destroying one of the main fallacies about African peoples, which is they have no history.

This groundbreaking work would lay the foundation for generations to come by providing some of the data on Africa necessary to establish a point of departure for the form of analysis that is Africancentricity. One of Du Bois' contemporaries, Marcus Garvey, with his United Negro Improvement Association (UNIA) and his push for "Africa for the Africans," George Padmore, C. L. R. James, Kwame Nkrumah, Frantz Fanon, Aimé Césaire, etc. helped to raise the awareness of Africans in the United States and throughout the Diaspora while laying the foundation for organization building as well as nationalism. These giants, along with people like Booker T. Washington and others during the 1920s, ushered in a period of heightened consciousness for people of African descent, even though they were not totally able to free themselves of the effects of Eurocentric education on the character of their analyses.

It is said that in the beginning there was the "Word," but in ancient Kemet, it was the thought or conceptualization that came before word or form. The second stage of Africancentricity was characterized by conceptualization. And during this period, scholars like Cheikh Anta Diop, Yosef A. A. ben-Jochannan, John Henrik Clarke and others contributed and shaped it. Diop, in particular, probably did more than anyone else, through his exhaustive and pluridisciplinary approach, to bring Kemet or Egypt back to its African home and even to a greater degree in proving the continuity of African culture despite the superficial differences. The most important aspect of this period, however, was that scholars like Diop recognized the distinct character of African sociocultural constructs and the inability of a Eurocentric mode of analysis to correctly interpret these constructs. This period covered from the early 1950s on to the transitional 1960s and 1970s. During this period, African scholars placed Africa at the center of their analysis, making it not only the object of study but also that which determined understanding not clouded with the ethnocentrism of some alien culture.

Unlike the first two stages, the period of Africancentricity's concretion saw the formation of the concept into a systematic mode of analysis. Even though many of those people who played a role in this period were contributing to the prior periods, especially the 1970s, their work had its most dramatic impact during the early 1980s. Scholars like Asante, Karenga, R. Marimba Ani, Jacob Carruthers, and many helped to usher in the period of Africancentricity's definition and formation into something tangible for use by generations to follow. In prior periods, it not only lacked the name Africancentricity, but many African scholars, although they saw the importance of Africa and a deep understanding of it, either lacked an Africancentric perspective or failed to present a clear definition of that which characterized their perspective. The scholars of this period not only looked from the proper perspective, they were also conscious of that perspective and could facilitate the understanding of it to African peoples through its definition. In Asante's own words, "Afrocentricity is the most complete philosophical totalization of the African being-at-the-center of his or her existence" (1987:125). Once Asante spoke the word, a thought was now given form: i.e. Africancentricity.

In the end, or should I say from that point forward, Africancentricity would continue to evolve as more information about Africa is brought to light. From preconception to conception to concretion, Africancentricity is a rebirth of the African being at its "center." In the first stage, the foundation was laid for the conception, which found its manifestation in Africancentricity. Even though those scholars during the preconception stage were not, they still made it possible for Africancentricity's rebirth. During the conception, those African scholars in their application of an African-centered perspective enabled those to follow to define and add structure to Africancentricity. Ironically, the history of Africancentricity can only be understood from an Africancentric perspective, which applies a circular character to history instead of the linear character given to history by Europeans.

Major Research Questions in Africancentricity

The foundation upon which Africancentricity rests is based on certain essential research questions. All forms of inquiry done in the name of Africancentricity must use these questions and their answers as their basis. Here are those questions according to Diop (1987:270):

a) What is the unique history of African peoples?

b) What is the unique cognitive style of African peoples?

c) What are the unique characteristics of African languages?

These questions facilitate an understanding of Africancentricity as well as establish a point of departure for Africancentric analysis.

Major Theories and Concepts in Africancentricity

In Africancentricity, there are major theories and concepts that form the basis of the discipline. Here are some of them:

a) *Matriarchal Origins Theory* is the theory that most of African civilizations prior to the incursion of Christianity and Islam were matriarchal in structure with women being shown high levels of respect (Diop, 1989).

b) *Two-Cradle Theory* is the theory that Blacks in Africa and Whites in Eurasia resided in two separate localities or cradles which characterized their physical features, cultures, histories, and cognitive styles (Diop, 1991). It should be noted here that this theory has been dismissed by many scientists, both Black/African and White, in light of the new pieces of evidence that point to Africa as the sole cradle of civilization.

c) *Analogical Symbols Theory* is the theory that African symbolism was based on finding the similarities between things and representing those things incomprehensible, abstraction, by that which is comprehensible (De Lubicz, 1978).

d) *Kawaida* was conceived and crafted as a philosophy in the midst of the liberation struggle of the 1960s as an emancipatory philosophy dedicated to Cultural Revolution, radical social change, and bringing good in the world. *Kawaida* was shaped by its focus on culture and community as the basis and building blocks for any real movement for the liberation of African people everywhere. This means that culture is conceived as the crucible in which the liberation struggle of African people takes form and the context in which it will ultimately succeed. *Kawaida* is an ongoing synthesis of African thought and practice in constant exchange with the world, asking questions and seeking answers to central and enduring concerns of the African and human community (Karenga, 2007).

e) *The African Code* is a concept within Pan-Africanism that stresses unity through diversity based upon the seven key principles of *Kwanzaa* delineated by Maulana Karenga: (1)

Maat—truth, justice and equality; (2) *Kujichagulia*—self-determination; (3) *Umoja*—unity; (4) *Mila*—culture; (5) *Kulea*—education; (6) *Jamii*—family; and (7) *Ujamaa*—economics. The African Code acts as an intersection of a global Pan-African ethos for unity through diversity. It has been translated into over 30 languages and functions as a non-political, non-religious cultural commonality for African people seeking self-determination everywhere. The African Code employs the Ge'ez alphabet and treats Kiswahili as the official Pan-African language, and subsequently Ge'ez as an African script to replace all forms of Latin to write all African languages (Karenga, 1993).

f) *Negritude* is a concept originated by Aimé Césaire as a cultural expression of Blackness. In a poem written during World War II, Césaire coined the phrase "African personality" as an expression of an original and unique personality peculiar to Africans that manifest the foundation of Africancentricity in the African Diaspora where French is spoken. Leopold Sedar Senghor of Senegal became a major proponent of Negritude, arguing that the African emotional quality to life is different from the materialism of Europeans. Negritude has usually been described as "passive" by many social critics, but one of the followers of the movement, Cheikh Anta Diop, used the idea of "Africanity" to add a historical component to Negritude to explain the African origin of Egyptian and Mesopotamian civilization (Winters, 1998).

g) *Analyses Acculturaliste,* or topological analysis, is Cheikh Anta Diop's approach used to study the origin and spread of African cultural features from the Nile Valley to West Africa through his analysis of toponyms (Diop, 1974).

h) *Linguistic Constancy* refers to the rate at which languages change. It appears that linguistic change is culture-specific; thus, the social organization and political culture of a particular speech community can affect the speed at which languages change. African languages have been found to change much slower than European languages (Winters, 1998).

These theories and concepts help to guide research in Africancentricity. They define that which is African; and by doing this, they establish a means of differentiating between that which is African and anything else.

In sum, Africancentricity will continue to evolve as more information on African peoples' past comes to light and as greater and differing demands are placed on African peoples. It can be defined by its history, the organizations and people who contributed to its development, and the body of ideas that make it up. Applied within its scientific framework, Africancentricity has greater value and respect.

Indeed, Africancentricity in regards to the wellbeing and overall development of African peoples is invaluable. It can ultimately give them a greater degree of understanding of themselves. Probably the most beneficial aspect of Africancentricity is its potential in uniting people of African descent for their collective wellbeing.

Organization of the Rest of the Book

The rest of this book is divided into 12 chapters. Chapter 2 is based on the premise and demonstrates that before identifying the objects of study of linguistics and mathematics and delineating which ones they study in common, it makes sense to begin with brief definitions of the two fields, with the caveat that, as many linguists and mathematicians have pointed out, providing concise and meaningful definitions of these fields on which everyone can agree is virtually impossible. Nonetheless, to avoid providing working definitions of the fields will cause more havoc than it solves.

Chapter 3 focuses on why sending and receiving African mathematical and other messages define the very nature of a system of communication. The means by which these messages are transmitted and received is referred to as *mode of communication.* Users of human languages employ voice (or oral language) and gesture (or body language) to transmit messages. These two modes of communication are imperative for the transmittal of the complex sorts of African mathematical messages that languages require.

Chapter 4 looks at why African mathematical and other *semanticity* call for all signals of communication to possess a meaning and function. It is vital that a speaker/writer and a hearer/reader share a similar idea about what is being talked or written in order to have successful linguistic mathematical communication. Even when a person hears a word s/he does not know, s/he nevertheless assumes that it must possess some meaning.

Chapter 5 is about how an African mathematical communication system can be shown to have a *pragmatic function* when it serves some useful purpose. Such functions include language used to help individuals to stay alive, to influence others' behaviors, to find out more about the universe, to ask for food, to yell for help, and to influence voting behavior. Some communicative acts such as gossip are usually questioned, but they do have a pragmatic function in helping us to understand our social environment.

Chapter 6 shows that when individuals are able to both transmit and receive African mathematical messages, they can engage in *interchangeability*. An individual can both produce messages by gesturing, speaking, or writing and understand the messages of others by looking, listening, or reading.

Chapter 7 examines how the *cultural transmission* of African mathematical ideas reflect the fact that there are aspects of language which people can acquire only through communicative interactions with other users of that language system. Even children, whose ability to learn languages may seem innate, must still learn all of the specific signals of those languages by interacting with other speakers. The language we acquire as children is not influenced whatsoever by our genetic or hereditary background.

Chapter 8 reveals how African mathematical linguistic features entail five aspects. First, the words of a language are generally recognized to represent a nexus between a group of sounds or signs that give the words their forms and the meanings, which the sounds are said to represent. Second, evidence of *arbitrariness* can be observed in the fact that the inner core of an object can be given a name. This evidence of arbitrariness can also be observed in cross-linguistic comparisons when words with the same meanings usually have different forms in different languages, and similar forms usually express different meanings. Third, *onomatopoeia* (i.e. words that imitate natural sounds or possess meanings that are associated with such sounds of nature) provides evidence of arbitrariness being the norm in languages, at least in terms of the connection between the forms of words and their meanings. Fourth, a counterexample of arbitrariness is *sound symbolism*: i.e. certain sounds occur in words not because they directly imitate some sounds but because they are evocative of particular meanings. Fifth, *non-arbitrariness* or *iconicity* plays a minimal role in language.

Chapter 9 concerns linguistic *discreteness* as it pertains to the feature of African mathematical linguistic systems that makes it possible for Africans to put together discrete units to make larger communicative units. This is imperative because every language has a limited number of sounds, between ten and 100. The sounds themselves, for the most part, lack meanings until they are combined together. The fact that humans can develop a large number of meaningful elements (words) from a few meaningless units (sounds) is known as *duality of patterning*.

Chapter 10 is an investigation of linguistic *displacement*. The analysis hinges upon the fact that certain African mathematical ideas about things, actions, people, etc. can be communicated without these being spatially present.

Chapter 11 discusses linguistic *productivity*, which augments linguistic *discreteness*, vis-à-vis an African language's capability to develop fresh mathematical meanings from discrete units. *Productivity* contrasts with *discreteness* in that the latter feature only requires re-combinable units in a fixed set of ways in which they could be combined while the former has no fixed set of ways in which units can be combined. Language *productivity*, therefore, makes it possible for humans to develop and comprehend any number of innovative statements that they may have never heard/read before, allowing them to express postulates that may never have been proffered before.

Chapter 12 demonstrates that mathematicians did play a significant role in the development of African societies during antiquity and that they are sorely needed now to help fulfill the *African Renaissance*. In this chapter, I suggest two ways mathematicians can work with experts in other disciplines to do so: (1) Mathematics of Sustainability and (2) Mathematics of Tipping Point.

Chapter 13 entails the general conclusion. Here, I present an interpretive overview of what is covered in the preceding chapters, thereby reinforcing the analyses in them. I then end by showing what unites all of them.

Each of the chapters dealing with the nine design features of language (i.e. Chapters 3 through 11) is divided into three major sections. The first section entails a general overview of the topic, the second section encompasses the African linguistic evidence, and the third section comprises the African mathematical evidence. The chapter ends with a conclusion drawn from lessons gleaned from Ancient Kemetian/Egyptian tenets on the topic.

Chapter 2

Objects of Study of Linguistics and Mathematics

This chapter identifies the objects of the fields of Linguistics and Mathematics investigated in common. In order to do so, it begins with brief general surveys of both fields, albeit separately for lucidity. Before doing all this, however, it makes sense to first provide brief definitions of the two fields, with the caveat that, as many linguists and mathematicians have pointed out, providing concise and meaningful definitions of these fields on which everyone can agree is virtually impossible. Nonetheless, to avoid providing working definitions of the fields will cause more havoc than it solves.

Linguistics, according to Francis Dinneen (1995), can be generally defined as the scientific study of *language*. Mathematics, following Keith Devlin (2003), can be generally defined as the systematic study of *change, quantity, relation, space, structure*, and other topics dealing with *entity, form*, and *pattern*. Consequently, these italicized aspects of the definitions constitute the objects of study for the two disciplines, whether they are dealing with numbers, other fields, hypothetical abstractions, or other things. Before discussing these aspects, however, it is necessary to end this section with a retelling of how the nexus between the fields of Linguistics and Mathematics was birthed, although very briefly.

For starters, as narrated by Keith Devlin in his book titled *The Language of Mathematics: Making the Invisible Visible* (2000), the development of mathematical logic in the 1930s and 1940s inspired that of *syntactic structure*: i.e. the arrangement of words and phrases to develop well-formed sentences in a language. The outgrowth of this endeavor is *synchronic linguistics*—the study of the elements and usage of a language as a communicative system at a particular point in time—with the focus on *grammar*, defined by Devlin as "a set of 'axioms' that describes" the structure of a language (Devlin, 2000:87).

Next, according to Devlin, a movement from *historical linguistics* or *philology* (i.e. the study of the historical features of languages in terms of their origins and evolutions) to synchronic linguistics took place toward the end of the 20th Century. This shift led to the birth of modern, mathematically based linguistics. The pioneers of this movement included Mongin-Ferdinand de

Saussure in Europe and Leonard Bloomfield and Frank Boas in the United States (Devlin, 2000:87).

Also, Devlin points out that the scientific study of linguistics was promoted by Bloomfield in particular, since he was an ardent advocate of philosophical positivism advanced by German-language philosopher Rudolph Carnap and the Vienna Circle of Logical Empiricism ("a group of philosophers and scientists drawn from the natural and social sciences, logic and mathematics who met regularly from 1924 to 1936 at the University of Vienna, chaired by Moritz Schlick"). As Devlin further states, scholars of logical positivism, motivated by work in logic and mathematical foundation such as that of the Hilbert program, sought to represent all meaningful statements to an amalgam of propositional logic and the five (traditional) sense data—i.e. feeling, hearing, seeing, smelling, and tasting. Devlin adds that certain linguists such as the American Zellig Harris went further than Bloomfield and called for the application of mathematical approaches to the scholarly investigations of languages (Devlin, 2000:87).

Furthermore, Devlin mentions that American linguist Noam Chomsky initiated the work on discovering axioms to illustrate the syntactic structure of a language, although Friedrich Wilhelm Christian Karl Ferdinand von Humboldt (Prussian philosopher, linguist, government functionary, diplomat, and founder of the Humboldt University of Berlin) had proposed the idea a century earlier. Devlin quotes Chomsky as asserting that "To write a grammar for a language is to formulate a set of generalizations, i.e., a theory, to account for one's observations of the language" (Devlin, 2000:87).

Finally, Devlin recounts that Chomsky's 102-page book titled *Syntactic Structures,* which was published in 1957, dramatically revolutionized the study of language in the United States, albeit less so in Europe. In just two years, this short disquisition had changed American linguistics from a subfield of anthropology into a mathematical science (Devlin, 2000:88).

A Brief General Survey of Linguistics

In this section, I briefly examine each of the following aspects of the Linguistics discipline: (a) subfields, (b) intellectual role, (c) applications, (d) technological importance, and (e) foci. I present these features sequentially for the purpose of cohesion. Discussions of the first four aspects are based on the pamphlet written by Geoffrey Nunberg for the Executive Committee of the Linguistic Society of America (1988), and the review of the last aspect is based on the works of William O'Grady et al. (1989) and Francis P. Dinneen (1995) because of their completeness and simplicity in the coverage of the topics.

Subfields of Linguistics

The following are the labels of the subfields of Linguistics with their attendant descriptions as presented by Nunberg:

a) "*Phonetics* deals with the sounds of language. Articulatory phonetics studies the way sounds are produced by the vocal organs; acoustic phonetics the physical properties of the sounds themselves" (Nunberg, 1988:4).

b) "*Phonology* studies the ways in which speech sounds are classified in the sound systems of particular languages and are combined to form meaningful units." For example, "English and German are alike in that both have the sounds 'p' and 'f,' but the phonologist observes that only in German can a word begin with the sequence 'pf'" (Nunberg, 1988:4).

c) "Morphology studies the way in which words are put together out of prefixes, roots, and suffixes. A morphologist asks how to determine in," for instance, "English whether an adjective is formed by the addition of -*al* (as in *regimental*), -*ar* (as in *polar*), or –*ic* (as in *parasitic*) (Nunberg, 1988:4).

d) "*Syntax* is concerned with the rules used to combine words into larger units of phrases and sentences." For example, "Why do we say a 'big red rooster' rather than 'a red big rooster' or 'a rooster red big'? (Nunberg, 1988:5).

e) "*Semantics* studies the meanings of linguistic units and how they are combined to form the meanings of sentences. A semanticist describes the difference in meaning between *recall* and *remember*, or between *I remembered going* and *I remembered to go*" (Nunberg, 1988:5).

f) "*Discourse Analysis* looks at the way in which sentences are combined in the larger stretches of narrative or conversation" (Nunberg, 1988:5).

g) "*Pragmatics* is the study of the way in which the meaning of a sentence is interpreted relative to the context in which it is spoken." For instance, "What makes a sentence like 'Can you pick up the package?' count as a question in one situation and a request in another?" (Nunberg, 1988:5).

h) "*Historical Linguistics* considers how languages change over time in pronunciation, syntax, and meanings" (Nunberg, 1988:6).

i) "*Stylistics and Poetics* study literary language and how it achieves its effects" (Nunberg, 1988:6).

j) "*Sociolinguistics* considers how languages are embedded in the surrounding frameworks of cultures and social institutions." It examines questions such as "How do the various features that make for different social groups correlate with the linguistic features that distinguish dialects and language varieties? How do social features influence the way in which linguistic changes spread throughout the larger community?" (Nunberg, 1988:6).

k) "*Dialectology* observes how language varies across geographical or social boundaries" (Nunberg, 1988:6).

l) "*Psycholinguistics* looks at the relation between language and mind and tries to uncover the perceptual mechanisms that people use in processing and producing utterances" (Nunberg, 1988:7).

m) "*Developmental Psycholinguistics* is concerned with the question of how children acquire their knowledge of language" (Nunberg, 1988:7).

n) "*Neurolinguistics* deals with the way in which language is represented in the brain" (Nunberg, 1988:7).

o) "*Mathematical and Computational Linguistics* look at languages as essentially formal systems, which could be mastered by (actual or hypothetical) machines. The goal of research in these areas is to uncover the logical and mathematical structures that underlie linguistic systems" (Nunberg, 1988:7).

p) "*Other Specialties* cross-cut the areas...already mentioned. 'Applied linguistics' is a heterogeneous area that has traditionally involved the application of linguistic theory to such practical issues as first- and second-language teaching, translation, language planning, and the development of

> computational systems to deal with natural language. Finally, many linguists specialize in another way, by concentrating on the study of one language or language family... (Nunberg, 1988:8).

Nonetheless, as Nunberg explains, the field of Linguistics is not as fractured as other intellectual disciplines despite its numerous subfields for a number of reasons. First, most linguists receive similar training since the professional community is somewhat small. Second, it is possible to keep up with the scholarship in several areas as the works in the specialized subfields are not so vast. Third, most linguists are familiar with the general developments in a number of specialties and have conducted research in several others. Fourth, none of the subareas are completely independent. For example, a sociolinguist needs to know about developments in phonetics and phonology to do his/her work; and, a semanticist must be familiar with developments in syntax and mathematical linguistics to do his/her work. Finally, despite the growth in the number of linguists, the community is still small enough to facilitate the development of professional and personal relationships (Nunberg, 1988:8).

Intellectual Role of Linguistics

As it pertains to the intellectual role of Linguistics, Nunberg begins with the observation that given the distinctive possessions of language among institutions and human capacities and the technical comprehension of language it provides, the field of Linguistics has throughout its history had an intellectual impact on contiguous disciplines that is larger vis-à-vis its size. For starters, by developing a "metalanguage"—i.e. a language for talking about language, linguists provide scholars in other disciplines a tool they can use to analyze phenomena, at the level of words, phrases, sounds, etc., from their disciplinary perspectives. For example, a literary critic can utilize this metalanguage to describe a particular author's style when, for instance, s/he says that the author privileges the passive voice over the active voice, or that a poem employs words beginning with labial consonants at particular major places. Also, before a neurologist can categorize different types of aphasia and correlates them with the types of physical lesions that made them occur, s/he would need a linguistic classification of speech defects (Nunberg, 1988:9).

Next, Nunberg points out that Linguistics has been much more than a fount of vocabulary and taxonomic devices for many human sciences. For instance, in their efforts to investigate earlier problems with thoroughness and comprehension, some scholars in the human sciences have utilized approaches in the field of Linguistics to develop new theoretical frameworks and methods. This has been possible because linguistics data are relatively more controllable compared to those in many other disciplines. For the literary theorist, it is easier to pinpoint words and phrases which are the distinct elements of a sentence compared to identifying episodes, events, etc. which comprise the elements of a narrative. Thus, some concepts from syntactic theory become very useful for the literary theorist. For the ethnographer interested in examining various cultural attitudes, the structural elements of linguistic systems are imperative for his/her analysis of the structure of cultural systems. It is therefore not surprising that a number of the general intellectual movements of the 20th Century emerged from developments in linguistic theory. Among the most known theoretical approaches from this development is "structuralism" which has at various times been dominant in the disciplines of Anthropology, Literary Theory, and Psychology. Contemporary developments of the fields of Philosophy, Psychology, and Sociology have also seen the employment of generative grammar in various ways. And, in the theories of the fields of Architecture and Genetics, linguistic terms have been used for more metaphorical developments (Nunberg, 1988:9-10).

In addition, according to Nunberg, the human sciences have found the Linguistics discipline as a source for developing and evaluating various perspectives about fundamental questions that arise from their human scientific inquiries. Citing the conceptual framework of "cultural relativism" as an example, he states that in order to justify the introduction of theoretical ideas such as "kinship," "the market," and "oligarchy" in their investigations, some anthropologists, economists, and political scientists have had to deal with how much of the conceptual framework is employed to classify an experience that is particular to one's culture and how much of that experience is universal. Nonetheless, he suggests that it is easier to use relatively real-world examples provided by language compared to metaphysical ideas from other contexts to elucidate general issues by asking, for instance, about the universal systems that exist in all systems of color names instead of the universal idea of justice. Another interesting issue alluded to by Nunberg is the debate over "innate ideas." On this matter, he begins with the question about whether we are born knowing certain things or we acquire all of our knowledge through experience. Thereafter, he relays that certain linguists have employed a formal mathematical approach of inductive learning and language structure to postulate that some linguistic information must have been included in children's genetic programming that makes it easier for them to learn a language. Despite the controversy generated by their postulate, philosophers and psychologists on both sides of the debate now better comprehend what is at stake on the matter (Nunberg, 1988:10-11).

Also, Nunberg apprises us that even though there are not many professionals in the field of Linguistics, their work, however, is a significant component of the wider intellectual environment. Even colleges and universities that do not have linguistics departments, as do major research institutions, have linguists to make their programs in fields ranging from Computer Science to Literature adequate. As long as an academic institution is engaged in advanced research and graduate training, it will need experts in Linguistics, even though this is less so the case in an academic institution that focuses on undergraduate education or lacks Linguistics as an independent subject in its curriculum. But even at undergraduate institutions and secondary or high schools, Nunberg mentions that Linguistics has had a significant impact on the manner a language is taught because the use of linguistic concepts is imperative for conveying an understanding of the language (Nunberg, 1988:11).

Applications of Linguistics

In regards to the applications of Linguistics, Nunberg maintains that in disciplines ranging from Education to Medicine to Public Policy, theories and methodologies developed in the field of Linguistics have found a home whenever language emerges as a practical concern. He recounts that (a) language teaching was the earliest area to employ the study of grammar by incorporating the categories of traditional school grammar such as "predicate grammar," etc. derived from classical linguistic theories; (b) the ways first- and second-language skills are taught have benefited from the application of modern linguistics understanding of language structure and how language learning takes place; and (c) in every endeavor pertaining to language education, from the business person's "immersion course" in a language to a first-grade reader, linguistic theory can be discerned (Nunberg, 1988:12).

Thereafter, Nunberg tells us about how the herculean challenge to learn and describe the thousands of unwritten languages across the globe and efforts to develop writing systems to record them engendered a different utilization of linguistic theory and method. He recounts that during earlier historical eras, such work in the Third World was the enterprise of Christian missionaries who sought to convert indigenous people to their religion. These missionaries called upon the expertise of linguists to help them with linguistic research and make the scriptures available to the indigenous people in both their own languages and those of the missionaries. In fact, many early linguists came from these missionary organizations. As Third World countries gained independence from the colonial powers, the work took on greater importance at the secular level since writing systems, grammars, and dictionaries had to be developed in the indigenous languages to facilitate wider literacy. Linguists with their tools have been pivotal in the development of traditional languages to meet the requirements for modernization by developing new alphabets and neologisms, and simplifying grammars. Other social scientists have found it advantageous to work with linguists in the larger area of language planning in order to develop the same linguistic tools to help speed up the modernization process. Also, the manner in which dictionaries are compiled and how they present their information has been greatly influenced by linguists (Nunberg, 1988:12-13).

Furthermore, as Nunberg informs us, areas where problems arise when language does not work the way it is expected have also enjoyed tools from the field of Linguistics. This is the case of language disorders such as aphasia, hearing and speech defects, learning disabilities, and schizophrenia that have seen linguistic research conducted for their diagnoses and treatments. When the language process is physiologically or psychologically distorted, it is impossible to comprehend what transpired without the knowledge of how language and speech function in normal human beings. Also, when communication breaks down for social reasons such as air accidents occurring due to conversational misunderstandings, when an average citizen does not understand his/her rights and obligations in a contract because of the convoluted way it was written, or when advertisements and product labels are written to mislead the average consumer, linguists have been invited to help resolve these types of problems. In addition, when there is a need for the precise interpretation of a recorded or written text in a court case, linguists have often been called upon to serve as expert witnesses (Nunberg, 1988:13).

Technological Importance of Linguistics

As it concerns the technological importance of the Linguistics discipline, Nunberg explains that the field will have a far greater impact on computational applications, even though it is a relatively new technological phenomenon for the field. He identifies two endeavors in which both Linguistics and Computer Science are benefitting tremendously via collaborative explorations. One endeavor is how the analyses of speech signals and the collection of data on usage for new dictionaries are being advanced by the use of computational techniques in many of the traditional research areas of Applied and Theoretical Linguistics. Another endeavor is how new computational systems are designed to enable people to interact efficiently and naturally made possible by the emergence of new problems when computer scientists sought to design these new systems. By developing "user interface" that behave the way other people do—i.e. by being able to employ natural language in a number of realistic modes, these new computational problems are being resolved through the collaborative work of linguists and computer scientists (Nunberg, 1988:14).

Next, Nunberg indicates that in one area, computer scientists continue to collaborate with phoneticians and phonologists to advance "speech recognition" and "speech synthesis" technology so that computers can understand spoken speech and also talk to users. In the other area, which has proven to be more challenging, these scientists are working to enhance the capabilities of computers to understand spoken input. In another area, experts in semantics and syntax are working on a number of natural-language understanding systems to ascertain the structure of input sentences of a natural language in order to assign some meanings to those sentences in the internal representation of computers (Nunberg, 1988:14).

Also, Nunberg concedes that we cannot expect an artificial system to be developed in the near future that will have everything that a human being possesses in terms of linguistic capacity because a person culls from a huge amount of background assumptions and information to interpret utterances produced by others. Nonetheless, as he points out, the partial successes of artificial systems show the great potential of this area. He cites the following examples: (a) the partial automation of telephone information services has been made possible by the development of speech synthesis programs, a process that is being extended as work on speech recognition continues; (b) "machine-assisted translation" programs are being used to produce adequate first drafts of articles that are then edited by human experts; (c) machine-user interactions are being changed by more sophisticated systems of natural-language generation and understanding as users are able to get information from a database system by asking a simple question, medical expert systems "talk" to clinicians looking for assistance to make accurate diagnoses, etc.; (d) the development of voice recognition programs and the improvement of various types of document preparation systems are being made possible by some linguistic expertise. The upshot of all these developments, according to Nunberg, is that the field of Linguistics is finding homes beyond academia as more linguists are launching their careers in the private sector, and corporations and public agencies are funding research dealing with applied and theoretical linguistic issues (Nunberg, 1988:14-15).

Foci of Linguistics

The foci of the field of Linguistics are numerous. For relative brevity, I briefly treat each of the following most frequently discussed foci in general linguistics textbooks: (a) the study of language and linguistics, (b) language as a system of sounds, (c) language as a system of forms, (d) etymology and historical linguistics, (e) transformational grammar, (f) non-transformational constants,

(g) after aspects, (h) language classification, (i) language and brain, (j) language acquisition, (k) second language acquisition, (l) language in social context, (m) writing systems, and (n) animal communication. These aspects are explored here one at a time for perspicuity.

Beginning with the *study of language and linguistics*, we learn from Dinneen that Linguistics is just one of the many disciplines that study language. He states that while it is evident that fields such as Grammar, traditional Logic, Poetry, and Rhetoric study language, it is also evident that these disciplines focus on different objects in their inquiries. He adds that a question is raised about whether the objects Linguistics study are not already covered by what these other disciplines study, making the discipline redundant. Thus, Dinneen asserts that in order to address this legitimate question, Linguistics transcends what these other disciplines investigate by demonstrating that "there are aspects of *Language* common to any *language*, and that one stage of the study called Linguistics" has involved making very clear what these aspects are (Dinneen, 1995:1-2).

For O'Grady et al., Linguistics is a discipline that focuses on how language is organized and utilized, which are processes through which all of us can be affected. Therefore, language, according to these scholars, denotes numerous aspects that include "a system of communication, a medium for thought, a vehicle for literary expression, a social institution, a matter for political controversy, a factor in nation building" (O'Grady et al., 1989:1). Thus, these scholars also state that since every normal person speaks at least one language, it is hard to envisage a social activity occurring without the use of a language (O'Grady et al., 1989:1).

Next, about *language as a system of sounds*, O'Grady and his colleagues inculcate for us that since language can be written, recorded mechanically, and even produced by computers, it is not even imperative for us to speak in order to use language. Nonetheless, they add, because speech is still the primary way to encode language, we are condemned to use it. Thus, linguists working in the subarea of the field called Phonetics use their relevant tools to study the structure of the sounds of language. As O'Grady et al. point out, there are two ways phoneticians approach their subject matter. One way is *articulatory phonetics*, which deals with the physiological aspects of speech production. The other way is *acoustic phonetics*, which addresses the physical aspects of speech sounds (O'Grady, 1989:13).

As Dinneen also instructs us, Phonetics is impartially employed to study three different "objects." The first object is *composition*, which refers to "the discrete units of sound." The second object is *distribution*, which denotes the positions in which sounds "occur relative to each other." The third object is *function*, which is about what "sounds do in the system of the language"

(Dinneen, 1995:25). On these aspects, Dineen further makes an important distinction between Phonetics and Phonology that is worth restating here, although briefly: i.e. "One definition of Phonology is *Functional Phonetics*. While the first focus of Phonetics is what a sound *is* (its composition), in phonology, it is precisely what a sound *does*, how it functions, that is relevant" (Dinneen, 1995:26).

In addition, as it relates to *language as a system of forms*, Dineen edifies for us that while the representation of speech is simplified by Phonemics, the approach has empirical limitations. First, Phonemics' ideal goal to extract all of the common, contrastive elements in a language is impossible to fulfill for a language that is rich in dialectical pronunciations. Second, grammatical systems may not be represented in this way. Third, since grammatical evaluations seem to be quite numerous, Phonemics cannot handle these since the approach's easily agreed-upon contrasts tend to be few. Fourth, we must depend mostly on techniques of contrastive substitution in a linear distribution, as the minimal-pair test for phonemes tends to reveal, should we consider phonemic results as a type of micro-model for describing other linguistic systems. Fifth and finally, for grammatical systems, the interlocking factors that undergird them are quite complex (Dinneen, 1995:79).

It behooves me to briefly explicate here what the aforementioned minimal-pair test is for those readers who may not be familiar with it. According to O'Grady and his colleagues, the *minimal-pair test* is "the basic test for a sound's distinctness" (O'Grady, 1989:57). They add that "The minimal pair consists of two forms with distinct meanings that differ by only a segment found in the position in each form. The examples [sɪp] *sip* and [zɪp] *zip* form a minimal pair and show that the sounds [s] and [z] contrast in English" (O'Grady et al., 1989:57).

Also, in reference to *etymology* (i.e. the study of the origins of words and the way in which their meanings have changed throughout history) *and historical linguistics*, Dinneen instructs us that the phenomena concern language changes. He mentions that some people believe that the "one correct way of speaking" is in jeopardy when languages change, making these individuals quite perturbed about the change. He raises the question about whether the "correct" etymological or historical stage can be restored by bringing to light the "errors" of today. He states that instead of some reconstructions informing us about what reformers claim to recreate, they instead apprise us about what is in the reformers' heads. He also raises the question about whether anyone can be an Analogist if s/he has a sense of order, even if it was anarchy, since "*order, cause, basic* and *real*," essentially, seem to define one another. He suggests that one must substitute one type of "*basic reality*" for another, if s/he is to demonstrate that a "cause" is a "non-cause," or to determine that the idea is not accurately applied. Consequently, he opines, since researchers perceive different things to be basic, or completely real, or just ideal in language, the goals, methods and results of language reconstruction are to provide examples of the phenomenon of language change (Dinneen, 1995:199).

O'Grady et al. tell us that there are many causes for language change which often interact with one another in complex manners. While this situation makes it quite challenging for an analyst to determine the exact cause of a particular linguistic change, these scholars, however, identify five specific triggers for such a change. The first trigger they discern comprises *physiological factors*: when articulation is made easier by modifying a sound so that it is similar or dissimilar to a neighboring sound. The second trigger is *analogy*: the proposition that when two elements are similar in certain cases, they should be similar in other cases as well. The third trigger is *folk etymology*: when the misanalysis of a word by speakers of a language leads to change in the meaning of the word. The fourth trigger is *borrowing*: acquiring linguistic rules, sounds, and words from another language. The fifth trigger is composed of *sociological factors*: when speakers intentionally or unintentionally alter the ways they speak in order to come close to what they believe are the more socially acceptable or prestigious speech varieties (O'Grady et al., 1989:191-192).

Thereafter, regarding *transformational grammar*, Dinneen illuminates that the approach can be traced to the adoption of Immediate Constituent Analysis' representations in Stratificational and Systemic Grammar. This development allows the field of Linguistics to deal with *Text* instead of just isolated sentences. Thus, a set of sentences, connected by *Constellation* (i.e. "a relation among variables"), does not constitute a *text*. Dinneen recalls that in order to demonstrate that the sentences of a language are connected via *Systemic Interdependence*, *Syntactic Structures* were developed so that grammarians could represent language explicitly as a *Determination of Process* between basic and generated sentences (Dinneen, 1995:367).

Thereupon, in terms of *non-transformational constants*, Dinneen teaches us that the *generative* component of *Transformational Generative Grammar* insists that the subcomponents of the structure of a language are related in a rule-governed manner, in addition to being related to higher categories. He notes that this postulate is easier to falsify compared to that of *langue as a pure form*, which treats "a pattern /structure / design as indifferent to semantics and phonology as numbers are to what we count" (Dinneen, 1995:403). To support his point, he cites the fact that there are no "5½ real people in the average [United States] family, despite the impeccable statistics which generates it" (Dinneen, 1995:403).

Subsequently, apropos to *after aspects*, we are enlightened by Dinneen that the notion focuses on the frontier between semantics and syntax. He ventilates the question about whether a grammar is a model of what a speaker plans to communicate, what a hearer understands, or the linguistic structure shared by both the speaker and the hearer. He proposes that what the Ideal Speaker-Hearer "knows" or of "language itself" will vary based on what an analyst selects to treat as constants and variables. He notes that analysts isolate aspects of language, even though we experience it holistically. He points out that just as machine-parts are designed to adjust to the changes they produce, so do aspects of language affect one another. He adds that clarity about their autonomy is proposed when the parts of a model are labeled *syntax, semantics,* and *phonology*. He also makes the point that the clarity of a model is enhanced when its modules designated as *surface, intermediate,* and *surface structures* are differentiated from one another. He concedes that we are bound to make hasty formalizations; nonetheless, he also believes that we can learn to remodel such formalizations (Dinneen, 1995:445).

Regarding *language classification,* we are told by O'Grady and his colleagues that the notion has to do with the fact that linguistic diversity is one of the most conspicuous realities of our world. Each of the thousands of different languages has its own lexicon, sounds, and syntactic rules. But, despite the significant diversity in these languages, linguists have capitalized on their shared features in order to group them into a relatively smaller number of types and families. O'Grady et al. identify two approaches linguists have utilized to classify languages. One approach is the use of the structural features of languages to classify them according to their grammatical rules and sound patterns. The other approach is the utilization of the genetic relationships of languages to classify them in terms of their ancestries (O'Grady et al., 1989:223).

As it concerns *language and brain,* O'Grady and his fellow scholars explain that the study of this aspect, which is referred to as *Neurolinguistics,* is concerned about how grammatical categories and rules are represented in the brain: i.e. the domain of all cognitive functions. They add that linguists working on this aspect have been able to identify the components of the brain that deal with language and how linguistic knowledge is represented in it. They also state that neurolinguists usually begin their discussions with descriptions of how the different parts of the brain that are used for language are structured and function. Neurolinguists then discuss the various language disorders that occur when different areas of the brain are damaged. They end with discussions about whether there exists a critical age for a person to acquire a language (O'Grady et al., 1989:253), an aspect that is examined in more depth by linguists focusing on language acquisition.

Vis-à-vis *language acquisition,* in addition to the aforementioned interest, we also learn from O'Grady and his colleagues that linguists studying this aspect have focused more on children's acquisition of a first language. But, as they also quickly caution, what is actually acquired is grammar: i.e. the set of conditions, elements, and rules that make it possible for people to speak and understand a language (O'Grady et al., 1989:269-270). They further briefly recount the early attempt to investigate language acquisition dating back to Ancient Kemet/Egypt that is worth retelling here verbatim as follows:

> Fascination with this issue dates back to at least the seventh century B.C., when the Egyptian Pharaoh Psammetichus had two infants brought up in complete isolation in an attempt to determine the type of language they would learn on their own. The Pharaoh had hoped that the children's utterances would provide some clues about the origin of language. The story is that the children were brought up by an old shepherd couple, who were instructed not to speak to them (or, who were mute, depending on which version you hear). After some years, the children were heard to utter *be*, and the Pharaoh concluded that the original language of humankind was Phrygian, since the Phrygian word for 'bread' is *bekos*. It has been pointed out that it is not surprising that children raised in an environment of sheep cries would produce the syllable *be* (O'Grady et al., 1989:269).

About *second language acquisition*, O'Grady et al. inculcate us that unlike the preceding language acquisition aspect that focuses on how children acquire a first language, second language acquisition focuses on how a learner who already has an established system of communication and, thus, cognitively more mature, acquires a second language. Since second language learners approach a second language differently, O'Grady and his colleagues argue that it is justification enough for the study of second language acquisition to be considered as a separate subarea. They add that because linguistic theory is employed to investigate second language learning and teaching, the concept *Applied Linguistics* is frequently employed to refer to the phenomenon (O'Grady et al., 1989:299).

Additionally, in relation to *language in social contexts*, we are edified by O'Grady and his coauthors that this aspect entails the analyses of the use of language and the effects of extralinguistic elements on the use in various social environments. With the objective of revealing that a speech community is highly structured and complex, linguists working in this subarea investigate topics ranging from language use in interaction to social variation in language and to regional variation in language. O'Grady et al. alert us that in actuality, language in social contexts is not about proper speech in opposition to all other speech; instead, it is about a set of reciprocal speech varieties that exist in a structured speech environment (O'Grady et al., 1989:326). I must add here the names of some of the relatively new foci that have emerged, the first three of which I am the pioneer, albeit which traditional linguists tend to ignore: (1) fractal linguistics, (2) language and peace, (3) accelerated language deaths, (4) linguistic forensics, (5) language and bilingual education, (6) language and politics, (7) language and testing, and (8) language and law.

Furthermore, as it pertains to *writing systems*, O'Grady et al. inform us that researchers engaged in the investigation of these systems focus on the symbolic representations of language in storage graphic forms. They also point out that (a) writing is a relatively newer cultural form of communication compared to speaking; (b) there are numerous communities in the world that do not have writing systems; (c) numerous people across the globe lack writing skills; and (d) while language comes naturally to people, writing does not. In addition, they describe the three types of writing: (1) *logographic*—a system of writing in which an entire word is represented by a sign; (2) *syllabic*—a system of writing in which syllables [i.e. units of pronunciation having one vowel sound, with or without surrounding consonants, forming the whole or a part of a word; e.g., there are two syllables in water and three in inferno] are represented by signs; and (3) *alphabetic*—a system of writing in which consonant [e.g., b, c, d, f, g in English] and vowel [a, e, i, o, u in English] segments are represented by signs (O'Grady et al., 1989:358-359).

Finally, in regards to *animal communication*, according to O'Grady and his colleagues, linguists working on this aspect ask at least two questions dealing with (1) whether animals do communicate among themselves as humans do and (2) whether animals can communicate with humans. Given the long-held belief that language is what sets humans apart from other animals, linguists investigating animal communication take these questions very seriously. Their major argument is that we may have to discard language as a variable that differentiates us humans from other animals should it be scientifically established that the other animals possess a system of communication that is structured just like ours (O'Grady et al., 1989:383).

A Brief General Survey of Mathematics

Drawing upon many sources, briefly examined in this section are the following frequently discussed aspects in the field of Mathematics: (a) stages of mathematics; (b) areas of mathematics; (c) general attributes of modern mathematics; (d) concepts of traditional mathematics; (e) concepts of modern mathematics; and (f) mathematical applications in the business, social and life sciences. These aspects are discussed individually for clarity.

Stages of Mathematics

After consulting more than 30 relevant mathematical sources (e.g., Bangura, 2000a, 2000b, 2000c & 2011a; Benacerraf and Putnam, 2004; Katz, 2008; Merzbach and Boyer, 2011; O'Connor and Robertson, 2019; and Smith, 2012), I was able to identify three major stages of the field of Mathematics. The first is the *Rhetorical Stage* during which mathematical calculations were performed by using words and no symbols. This era began in about 70000 BC with South Africans adorning ochre rocks with scratched geometric patterns and lasted until 256 BC with China's *Chou Pei Suan Ching*—one of the oldest Chinses texts on arithmetic and geometric algorithms and proofs.

The second is the *Syncopated Stage* during which frequently used mathematical operations and quantities were represented by symbolic syntactical abbreviations. This stage is subdivided into two sub-stages. The first sub-stage covered the 1st Millennium BC; it began with the Ancient Kemetians/Egyptians inventing vulgar fractions in 1000 BC and lasted until the 1st Century BC with the mathematical work of the Greek astronomer and mathematician Geminus of Rhodes. The second sub-stage ranged from the 1st Century AD with the invention of the square roots of negative numbers by Heron (Hero) of Alexandria, Egypt to 975 AD with the Turkish-born Arab astronomer, astrologer, and mathematician Abū ʿAbd Allāh Muḥammad ibn Jābir ibn Sinān al-Raqqī al-Ḥarrānī aṣ-Ṣābiʾ al-Battānī extending the Indian concepts of sine and cosine to other trigonometrical ratios.

The third is the *Symbolic Stage* during which numbers, notations and mathematical symbols have and continue to be used. There are three sub-stages of this phase. The first sub-stage began in 1000 AD with the Persian astronomer, mathematician and physicist Abū Sahl Wayjan ibn Rustam al-Qūhī solving equations higher than the second degree and ended in 1494 AD with the Italian Fra Luca Bartolomeo de Pacioli writing the *Suma de arithmetica, geonetrica, proportioni et proportionalità* and introducing primitive symbolic algebra using "co" (cosa) for the unknown.

The second sub-stage or modern era of the *Symbolic Stage* covered (a) the 16th Century, beginning in 1501 with the Indian astronomer and mathematician Nilakantha Somayaji writing the *Tantrasamgraha*—an astronomical and mathematical work comprising 432 Sanskrit verses divided into eight chapters—and ending in 1596 with the German-Dutch Ludolf van Ceulen computing π to 20 decimal places by employing inscribed and circumscribed polygons; (b) the 17th Century, beginning in 1614 with the Scottish astronomer, mathematician and physicist John Napier's discussion of Napierian logarithms in his *Mirifici Logarithmorum Canonis Descriptio* and ending in 1699 with the English astronomer and mathematician Abraham

Sharp calculating π to 72 digits, although 71 are correct; (c) the 18th Century, starting in 1706 with John Machin developing a quickly converging inverse-tangent series for π and computing π to 100 decimal places and ending in 1799 with the Italian mathematician and philosopher Paolo Ruffini partially proving the Abel-Ruffini theorem that quintic or higher equations cannot be solved by a general formula; and (d) the 19th Century, beginning in 1801 with the German mathematician and physicist Johann Carl Friedrich Gauss' number theory treatise published in Latin as *Disquisitiones Arithmeticae* and ending in 1900 with the German mathematician David Hilbert stating his list of 23 problems showing that some further mathematical work is needed.

The third sub-stage or contemporary epoch of the *Symbolic Stage* which began in the 20th Century started in 1901 with the French mathematician Élie Joseph Cartan developing the exterior derivative and ended in 2000 with the Clay Mathematics Institute based in Peterborough, New Hampshire, United States proposing the seven Millennium Prize Problems of unsolved important classic mathematical questions. As of this writing, only one of the seven problems has been solved. The contemporary epoch continues into the 21st Century, beginning in 2002 with Indian computer scientists Manindra Agrawal, Nitin Saxena, and Beeraj Kayal of the Indian Institute of Technology in Kanpur presenting an unconditional deterministic polynomial-time algorithm to calculate whether a specified number is prime and ending in 2019 with the Japanese computer scientist Emma Haruka Iwao using y-cruncher v.0.7.6 to successfully calculate π to 31.4 trillion digits.

Areas of Mathematics

Just as there is no one agreed-upon definition for Mathematics, there also is none for the classification of the various areas of the field. For example, the *Encyclopædia Britanica* (2019) classifies mathematical articles into three areas—(1) Pure, (2) Applied, and (3) Biographies—and the Dewey Decimal Classification (2019) system has seven categories—(1) Applied and Number Theory, (2) Arithmetic, (3) Topology, (4) Analysis, (5) Geometry, (6) Numerical Analysis, and (7) Probabilities and Applied Mathematics. In fact, a March 16, 2019 Internet search I conducted for terms that have been considered as being areas of study in Mathematics yielded almost 1,000. For the sake of conciseness, the traditional classification of the areas of Mathematics as *Pure Mathematics* and *Applied Mathematics* is employed here. While the division is sometimes opaque and new mathematical discoveries crosscut the two areas, the traditional classification system is still useful for subsuming the various subareas into the two major areas.

Beginning with the area of *Pure Mathematics*, it concerns the study of mathematical ideas in their own right. There are at least six subareas of *Pure*

Mathematics. The first subarea deals with the *foundations* of the field, including (a) recreational mathematics; (b) history and biography; (c) mathematical logic, including set theory; and (d) proof theory and constructive mathematics. The second subarea concerns *algebra,* which encompasses (a) order theory, (b) general algebraic systems, (c) field theory and polynomials, and (d) commutative rings. The third subarea is composed of *calculus and analysis,* which foci on the properties and relations of points, lines, surfaces, solids, higher dimensional analogs, the theory of functions and the use of limits, and continuity. The fourth subarea is *combinatorics,* which deals with combinations of objects belonging to a finite set in accordance with certain constraints, such as those of graph theory. The fifth subarea is composed of *geometry and topology,* which focus on (a) convex geometry and discrete geometry, (b) discrete and combinatorial geometry, (c) differential geometry, (d) algebraic geometry, (e) general topology, (f) algebraic topology, and (g) manifolds. The sixth and final subarea is *number theory,* which incorporates (a) arithmetic, (b) elementary number theory, (c) analytic number theory, (c) algebraic number theory, and (d) numerous other number theory subareas.

Next, the area of *Applied Mathematics* concerns the study of mathematical ideas as they apply to other disciplines such as biology, chemistry, engineering, physics, and the behavioral and social sciences. There exist at least four subareas of *Applied Mathematics.* The first subarea is comprised of *probability and statistics,* which entail (a) probability theory, (b) stochastic processes, and (c) the use of numerical data in large quantities for the purpose of inferring proportions in a whole from those in a representative sample. The second subarea involves the *computational sciences,* which deals with (a) numerical analysis and (b) computer algebra. The third subarea concerns the *physical sciences,* which address (a) mechanics of structure, (b) mechanics of deformable solids, (c) fluid mechanics, and (d) particle mechanics. The fourth and final subarea is composed of *other applied mathematics,* which include (a) operations research and (b) mathematical programming.

General Attributes of Modern Mathematics

From the work of Ian Stewart, who has made the study of modern mathematics his passion, one can discern several attributes of the subject. One attribute is in regards to *abstractness and generality.* To start with, Stewart points out that the propensity to become progressively abstract is one of the more discernible features of modern mathematics. Every key notion includes many different objects which have some property in common. The results of this general property are developed by an abstract theory which could then be executed on any of the different objects (Stewart, 1995:1).

Consequently, according to Stewart, aspects such as the "rigid motions in space, symmetries of geometrical figures, the additive structure of whole numbers, or the deformation of curves in a topological space" have enjoyed the application of the notion of "group" (1995:1). He adds that the possibility of amalgamating two objects of a particular kind to derive another object is what the common property is. He provides the following examples: (a) a rigid motion is formed when two rigid motions are performed successively, (b) a number results when two numbers are added, and (c) another curve is produced when two curves are stuck end to end (Stewart, 1995:1).

Next, Stewart states that abstraction and generality are closely connected. He also notes that a key upper hand of generality is that it minimizes work. Instead of trying to prove a certain theorem four different times, generality allows one to prove it only one time (Stewart, 1995:1-2).

In addition, Stewart mentions that dependence on the language of set theory is another principal feature of modern mathematics. For him, this feature is commonsensical since, at the generality level, a mathematical idea is more appreciated when it deals with whole collections of objects than when it deals with a particular object. He offers the following example and explanation: "That $5 = 1 + 4$ is not terribly significant. That every prime number for form '$4n + 1$' is a sum of two squares is significant. The latter is an observation about the collection of *all* prime numbers rather than about any particular prime number" (Stewart, 1995:2). Stewart continues:

> A *set* is just a collection: we use a different word to avoid certain psychological overtones associated with the word 'collection'. Sets can be combined in various ways to give other sets in the same way that numbers can be combined (by addition, subtraction, multiplication...) to give other numbers. The general theory of arithmetical operations is *algebra*: so we also can develop an algebra of set theory. Sets have certain advantages over numbers, particularly from the point of view of teaching. They are more concrete than numbers. You cannot show a child a number ('I am holding in my hand the number 3'), you can show him a number of things: 3 lollipops, 3 ping-pong balls. You will be showing him a *set* of lollipops or ping-pong balls. Although the sets of interest in mathematics are not concrete—they tend to be sets of numbers, or functions—the basic operations of set theory can be demonstrated by means of concrete material (Stewart, 1995:2).

Another attribute of modern mathematics concerns *intuition and formalism*. Stewart provides a number of rationales for this feature. First, he alludes to the fact that a growing degree of logical thoroughness has followed the movement toward greater generality. He cites the following examples of this phenomenon: (a) since Euclid did not proffer an axiom that a line passing through a point inside a triangle must cut the triangle somewhere, he is being criticized for it; and (b) mathematicians cannot play with fluctuations because Euler's denotation of a function as "a curve drawn by freely leading the hand" is too vague, and even the term "curve" is not defined. Stewart argues that this sort of things can be overdone, as "verbal arguments can be replaced by a profusion of symbolic logic and checked for validity by a blind application of a standard technique. If carried too far (and in this case, enough is too much) this destroys understanding, instead of aiding it" (Stewart, 1995:3).

Second, Stewart points out that the dictates for greater rigor are not a mere impulse. This is because it is imperative to embrace an evaluative orientation toward a subject as it becomes complex and immense. Stewart makes the point that just as a sociologist would throw out masses of experimental data whose experiments are found to be badly performed or conclusions are uncertain, so, too, must a mathematician act when faced with the same situation. On many occasions, the "obvious" has been incorrect. For instance, there exist geometrical figures that lack an area (Stewart, 1995:3).

Third, Stewart suggests that when addressing subtleties or confronted with perilous situations, a moderating effect is provided by logical rigor. Many mathematicians believe that certain theorems are true; but, these theorems will remain untenable assumptions until someone proves that they are valid (Stewart, 1995:3).

Fourth, Stewart proposes that proving the impossible is another aspect that calls for great caution. Concomitantly, he advocates that very careful specializations are needed, as what can be easily performed by using a particular method may not be performed by utilizing another method. Stewart provides the following example and counsel for this aspect: "There exist proofs that quintic equations cannot be solved by radicals, or that angles cannot be trisected with ruler and compass. These are important theorems, because they close off unproductive paths. But if we are to be certain that the paths really are unproductive, we must be very cautious with our logic" (Stewart, 1995:3-4).

Fifth, Stewart alludes to the fact that a growing characteristic in the Mathematics discipline is the notion of impossibility proofs. The idea has emerged as the singular subject that can assure its own limitations. The subject has become so much obsessed with its own limitations that some mathematicians are simply content with proving that something cannot be done instead of seeking ways to do it. As Stewart euphemistically puts it, "If self-knowledge be a virtue, then mathematicians as a breed are saints" (Stewart, 1995:4).

Sixth, Stewart argues that logic cannot be everything. Just as a formula cannot suggest anything on its own, so can logic not suggest which problems to solve; instead, logic can only be utilized to solve problems. Similarly, *significance* has not been formalized by anyone. Some amount of experience and intuition are required to perceive something that is significant (Stewart, 1995:4).

Seventh, Stewart broaches the truism that virtually every human being has some level of mathematical intuition. Children solving jig-saw puzzles possess mathematical intuition, people packing families' holiday luggage in the trunks of automobiles possess it, and the list goes on. For training mathematicians, however, the key objective is to get them to be able to use their intuitions as controllable tools (Stewart, 1995:4).

And, eighth, Stewart deliberates on the fact that a virulent debate is taking place on whether rigor should be privileged over intuition or vice versa. He argues that proponents of both extreme positions miss the point that both rigor and intuition make Mathematics a powerful discipline. He therefore calls for the avoidance of these extreme polemics because, as he succinctly articulates it, "Controlled genius, inspired logic. We all know the brilliant person whose ideas never quite work, and the tidy, organized person who never achieves anything worthwhile because he is too busy getting tidy and organized" (Stewart, 1995:4).

The other attribute of modern mathematics relates to *pictures.* Stewart estimates that while about ten percent of mathematicians think in formulae, the rest (or about 90 percent) think in pictures, since their intuition is geometrical. This characteristic is an advantage because pictures convey a great deal more information compared to words. Stewart recalls that for a long time schoolchildren were dissuaded from drawing pictures because "they aren't rigorous." He believes that this action was a bad mistake because while pictures may not be rigorous, they are, nonetheless, "an essential aid to thought and no one should reject anything that can help him to think better" (Stewart, 1995:5).

Concepts of Traditional Mathematics

For a discipline that dates back to 70000 BC, it is only to be expected that it contains a plethora of traditional concepts that are used to this day. Presented in this section are just some of the more important concepts with their definitions, which are generated by me.

a) *Number*—an arithmetical value, expressed by a word, symbol, or figure, representing a particular quantity and used in counting and making calculations and for showing order in a series or for identification.

b) *Angle*—the space, usually measured in degrees, between two intersecting lines or surfaces at or close to the point where they meet.

c) *Fraction*—a numerical quantity that is not a whole number: e.g. ½, 0.5.

d) *Two-dimensional shape*—a flat shape such as a circle, a square, a triangle, etc. which has or appears to have length and breadth but no depth.

e) *Three-dimensional shape*—a solid such as a cube, a prism, a pyramid, a sphere, a cone, a cylinder, etc. which has or appears to have length, breadth, and depth.

f) *Decimal*—a fraction whose denominator is a power of ten and whose numerator is expressed by figures placed to the right of a decimal point.

g) *Statistics*—the practice or science of collecting and analyzing numerical data in large quantities, especially for the purpose of inferring proportions in a whole from those in a representative sample.

h) *Graph*—a diagram showing the relation between variable quantities, typically of two variables, each measured along one of a pair of axes at right angles.

i) *Measurement*—the size, length, or amount of something, as established by measuring: i.e. ascertaining the size, amount, or degree of something by using an instrument or device marked in standard units or by comparing it with an object of known size.

j) *Perimeter and area*—the continuous line forming the boundary of a closed geometric figure and the extent or measurement of a surface or piece of land.

k) *Algebraic expression*—a collection of symbols that jointly indicate a quantity or mathematical equation in which a finite number of symbols is combined using only the operations of addition, subtraction, multiplication, division, and exponentiation with constant rational exponents.

l) *Approximation*—a value or quantity that is nearly but not exactly correct.

m) *Equation*—a statement that the values of two mathematical expressions are equal (indicated by the sign =).

n) *Percentage*—a rate, number, or amount in each hundred.

o) *Formula*—a mathematical relationship or rule expressed in symbols.

p) *Circle*—a round plane figure whose boundary (the circumference) consists of points equidistant from a fixed point (the center).

q) *Probability*—the extent to which an event is likely to occur, measured by the ratio of the favorable cases to the whole number of cases possible.

r) *Ratio and proportion*—the quantitative relation between two amounts showing the number of times one value contains or is contained within the other; and a part, share, or number considered in comparative relation to a whole.

s) *Pythagoras' theorem and trigonometry*—the general proposition, not self-evident but proved by a chain of reasoning, that the square of the hypotenuse of a right triangle is equal to the sum of the squares of the other two sides; and the branch of mathematics dealing with the relations of the sides and angles of triangles and with the relevant functions of any angles.

t) *Indices and standard form*—exponents or other superscript or subscript numbers appended to quantities and, depending on the system or subject, written down in the most commonly accepted way.

The reader interested in detailed descriptions, examples and exercises of the preceding concepts can consult Trevor Johnson and Hugh Neill's *Complete Mathematics* (2010).

Concepts of Modern Mathematics

Similar to the traditional mathematical concepts, numerous modern ones have been developed over the years. Therefore, just as I did in the preceding section, I present here only some of the more important modern mathematical concepts with their definitions. Here also, the definitions are generated by me.

a) *Set*—a collection of distinct entities regarded as a unit, being either individually specified or, more usually, satisfying specified conditions.

b) *Function*—a relationship or expression involving one or more variables.

c) *Abstract Algebra*—the type of algebra that is concerned with various types of theoretical units like functions, mathematical spaces, operations, and sets which are more multiplex and copious than usual numbers.

d) *Symmetry*—a law or operation in which a physical property or process has an equivalence in two or more directions.

e) *Axiomatics*—relating to or containing axioms: i.e. statements or propositions on which abstractly defined structures are based.

f) *Counting: Finite and Infinite*—determining the total number of a collection of items, either having or not having limits or bounds.

g) *Topology*—the study of geometric properties and spatial relations unaffected by the continuous change of shape or size of figures; a family of open subsets of an abstract space such that the union and the intersection of any two of them are members of the family, and which includes the space itself and the empty set.

h) *Hyperspace*—a space of more than three dimensions.

i) *Linear Algebra*—a type of algebra that is concerned with structures that can be described by systems of equations containing only the first powers of any variables.

j) *Real Analysis*—the study of mathematical notions such as continuity, differentiation, integration and sequences of functions, and sequences and their limits; the focus is on the real numbers, including positive and negative infinity, to delineate the extended real line.

k) *Probability Theory*—the branch of mathematics that deals with quantities having random distributions.

l) *Voting Systems*—a mathematical procedure employed to determine who wins and who loses.

m) *Game Theory*—the branch of mathematics concerned with the evaluation of strategies for dealing with competitive situations where the outcome of a participant's choice of action depends critically on the actions of other participants.

n) *Linear Programming*—a mathematical technique for maximizing or minimizing a linear function of several variables, such as output or cost.

o) *Social Choice*—a theoretical procedure used to analyze the combination of individual interests, opinions, preferences, or welfares to reach a collective decision.

p) *Shape and form*—the outline of an area or figure and the visible shape or configuration of something.

q) *Normality* (normal curve)—bell-shaped distribution of data used for statistical analysis.

For detailed descriptions, examples and exercises on the aforementioned concepts, the interested reader can consult the works of Saul Stahl and Paul E. Johnson (2007), Peter Tannenbaum (2014), and Ian Stewart (1995), to name but a few.

Mathematical Applications in the Business, Social and Life Sciences

To assert that there has been an exponential growth in the utility of mathematical ideas in the business, social and life sciences is hardly a matter of dispute. This can be seen in the numerous articles and books that have been published in these fields employing mathematical ideas. To cover all of this material here is impossible and unnecessary. What I present in this section, instead, are brief synopses of a sample of these ideas from the works of Will H. Moore and David A. Siegel (2013), Peter Tannenbaum (2014), and Bill Armstrong and Don Davis (2014).

To begin with, Moore and Siegel in their exploration of mathematical ideas for political and social research examine five topics. The first topic deals with the *building blocks*, which include (a) preliminaries—key concepts such as variables and constants, sets, operators, relations, level of measurement, notation, and proofs; (b) algebra review, encompassing basic properties of arithmetic and algebra, and computational aids; (c) functions, relations and utility, focusing on examples of functions of one variable, preference relations and utility functions; and (d) limits and continuity, sequences and series, and more on sets, with emphases on open, closed, compact and convex sets, and continuous functions (Moore and Siegel, 2013).

The second topic covered by Moore and Siegel is *calculus in one dimension*. For this, the following mathematical ideas are explored: (a) introduction to calculus and the derivative, with a brief introduction to calculus, definition of the derivative, and a formal discussion on derivatives; (b) the basic rules of differentiations, entailing discussions of the rules and when to apply them, and derivatives of functions; (c) the integral, highlighting the definite integral as a limit of sums, indefinite integrals and the fundamental theorem of calculus, computing integrals, and the rules of integration; and (d) extrema in one dimension, with a particular attention on higher-order derivatives, concavity and convexity, and finding extrema (Moore and Siegel, 2013).

Moore and Siegel's third topic is *probability*. Their investigation here centers on (a) an introduction to probability, which includes discussions of the basic probability theory, computing probabilities, and some specific measures of probabilities; (b) an introduction to discrete distributions, encompassing the distribution of a single concept or variables, sample distributions, empirical joint and marginal distributions, the probability mass function, the cumulative distribution function, probability distributions and statistical modeling, and expectations of random variables; and (c) continuous distributions, looking at continuous random variables, expectations of continuous random variables, and important continuous distributions for statistical modeling (Moore and Spiegel, 2013).

The fourth topic interrogated by Moore and Siegel deals with *linear algebra*. Studied here are (a) vectors and matrices, with reviews on scalars, properties of vectors and matrices, and matrix illustration of ordinary least squares estimation; (b) vector spaces and systems of equations, focusing on what vector spaces are and their applications, and solving systems of equations; and (c) eigenvalues and Markov chains, emphasizing eigenvectors and matrix decomposition, and stochastic processes (Moore and Spiegel, 2013).

The fifth and final topic Moore and Spiegel treat entails *multivariate calculus and optimization*. Examined in relation to this subject are (a) the fundamentals of calculus, with attention paid to functions of several variables, calculus in several dimensions, and concavity and convexity redux; (b) multivariate operations, focusing on unconstrained optimization, equality and inequality constraints; and (c) comparative statics and implicit differentiation, dealing with properties of maximum and minimum, and implicit differentiation (Moore and Spiegel, 2013).

Next, Tannenbaum in his expeditions in modern mathematics also investigates five topics. His first topic is about *social choice*. Here, he deliberates on (a) the mathematics of elections by examining aspects of the paradoxes of democracy—basic elements of an election, plurality method, Borda count method, plurality-with-elimination method, method of pervasive comparisons, and fairness criteria and Arrow's impossibility theorem; (b) the mathematics of power by focusing on characteristic of weighted voting—Banzhaf power, Shapley-Shubik power, and subsets and permutations; (c) the mathematics of sharing by investigating fair-division games—divider-chooser method, lone-divider method, lone-chooser method, method of sealed bids, and method of markers; and (d) the mathematics of apportionments by analyzing making the rounds—apportionment problems and methods, Hamilton's method, Jefferson's method, Adam's and Webster's methods, Huntington-Hill method, and quota rules and apportionment paradoxes (Tannenbaum, 2014).

Tannenbaum's second topic deals with *management science*. On this subject, he investigates (a) the mathematics of getting around by exploring Euler paths and circuits—street-routing problems, introduction to graphs, Euler's theorem and Fleury's algorithm, and Eulerizing and semi-Eulerizing graphs; (b) the mathematics of touring dealing with the traveling salesman problems—Hamilton paths and circuits, brute-force algorithm, nearest-neighbor and repetitive nearest-neighbor algorithms, and cheapest-link algorithm; (c) the mathematics of networks pertaining to the cost of being connected—networks and trees, minimum and maximum spanning trees, and Kruskal's algorithm; and (d) the mathematics of scheduling related to chasing the critical path—introduction to scheduling, directed graphs, priority-list scheduling, decreasing-time algorithm, critical paths and the critical-path algorithm (Tannenbaum, 2014).

The third topic examined by Tannenbaum concerns *growth*. The foci on this subject include (a) population growth models vis-à-vis the notion that there is strength in numbers, thereby presenting population sequences, linear growth model, exponential growth model, and logistic growth model; and (b) financial mathematics—percentages, simple interest, compound interest, and consumer debt (Tannenbaum, 2014).

Tannenbaum's fourth topic has to do with *shape and form*. For this subject, he broaches (a) the mathematics of symmetry by looking at attributes that are beyond reflection—rigid motions, rotations, translations, glide reflections, symmetry types, and patterns; (b) fractal geometry, focusing on techniques used to examine the kinky nature of nature—Koch snowflakes and self-similarity, Sierpinsky gasket and the chaos games, twisted Sierpinski gasket, and the Mandelbrot set; and (c) Fibonacci numbers and the golden ratio with a focus on rabbits and gnomons, thereby delineating Fibonacci numbers, the golden rule, gnomons, and spiral growth in nature (Tannenbaum, 2014).

The fifth and final topic analyzed by Tannenbaum relates to *statistics*. On this subject, he interrogates (a) census, surveys, polls, and studies by examining aspects of the joys of collecting data—enumeration, measurement, and cause and effect; (b) graphs, charts, and numbers as features that data show and tell—the different types of graphs and charts, means, medians, percentiles, ranges and standard deviations; (c) probabilities, odds, and expectations in terms of the mathematical tools utilized for measuring uncertainty and risk—simple spaces and events, multiplication rule, permutations, combinations, probabilities and odds, expectations, and measuring risk; and (d) the mathematics of normality dealing with aspects of the bell curve—approximately normal data sets, normal curves and normal distributions, modeling approximately normal distributions, and normality in random events (Tannenbaum, 2014).

In addition, Armstrong and Davis in their work on the use of calculus in the business, social, and life sciences examine eight topics. The first topic deals with *functions, modeling, and average rate of change*. On this subject, the two scholars discuss (a) the coordinate system and function, with a focus on plotting points, function notation and evaluating functions; (b) linear functions and average of change, encompassing a review of linear functions, equations of lines, intercepts, marginal cost, piecewise-defined functions, and absolute value function; (c) quadratic functions and the secant line slope, comprising an examination of properties of quadratic functions, determining zeros of quadratic functions, and secant line slope and the difference quotient; (d) operations on functions and composite functions, entailing treatments of functions of business, polynomial functions, and end behavior of functions; and (e) rational, radical, and power functions, with an emphasis on end behavior of rational functions (Armstrong and Davis, 2014).

The second topic Armstrong and Davis analyze pertains to *limits and the derivative.* Here, the authors deal with (a) limits and continuity, with an examination of determining limits numerically and graphically, left-hand and right-hand limits, algebraic determining limits, continuity, and continuity for specific functions; (b) limits and asymptotes by considering infinite limits and vertical asymptotes, and limits at infinity and horizontal asymptotes; (c) approximating instantaneous rates of change, alluding to thinking about rate of change, thinking about approximating and applying instantaneous rates of change; and (d) the derivative of a function regarding difference quotient and the derivative, derivatives in general, and non-differentiable functions (Armstrong and Davis, 2014).

Armstrong and Davis' third topic is in regards to *differentiation techniques, the differential and marginal analysis.* Investigated on this subject are the following mathematical ideas: (a) derivatives of constants, powers, and sums, with attention paid to derivative of a constant function, power rule, constant multiple rule, and sum and difference rule; (b) derivatives of products and quotients, focusing on product rule and quotient rule; (c) the chain rule; (d) the differential and linear approximations; and (e) marginal analysis, encompassing marginal business functions and average business functions (Armstrong and Davis, 2014).

The fourth topic covered by Armstrong and Davis concerns *exponential and logarithmic functions.* On this topic, the coauthors analyze (a) exponential functions, with lenses on general exponential functions, compound interest, the number e (the Euler number often referred to as the natural exponential function), and logistic curves; (b) logarithmic functions, with an emphasis on inverse functions; (c) derivatives of exponential functions with base e and of general exponential functions; and (d) derivatives of natural logarithmic functions and of general logarithmic functions (Armstrong and Davis, 2014).

Armstrong and Davis' fifth topic is about *applications of the derivative,* with the discussion focused on the following mathematical notions: (a) first derivatives and graphs, entailing discourses on intervals of increase and decrease for functions, relative extrema and critical values, and the first derivative test; (b) optimizing functions on a closed interval, with a focus on absolute extrema on a closed interval; (c) second derivatives and graphs dealing with higher-order derivatives, concavity, inflection points, and rates of change; (d) the second derivative test and optimization, with considerations on absolute extrema on open intervals, sustainable harvest, and inventory cost; (e) graphical analysis and curve sketching; (f) elasticity of demand, with explorations on arch elasticity and point elasticity; and (g) implicit differentiation and related rates (Armstrong and Davis, 2014).

The sixth topic perused by Armstrong and Davis relates to *integration*. The dissection of this topic includes (a) the indefinite integral, with foci on recovering functions from rate functions and recovering business functions; (b) area, the definite integral, and the fundamental theorem of calculus encompassing scrutiny of finding the area under a curve numerically, the left sum method, the right sum method, the definite integral, and the fundamental theorem of calculus; (c) integral calculus and total accumulation, with an emphasis on the integral as a continuous sum; (d) integration by u-substitution (also known as "integration by substitution" or the "reverse chain rule") with indefinite and definite integrals; and (e) integrals that yield logarithmic and exponential functions, paying particular attention to integrating general exponential functions (Armstrong and Davis, 2014).

In reference to their seventh topic, *applications of integral calculus*, Armstrong and Davis review these mathematical ideas: (a) average value of a function and the definite integral in finance, looking at the average value of a function, compound interest, continuous stream of income, and annuities; (b) area between curves and applications vis-à-vis a closed interval *[a, b]*, cross on *[a, b]*, and area bounded by two curves; (c) economic applications of an area between two curves, with analyses of price-demand function, consumers' surplus, supply function, equilibrium point, and Lorenz curve and Gini index; (d) integration by parts with applications on present value of a continuous income stream; (e) numerical integration, with explorations on the trapezoidal rule, discrete data and Simpson's rule; (f) improper integrals entailing a study of limits at infinity, capital value, endowments and perpetuities, and probability density functions; (g) differential equations pertaining to separation of values; and (h) differential equations dealing with growth and decay, with treatments of unlimited growths, exponential decay, limited growth models, and logistic growth model (Armstrong and Davis, 2014).

Armstrong and Davis' eighth and final topic is apropos *calculus of several variables*. On this subject, they investigate (a) functions of several variables by focusing on two or more independent variables, Cartesian coordinates in 3-space, graphing in 3-space, and Cobb-Douglas production function; (b) level curves, contour maps, and cross-sectional analysis, with foci on horizontal cross sections, isoquants, indifference curves, and vertical cross sections; (c) partial derivatives and second–order partial derivatives, with special attention paid to marginal productivity of labor and capital; (d) maxima and minima by analyzing relative extrema, critical points, and the second derivative test; (e) Lagrange multipliers by discussing the method, geometric analysis, and marginal productivity of money; and (f) double integrals, with delineations of partial anti-differentiation and iterated integrals, double integrals, average value over rectangular regions, and volume (Armstrong and Davis, 2014).

Objects Linguistics and Mathematics Study in Common

With the preceding backdrops of what linguistics and mathematics are and their objects of study, the issue probed here then is about those objects the two fields study in common. The discussion should show that the two disciplines are parallel and must interact where they happen to coincide, especially since there is even a sub-discipline in linguistics called mathematical and computational linguistics.

In my article, "Domesticating Mathematics in the African Mother Tongues" (2014), I explore the subject examined in this section: i.e. the objects Linguistics and Mathematics study in common. I begin by stating that "the most common answer one usually finds in many textbooks and articles is that mathematics is a universal language among the numerous different languages spoken all around the world and that these languages are constantly changing just as mathematics; thus, it is only natural that some linguists use the tools of their discipline to study mathematics. Scholars specializing in mathematical linguistics, however, go further than that. For example, Geoffrey Pullum and András Kornai (2003) provide a definition of mathematical linguistics and then survey the following traditional subfields of linguistics and how they study and employ mathematics: phonetics, phonology and morphology, syntax, semantics, and model-theoretic syntax" (Bangura, 2014:16).

Next, I mention that "Pullum and Kornai define mathematical linguistics as the reverie of mathematical structure and methods that are relevant to linguistics. They note that the agency of the empirical subject matter of mathematical linguistics is relatively indirect as it is another subarea of applied mathematics, since theorems are proved not for their applicability but for their mathematical value. Nonetheless, they add, the internal subdivisions of mathematical linguistics are still best guided by the internal organization of linguistics (Pullum and Kornai, 2003:17)" (Bangura, 2014:16).

In addition, I proffer that "for phonetics, Pullum and Kornai state that Hidden Markov Models (HMMs) are the key structures of both phonetics and mathematics. They note that the way the structure of HMMs is set up as a 'discrete, psychologically relevant underlying units as hidden states coupled with conditions, physically relevant output' makes the models important to the study of phonetics and mathematics (Pullum and Kornai, 2003:17). They add that if a modality is changed from spoken to written language, HMMs will be applicable because the models assume a Markovian underlying structure which is ideal for modeling the succession of linguistic units (Pullum and Karnai, 2003:17)" (Bangura, 2014:16).

Also, I explain that "as it pertains to phonology and morphology, Pullum and Kornai point out that Bloomfieldian postulates provide the fundamental apparatuses of mathematical linguistics, particularly the notion of hierarchical structures made up of relatively stable recurrent items developed principally on the basis of phonological and morphological attributes. They mention that the three Chomskian theoretical models—finite state automata (FSA), context-free grammars (CFGs), and context-sensitive grammars (CSGs)—and/or the even more powerful unrestricted rewriting systems (URSs) are critical for describing linguistic structure and emerged as the basis for work on formal language structure theory within computer science. They note that finite state tranducers (FSTs), which are very useful for mathematical work on phonology, are employed to describe the definitive formalization of theoretical phonology and morphology that was proposed by Noam Chomsky and Morris Halle (1968). Pullum and Kornai add that given the near-phonetic nature of Russian orthography, HMMs contributed to phonology and statistical linguistics. They conclude that 'today the mathematical apparatus of phonology and morphology is centered on the study of deterministic, nondeterministic, and probabilistic FSTs' (Pullum and Kornai, 2003:17-18)" (Bangura, 2014:16).

Thereafter, I note that "in terms of syntax, Pullum and Kornai state that Noam Chomsky's formalization of immediate constituent analysis utilizing context-free grammars (CFGs) was his first significant technical contribution to the field of linguistics. They note that to this day, CFGs maintain a key position in the design of programming languages, even though some of the widely-used programming languages transcend CFGs in some ways. They add that compiler designs emerged as the home of much of the early work in mathematical linguistics concerned with developing efficient parsing techniques (Pullum and Karnai, 2003:18)" (Bangura, 2014:17).

Furthermore, I delineate that "in the case of semantics, Pullum and Kornai inform us that there was insistence during early efforts dealing with the meaning of expressions in linguistic semantics to provide translations of those expressions in a representational system of some sorts within generative grammar. They point out that while philosophers objected that no such representation in any vocabulary can yield a specification of meaning, the work of Richard Montague (1974) led to providing natural language expressions with actual model-theoretic interpretations just as is done with formal languages in logic. Nonetheless, Pullum and Kornai, argue, 'On the whole, approaches to semantics based on information theory are still largely restricted to lexical semantics, though many tasks such as machine translation that were originally believed to require sophisticated semantic analysis are now often performed by purely statistical models' (Pullum and Karnai, 2003:18-19)" (Bangura, 2014:17).

Finally, I express the fact that "relating to model-theoretic syntax, Pullum and Kornai state that logical theory with well-formed structures in the language as its models have emerged as the recent line of research that connects model theory to syntax. They point out that CFG is used to devise monadic second-order logic that characterizes the sets of trees that can be generated. They also mention that phrase structure rules are employed to make directly the set of trees CFGs make, and that modal logic is utilized to extend these notions to structures that are more complex than trees (Pullum and Kormai, 2003:18-19)" (Bangura, 2014:17).

Conclusion

The Ancient Kemetian/Egyptian tongue has several denotations for the word *language*. One denotation is *tchet-t* ("language"), which also means "word," "speech," "sayings," "proverbs," "aphorisms," "Sayings of the Fathers," and "Sûdânî language." Another denotation is *tepi hesb met-t nefer-t*, meaning "the finest language imaginable," in reference to the Ancient Kemetian language. The other denotation is *metura en Kamt*, which means "word of the mouth of Kemet/Egypt," or *re en Kam*, meaning "speech of Egypt," for short (for these translations, see Budge, 1978). I must mention here that "Sayings of the Fathers" refer to the teachings of the great desert monks (the "elders") whose primary responsibility was to train disciples living in the semianchoretic communities of Egypt during the 4th and 5th Centuries AD. Given the discrepancy in terms of time, I am skeptical about E. A. Wallis Budge ascribing this aspect to an Ancient Kemetian meaning of language.

Also, as we are informed by Moustafa Gadalla, from an Ancient Kemetian linguistics perspective, the Eurocentric description of the Kemetian Hieroglyphics as a primitive form of writing with individual sound values is a misrepresentation. This is because the Kemetian Hieroglyphics are ideograms that are distinct from the Kemetian alphabetic system. As Gadalla also points out, the pictorial writings constitute a collection of images representing conceptual denotations and not independent sound values and, thus, pictures convey information more efficiently compared to letters or words. Add to this the verity that the Kemetian Hieroglyph is a metaphysical language of images or pictures, which makes it a language of the divine, the intellect, and the mind (Gadalla, 2016:vii-2).

For Mathematics, the Ancient Kemetians employed the word *s-qet*, which means "mathematical term" and also "relationship," "disposition," and "inclination" (for this translation, see Budge, 1978). I have already demonstrated in the introductory chapter and in several others that the Eurocentric myth of Ancient Kemetian Mathematics being only practical arithmetic and simple geometry is pernicious and that, in fact, Kemet taught many of Ancient Greece's mathematicians and other scholars who were the precursors of later Western mathematicians and other thinkers. Thus, there is no need to belabor the point here. Kim R. W. Zahrt, after analyzing a significant number of Kemetian mathematical documents, makes the point quite succinctly when he states the following: "the Egyptians solved the truncated pyramid problem which the Greeks did not. This problem could be a clue that the Egyptians had the ability to assess universal formulae" (Zahrt, 2000:93).

Chapter 3

Mode of Communication

Sending and receiving messages define the very nature of a system of communication. The means by which these messages are transmitted and received is referred to as *Mode of Communication.* Users of human languages employ voice (or oral language) and gesture (or body language) to transmit messages. These two modes of communication are imperative for the transmittal of the complex sorts of messages that languages require (Bergmann et al., 2007:17). Before going any further, it would be helpful to provide a definition of the concept *communication* from a linguistic point of view. The following denotation by David Crystal is quite instructive:

> Communication [is] the transmission and reception of information between a signaller and a receiver. Various steps in this process can be recognized. A message is formulated in the signaller's brain, and is then encoded in the nervous and muscular systems. It leaves the signaller (typically, via the vocal tract or hands), and is transmitted through air, paper, electrical system, or other medium to the brain of the receiver (typically, via the eye or ear), where is it decoded. The receiver may influence the nature of the message at any time by sending feedback to the signaller. In principle, any of the five [traditional] senses can be involved, but humans tend to use only the auditory/vocal, visual, and tactile modes for active communication (the other two modes, smell and taste, being widely employed among certain animal species) (Crystal, 1992:72-73).

In order to investigate the topic of language as a mode of communication, this chapter is divided into three sections. The first section is composed of linguistic expositions on the subject. The second section entails African linguistic evidence. And, the third and final section comprises African mathematical evidence. In the end, conclusion is drawn.

Linguistic Expositions on Mode of Communication

At least the following ten themes can be recognized in linguistics works that have discussed language as a mode of communication: (1) system of communication, (2) communication theory, (3) communicative grammar, (4) types of communication, (5) communicative effect taxonomy, (6) indices of communication and communicativity, (7) communicative act and strategies, (8) communicative function, (9) communicative power, and (10) language development. For brevity and clarity, these themes are each examined briefly and in sequential order in the ensuing subsections.

System of Communication

"To say that language serves as an instrument of communication," asserts John Lyons, "is to utter a truism" (Lyons, 1977:32). He proffers that there is hardly any definition of language that does not include the notion of communication in it. He mentions that for semanticists, meaning and communication go hand in hand. He ponders what communication is and states that in their pre-theoretical sense the terms "communicate" and "communication" are employed in numerous everyday domains. He states that we discuss the communication of factual information just as we do about attitudes, feelings, and moods. He argues that there is an interconnection among the various senses of these terms, leading some linguists to suggest various definitions for and subsume them under a general theoretical concept denoted vis-à-vis social interaction or a stimulus to which an organism responds. He then proposes the concept "signalling-system" as a narrower interpretation for the terms "communicate" and "communication" to distinguish them from their everyday employment. He further calls for the restrictive use of the terms to the "intentional transmission of factual, or propositional, information," and asserts that the major "signalling-systems employed by human beings for the transmission of information, although not the only ones, are languages" (Lyons, 1977:32).

Also, Paul A. Gaeng elucidates a couple of aspects pertaining to language as a system of communication. First, he states that broadly speaking, through gestures, signals, vocal sounds, written symbols, etc., language is a means of communicating meaning from one human mind to another. Speech, he points out, is obviously the most common and widely used language activity. It is not just the capacity to utter sounds, but humans' mental ability to connect sounds with meanings that are acceptable to one another so that there is a real transfer of thought from one human mind to the other and produce mutual comprehension that separates humans from other animals (Gaeng, 1971:2). Gaeng provides the following example to buttress his point:

> As a result of patient training, a parakeet may succeed in producing a sentence like "birds can't talk" with proper pitch and intonation. However, he will never be able to say "birds can't write" or "dongs can't talk" (unless we specifically teach him to produce these sentences) because animals are incapable of carrying on language activity, or the

> kind of word substitution that would allow complex sequences of "organized noises" that add up to definite meanings. The humanlike chatter of a parakeet is *imitation* of language, rather than *use*, because it is spontaneous and it is not, above all, motivated by organized thought (Gaeng, 1971:2).

Second, Gaeng mentions that human language is fundamentally a spoken method of communication. Highlighting the imperative of the speech organ "tongue" is the Latin word *lingua*, which is translated into English as "tongue," from which the word language is derived. Thus, the word "tonguing" and the phrase "wagging of the tongue" denote language. Additional proof for the essentiality of the role language plays in speech production is the use of the word *tongue* for language (Gaeng, 1971:12).

Communication Theory

For some linguists, the idea of language as a mode of communication is best treated theoretically and, thus, it is a scientific notion. Consequently, *communication science*, according to Crystal, involves the scientific investigation of every feature of communication. Also called "communication sciences," this area of study includes linguistics and phonetics, with their various subfields such as psycholinguistics and sociolinguistics and the pertinent utilizations of related subjects such as acoustics, anatomy, and neurology. Also involved in this area of scientific inquiry are all communication modes: signed, spoken, and written (Crystal, 1992:74).

Correspondingly, *communication theory*, following Lyons, is projected in the wider nomenclature of *communications-engineering* and transcends the electrical, electronic, or mechanical system of signal-transmission. Therefore, also included in the linguistic theoretical conceptualization of communications-engineering are the various language aspects or functions and the various types of information that are relayed in language-utterances (Lyons, 1977:36). Lyons goes on to describe the communication model, whose diagrammatical representation and description appear in numerous books and articles dealing with the topic, as follows:

> X sends some information to Y. But 'send' is ambiguous...as between the origination of what in communications-engineering is generally referred to as the message and the actual transmission; and we must distinguish, in principle, between the source of the message and the transmitter. But we can continue to use the word 'sender' when, as is commonly the case, the same machine or organism is both source and transmitter, or incorporates both mechanisms within it, or when it is

> not important to distinguish between origination and transmission. The message originated by X is encoded by the transmitter into a signal. The signal is sent over a particular communication channel to the receiver. The receiver decodes the signal into a message and passes the message on to the 'receiver' and 'destination' as 'sender' subsumes 'source' and 'transmitter'...the signal transmitted may differ from the signal that is received due to distortions introduced by noise in the channel. This may, but does not necessarily, lead to a failure in communication (Lyons, 1977:36-37).

Lyons also points out that in addition to the preceding X-Y, or one-to-one, communication model, the following two models also exist: (1) one-to-many communication, whereby X sends the same message simultaneously to Y, Y', Y", etc.; and (2) many-to-one communication, whereby X, X', X", etc. send the same message simultaneously to Y (Lyons, 1977:38). I would add here the X-only variety whereby someone communicates with himself/herself, which has been labeled intrapersonal communication. An example is a person using self-talk to calm himself/herself during a stressful event.

In an earlier work, Lyons discusses the notions of "information content" and "probability of occurrence" in terms of their connection to communication theory. According to him, the nexus pertains to the amount and frequency of *information* (in the specialized sense of communication theory) contained in a linguistic unit within a particular context. He defines the *information-content* of a particular unit as the function of that unit's *probability*, which is the frequency in the following manner, if, for example, two or more units are "equiprobable" in a particular context, with each having the same information-content in that context (Lyons, 1971:84):

> If two, and only two, equiprobable units may occur, x and y, each of them will occur (on average) in just half the instances of the context in question: each has a probability, *a priori*, of ½. The probability of a particular unit x is denoted by p_x. Thus, in the present instance, $p_x = ½$ and $p_y = ½$. More generally, each of n equiprobable units ($x_1, x_2, x_3 \ldots x_n$) has a probability of $\frac{1}{n}$. (It will be observed that the sum of all the probabilities in the set of units is 1. A special case of probability is 'certainty'. Units which cannot but occur in a given context have a probability of 1.) If the units are equiprobable, each of them carries the same amount of information (Lyons, 1971:84).

It is imperative to also mention here what Lyons states about the case of unequal probabilities, which are more typical of language. As he explains by way of an example, where two, and only two, units, x and y, occur, and x occurs twice as much, on average, than y, $p_x = \frac{2}{3}$ and $p_y = \frac{1}{3}$. In essence, the information-content of x is half of that of y. In other words, information-content is *inversely proportionate* to probability, which is the basic postulate of information theory (Lyons, 1971:84).

Communicative Grammar

To begin with, according to Geoffrey N. Leech, *communicative grammar* involves the use of the three levels of description to analyze any grammatical category, whether it is interrogation, modality, or negation. The first level is the *syntactic* (sign vs sign: the arrangement of words and phrases to create well-formed sentences in a language), which can be used, for instance, to generate clauses or sentences in English. The second level is the *semantic* (sign vs concept: relating to meaning in language or logic), which focuses on sense instead of force, as, for instance, determining the negative operator in reference to propositional logic. The third level is the *pragmatic* (sign vs context: language in use and the contexts in which it is used), which focuses on negative "uninformativeness" (Leech, 1983:152).

Next, Leech raises the question about how to conduct an analysis in terms of communicative grammar, whether from the syntactic to the semantic and then to the pragmatic level, or vice versa. His answer is that since syntax is clear-cut and relatively well known to us, it is only natural we begin from it and then move to semantics and pragmatics—which is relatively context-variable and the least clear to us. This approach implies that an analyst is taking the position of the addressee, who discerns "the force of an utterance by decoding its sense and then, by heuristic problem-solving, works out its force, or pragmatic interpretation" (Leech, 1983:152-153).

In addition, from William Labov, we learn that the essentiality in demonstrating how grammatical functions can directly alter a sound change for communicative necessities hinges upon the "compensating differentiation." This notion, according to him, suggests that language learners will gain enough information to maintain the fundamental distinctions and the underlying forms of the language when the variable rule is well ordered (Labov, 1984:224-225).

Also, from Kenji Hakuta's work on bilingualism, we glean a number of aspects apropos to a communicative approach to language development and communicative grammatical structure. First, he argues that the metamorphosis in the study of language as habits from interest in grammar to meaning and then to communication is more a change in approach, not in scientific progress. The change from grammar to meaning, for instance, was not precipitated due to the failure of grammar as a theory of language acquisition; instead, it was due to researchers' desire to chart a novel course on the theory (Hakuta, 1986:113).

Second, Hakuta propounds the idea that the communicative approach is the most percipient method for analyzing the earliest stages of language development. This is due to the fact that it is at this stage that the combining of words begin, thereby facilitating the description of child language in terms of meaning. The later stages of language development are absorbing due to the advanced forms of grammar that ensue. Nonetheless, Hakuta cautions us not to be "fooled, however, by the developmental order in which these characterizations of language emerge, from communication to meaning to grammar," since "it would be rash to conclude that there is a progression, with grammar developing out of meaning and communication" (Hakuta, 1986:113). He adds that "Grammar, meaning, and communication are distinct levels of language and cannot be interchanged. Each has its own course of development, and it is still unclear how they are related to each other" (Hakuta, 1986:113).

Third, according to Hakuta, both first and second language acquisitions are inspired by prefabricated language due to its communicative potential. He argues, however, that such an explanation is equivalent to stating that hunger inspires cooking. Thus, for him, more needs to be explained because just as hunger cannot explain the quintessence of the art of cooking and consumption of what is cooked, so can communication not explain all of the features of grammatical structure; communication can only furnish the common raison d'être of language (Hakuta, 1986:128).

Fourth, as Hakuta points out, some linguists argue persuasively that grammatical structures can be understood by analyzing communication. He quotes a general form of this proposition offered by Evelyn Hatch as follows: "One learns how to do conversation, one learns how to interact verbally, and out of this interaction syntactic structures are developed" (Hatch quoted by Hakuta, 1986:128-129). Hakuta adds that the cardinal position of these scholars is that linguistic structures could be delineated in the give-and-take context of conversation. This perspective deviates from the notion that second language acquisition is inspired by the social interactional nature of language. Instead, it is a point of view that suggests a sort of conventional correspondence or clearly identifiable nexus between linguistic structure and social interaction. Hakuta, nonetheless, is not convinced that it is easy to generate specific rules of grammar from conversational analysis such as in the case of number agreement between subject and verb in English (Hakuta, 1986:129)—e.g., *Malcolm X lives on; Thomas Sankara and Patrice Lumumba live on.*

Fifth, Hakuta opines that essentially, an analyst's definition of language will dictate his/her interest in language learners' elaborate employment of prefabrications. This is because, as he repeats his position, stated earlier, the analysis of language acquisition can be conducted at the levels of communication, grammar, and meaning, albeit they should not be treated as clear-cut analytical levels. He therefore also iterates his insistence that these levels should not be mixed since they have not been shown to be analytically analogous (Hakuta, 1986:129).

Finally, Hakuta suggests that prefabricated utterances would not be exceptionally appealing to a linguist with a penchant for analyzing the capability of obtaining fully-fledged linguistic structure. In order to analyze linguistic structure, this linguist would want to ascertain what a learner must know. Thus, of interest to this linguist is *how* a learner engages in the analysis of prefabrications. Hence, according to Hakuta, from the grammatical point of view, prefabrications simply provide some effortlessly accessible raw data for analysis and, consequently, might be only minimally enthralling (Hakuta, 1986:129).

Types of Communication

Communication has been categorized as either vocal or non-vocal (verbal/non-verbal or oral/non-oral). While the former refers to spoken and written language or the use of words, the latter refers to facial expressions, the tone and pitch of the voice, gestures displayed through body language (kinesics) and the physical distance between the communicators (proxemics).

According to Gaeng, while cats have been meowing and sheep have been bleating since time immemorial, only humans have used vocal symbols to develop a complex language communication system. He also expresses the fact that the roots of language as a medium of communication go far back to the prehistory of the human race when homo sapiens employed gestures, crude markings, and uncertain vocal cries to communicate (Gaeng, 1971:3). He then briefly reviews the following theories that have been suggested for the origins of this linguistic development:

a) Richard Paget's theory—it suggests that "speech was preceded by non-vocal communication...human speech originated in verbal imitations of such movements as gestures" (Gaeng, 1971:3).

b) *Bow-Wow Theory*— it postulates that "the first words ever uttered by man were imitative of natural sounds, as when a child calls a cow a *moo* or a locomotive a *choo-choo*...in accordance with view, primitive man copied the barking of dogs and thereby obtained a natural word with the meaning 'dog'; hence, the name Bow-Wow Theory" (Gaeng, 1971:4).

c) *Pooh-Pooh Theory*—"the theory...makes interjection the original speech...it claims that language is derived from instinctive emotional cries called forth by intense sensations and feelings, for example, utterances like *ouch* for mild pain and *oh* for surprise" (Gaeng, 1971:4).

d) *Ding-Dong Theory*—it "holds that there is a mystical harmony between sound and sense in language...the thought that language must have arisen by necessity from laws of nature and that there existed a mystical correspondence between the names of things and the objects they stood for" (Gaeng, 1971:4-5).

e) *YO-HE-YO Theory*—it "seeks the source of speech in the rhythmic chants of a group of men engaged in a common physical task, such as lifting a rock or dragging a heavy log...such noises would then develop into words associated with communal labor something like *heave* or *haul*" (Gaeng, 1971:5).

f) D. S. Diamond's theory—"it attempts to find the beginnings of human speech in commands involuntarily uttered in the course of violent or strenuous efforts of the arm, at first, were calls for assistance accompanied by gestures mimicking the desired action, such as cutting, breaking, crushing, or killing...inperatives like cut, break, crush, strike are most likely to have been the first authentic words, and certainly indicate the commonest forms of maximum arm effort" (Gaeng, 1971:5).

g) Otto Jespersen's theory—"sought the genesis of language not in the prosaic but in the poetic side of life, its source not in 'gloomy seriousness' but 'merry play and youthful hilarity'...the earliest language was expressive rather than communicative, poetic and emotional rather than matter-of-fact and practical" (Gaeng, 1971:5-6).

h) G. Révész's CONTACT Theory—it "argues that language arose out of man's instinctive need for contact with his companions...starting with the contact sound to identify

> himself with other members of his species...the next step is the *cry*, the first attempt at communication...next comes the *call*, a deliberate message addressed to a particular individual...and finally, the *word*, a specialized development of the call, which has symbolic function and hence is proper to man only" (Gaeng, 1971:6).

The general limitation of all of these theories, as Gaeng argues, is that they do not explicate how these basic speech forms evolved into expressed and wholly functioning language (Gaeng, 1971:6).

While the distinction between vocal and non-vocal (verbal/non-verbal or oral/non-oral) is necessary for theoretical discussion, it is the case, however, that both types of communication can and often do go hand in hand. Donn Byrne makes this same point when he broaches the notion of oral communication being a two-way activity between a speaker with a *productive* speaking skill and a listener (or listeners) with the *receptive* skill of listening and comprehending a message. He indicates that the speaker and the listener must perform a positive function, as the speaker must *encode* the message to be communicated in an appropriate language while the listener (or listeners) must *decode*, or interpret, that message. He adds that even though the message itself comprises a significant amount of "information," the listener gains additional comprehension from prosodic attributes such as body and facial movements, intonation, and stress, all of which escort the spoken utterance and comprise a component of its meaning (Byrne, 1976:8).

Communicative Effect Taxonomy

To start with, William O'Grady, Michael Dobrovolsky and Mark Aronoff identify several communicative effect taxonomies based on native English speakers' assessments of the comprehensibility of sentences containing various types of errors. They also suggest that overall, errors involving word order, connectors, and other features of sentence organization appreciably hamper communication. Nonetheless, O'Grady et al. further indicate that there are sentences in which communication is not hindered when any of the aforementioned features, say a verb, is missing. They provide the following two examples: (1) "in the sentence *My husband not here*, communication is not impeded even though the verb (*is*) is missing"; and (2) "*My doctor suggests me to take a vacation*...is understood that the speaker meant 'My doctor suggests that I should take a vacation'" (O'Grady et al., 1989:312).

Next, O'Grady and his colleagues apprise us that the undertaking of identifying learners' errors can be quite intricate and taxing. A native speaker's age, educational level, and extent of communication with foreigners will dictate how s/he reacts to leaners' errors. In addition, whether learners' errors are being heard or read and whether they are presented formally or informally are also factors that will influence how a native speaker of the language perceives those errors (O'Grady et al., 1989:312).

Furthermore, O'Grady and his fellow scholars contend that despite the fact that the development of taxonomies is vital in error analysis, not much knowledge is gained to significantly explicate the strategies a second language learner is utilizing by simply describing an error. Also, even though these scholars welcome the movement toward developing explanations by error analysts, they are, however, not convinced that these analysts have achieved much success in this movement (O'Grady et al., 1989:312).

Indices of Communication and Communicativity

Based on Ralph Fasold's exploration of the subject, the following three indices of communication and communicativity can be espied: (1) Joseph Greenberg's Index of Communication; (2) Stanley Lieberson's extensions of Greenberg's Index; and (3) Eddie C. Y. Kuo's Index of Communicativity, which is an extension of the Greenberg-Lieberson approach. Drawing upon Fasold's work, these indices are discussed sequentially in the rest of this section for the sake of cohesion.

Beginning with Greenberg's Index of Communication, Fasold inculcates that Greenberg's Formula H is a gauge of communication instead of diversity. The formula is computed as follows: "(1) drawing pairs of speakers at random…(2) taking note of those pairs that can communicate because they share at least one language; (3) calculating the probability of drawing each of those 'good' pairs; and (4) summing these possibilities" (Fasold, 1984:128). Fasold then provides the following hypothetical data as an example of how Formula H operates:

> In a given country, there are three languages, M, N, and O. Of the total population, 0.15 are M monolinguals, 0.20 are N monolinguals, 0.05 are O monolinguals, 0.25 are MN bilinguals, 0.30 are NO bilinguals, and 0.05 are MNO trilinguals. There are no MO bilinguals (Fasold, 1984:129).

Next, as it pertains to Lieberson's extensions, Fasold points out that Lieberson makes two observations regarding Greenberg's formulae. The first observation is that Greenberg's formulae should be capable of measuring communication or diversity for any definable population, not just limited to geographic areas such as cities, nations, and states. For Lieberson, Fasold states, geography should not even qualify as a factor for computing communication or diversity. He cites the example of Roman Catholics who are scattered across the globe that any two of them even randomly selected will be able to communicate in a common language. The second observation made by Lieberson is that Greenberg formulae only account for communication or diversity *within* a population, not *between* two spatially defined populations (Fasold, 1984:130).

Thereafter, in regards to Kuo's extension, Fasold explains that his (i.e. Kuo's) Index of Communicativity seeks to compute how a specific *language* can perform as a means of communication in a society. In essence, Kuo's index calculates communication in terms of a particular language whereas the methods of Greenberg and Lieberson calculate communication or diversity. Thus, Kuo provides the following formula for his index:

$$1_{am} = (P_{am})^2$$

where P_{am} is a proportion of a particular population m who can understand a specific language (Fasold, 1984:133). Fasold adds that in order for Kuo to extend his index to compute the value of a language as a medium for *interethnic* communication in the manner that Lieberson extended Greenberg's original formulae, Kuo adjusted his formula as follows:

$$1_{amn} = (P_{am})(P_{an})$$

where P_{am} is the proportion of people in group m who understand a language a, and P_{an} is the proportion of people in group n who understand the same language. Fasold concludes that Kuo's index is a worthwhile tool for computing wider communication function and can be employed to gauge the standing of a language as a full-fledged national language (Fasold, 1984:133).

Communicative Act and Strategies

Taking the action to communicate and developing plans designed to achieve a major or overall aim of that communicative act have received a significant amount of attention from certain linguists. Some of the perspectives that have been offered by these scholars are presented in this section.

For starters, Greenberg writes that a communicative act involves the complete effect of all external functions that include the total effect of conveying information and emotional expressions as well. Thus, according to him, this perspective proffers the notion that the function of language is communication (Greenberg, 1957:84).

Next, Gaeng deliberates on the fact that a communicative act, which requires the participation of at least two people, functions on three levels. One level is the *psychological*, which involves a speaker receiving the concept s/he wants to communicate from the billions of cells in his/her brain, translating the concept into the suitable sound image, comprising the set of sounds utilized to symbolize the concept, the sort of non-vocal sound symbol we utilize when we speak to ourselves without moving the lips or tongue or when we undertake some sort of mental recital. The other level is the *physiological*, which is a response to a nerve impulse from the brain which makes the speech organs to produce the sound sequence for the concept the speaker desires to convey to the hearer. Another level is the *physical*, which is the process whereby sound waves for the concept are sent through the air to reach the ears the hearer. Gaeng adds that "By a physiological process the vibration caused by the sound sequence...are transmitted to [the hearer's] brain, which, in turn, translates the audible sound symbols into the appropriate sound image. Finally, the message is decoded in terms of the psychological association of this sound image with the corresponding concept. Since the speaker and hearer speak the same language, [the speaker's] concept will more or less coincide with that of [the hearer] (Gaeng, 1971:24-26).

In addition, Stephen C. Levinson argues that a communicative act cannot be limited to the standard content of what is spoken, since the implicit, ironic and metaphoric content of an utterance is also communicated. He then queries whether the conventional content of what is spoken incorporates *all* that can be delineated from what is spoken in lieu of all that the speaker and hearer know about the world. He further states that since communication encompasses the ideas of agency and intention, only the inferences that are publicly meant to be relayed can appropriately be declared to have been communicated (Levinson, 1983:15-16). Given this backdrop, Levinson provides the following description of what a communicative act embodies:

> ...communication consists of the 'sender' intending to cause the 'receiver' to think or do something, just by getting the 'receiver' to recognize that the 'sender' is trying to cause that thought or action. So communication is a complex kind of intention that is achieved or satisfied just by being recognized. In the process of communication, the 'sender's' communicative intention becomes *mutual knowledge* to 'sender' (S) and 'receiver' (H), i.e. S knows that H knows that S knows that H knows (and so ad infinitum) that S has this particular intention. Attaining this state of mutual knowledge of a communicative intention is to have successfully communicated (Levinson, 1983:16).

Also, O'Grady et al. postulate that communication needs are served by communicative strategies, just like production strategies do. They also propose that communicative strategies are generally conscious efforts precipitated to meet immediate communication necessities. They give this example: a second language learner "may want to say *The teacher made the child go home*, but he or she does not know the causative form *make* and therefore uses his or her available knowledge to say *She asked the child to go home*. In many cases, the result may be a slight modification of the original intent of the message, but the central idea is conveyed" (O'Grady et al., 1989:313).

Furthermore, as O'Grady and his colleagues note that a number of sub-strategies are embedded in communication strategies. They include avoiding particular linguistic structures by changing the subject being discussed, borrowing from the native language, and word substitution. By utilizing such strategies, second language learners, for instance, are able to get a conversation to continue. The utilization of communicative strategies tends to be more ubiquitous in a natural environment than in a formal one. A well-known example of this is a tourist memorizing a "survival language" to communicate on the spot (O'Grady et al., 1989:313-314).

Communicative Function

It is not farfetched to assert that the topic of communicative function (i.e. reason to communicate) has received the most amount of attention from linguists who have investigated language as a mode of communication. Even the relatively brief review in this section, while not conclusive, lends credence to the assertion.

First, Greenberg asserts that the fundamental function of language is communication—which he says "is meant not only the conveying of information but all those effects on the society and satisfactions of societal needs" (Greenberg, 1957:61). He suggests that this perspective puts language in the complete frame of how the means of communication have evolved. The notion of language evolution, in turn, alludes to the standing of language among other means of communication and whether we can find a line of evolutionary advancement within this larger context (Greenberg, 1957:61).

Second, with a focus on teaching oral English, Byrne discusses two functions of language. The first function deals with meeting the communicative needs of learners. He indicates that the objective of teaching a foreign language has been denoted as helping a learner to conduct himself/herself in such a manner that s/he can engage to a certain degree in particular purposes as a member of another community besides his/her own. Given this objective, he states, focus is placed on the social function of language, which insists that learners must be taught a language in a way that will allow them to play some specific, although variable, roles as utilizers of the language. He adds that the more narrow linguistic definition of language learning puts more emphasis on mastering the language's *system* (Byrne, 1976:4).

The second function of language examined by Byrne concerns the use of dialogue as an instrument of communication to teach the spoken language. On this aspect, he begins by asserting that dialogue is undeniably the best suited approach for teaching a foreign language since the major concern in the preliminary phases is usually to teach the spoken facet of the language. Next, he points out that dialogue constructs a solid connection between language and situation because it presents the language directly in the contexts in which it is commonly employed and allows the learners to practice it in the same manner. In addition, he notes that dialogue easily accommodates many attributes of the spoken language and motivates learners to participate actively in the learning process. Also, he maintains that while the prose passage (descriptive, narrative, etc.) does not have the same advantages as the dialogue, the propose passage can be utilized occasionally to provide variety, albeit its language is usually less relevant to the immediate needs of the learners (Byrne, 1976:21).

Third, while broaching the subject of communicative function, Leech starts with the question about what functional explanation denotes. His response is that functional explanation refers to the occurrence of a particular phenomenon by demonstrating what it contributes to a larger structure of which it is a substructure. He proclaims that as it pertains to language, a functional theory defines it as a type of communication and, concomitantly, the theory is concerned about how language operates within society's larger structure. He posits that functionalism also presumes ends, goals, plans, and purposes. He mentions that we are engaging in functional explanation when we talk about illocutions or meanings in relationship to intentions or vis-à-vis goals. He urges us to employ the concept "function" when we discuss the properties of language, since the term allows us to know how much the achievement of a goal is a result of a person's conscious state or whether the goal is a feature of the individual, the community, or the humans in general (Leech, 1983:48).

Fourth, Gillian Brown and George Yule provide a cogent disquisition on computing communicative function; a discussion of its entirety is beyond the scope of this section. Instead, some salient aspects discussed by these scholars are briefly presented. To start with, Brown and Yule express the fact that anthropologists and ethnographers investigating the language of speech communities have for a long time made the assumption that when speakers produce certain forms of utterances in specific contexts, they convey both propositional and social meanings. They add that research conducted on this issue by sociolinguists has focused on the "social interaction" attribute of language usage, allowing them to inform us about how an utterance can be a social action such as a greeting, promise, warning, etc. (Brown and Yule, 1983:226).

Next, according to Brown and Yule, it has been articulated by some linguists that rules of interpretation exist that make it possible for an analyst to connect what is said to what is done. These social, although not linguistic, rules allow the analyst to determine which conversational sequences are coherent and which are not coherent. Brown and Yule add that it is the relationship between the actions performed with the utterances that determine coherence and incoherence in conversational sequences, not the relationship between the utterances themselves (Brown and Yule, 1983:226).

In addition, as Brown and Yule indicate, the employment of certain linguistics features such as the conjunction *because* can be interpreted only within an "utterance-as-action" analytical framework. This is due to the fact that in addition to being utilized to connect two clauses in a complex sentence, a conjunction like *because* can also be employed to introduce the reason for asking a question and for introducing a certain topic into a conversation. Such a conventional system, Brown and Yule suggest, yields "an account of how some utterances which are apparently unconnected in formal terms (lack cohesion) may be interpreted within a particular genre of spoken interaction, say conversation, as forming a coherent sequence" (Brown and Yule, 1983:227-228).

Furthermore, Brown and Yule argue that *interaction* can be analyzed without considering the language speakers in a conversation utilize. This is because little attention is given to the linguistic attributes of the realizations of sequences of actions in many of the analyses of discourse structure that depend on these actions. Thus, in analyzing discourse structure, the structure itself cannot be stated in grammatical terms, since what is of interest is the level of the function of a specific utterance within a specific social context and at a specific spatial context in a sequence, as a particular input into an evolving discourse (Brown and Yule, 1983:228-229).

Fifth, in his exposition of the place of social variation, as it pertains to the social setting of linguistic change, Labov questions what he characterizes as André Martinet's "strong negative position" that "communication alone shapes language." He first quotes Martinet as follows:

> It is, therefore, the communicative uses of language which must engage our attention if we wish to discover the causation of linguistic changes. What we shall establish and be able to formulate will not necessarily hold good for those linguistic utterances which do not serve the purpose of communication. But we shall be disposed to disregard these since

> they are modelled on communicative utterances and so offer nothing which will not be found in these (Martinet quoted by Labov, 1984:271).

Labov then comments that while "communication" could encompass all types of expressive communication, a consideration of the general context makes is evident that communication is geared toward excluding information in the linguistic form pertaining to the speaker and also excludes "phatic" expression (i.e. an expression denoting or relating to language used for general purposes of social interaction, rather than to convey information or ask questions; utterances such as "hello," "how are you?" and "nice morning, isn't it?" are phatic). He adds that it is a matter of neglecting one aspect and privileging others, since the information not represented is usually conveyed simultaneously with the one considered (Labov, 1984:271).

Sixth and finally, O'Grady and his colleagues explore the function of communicative language teaching methods geared toward producing communicatively competent language learners. According to these scholars, this subject is being touted as the way forward for second language teachers and researchers. Nonetheless, they point out that the journey from the foundations of audiolingualism and grammar to the ideas supporting communicative competence has not been a leveled one. A part of the problem, they point out, stems from the misinterpretation of the concept communicative competence itself, since simply having the ability for spontaneous self-expression is not sufficient to make one communicatively competent. For one to be communicatively competent requires him/her to have the grammatical knowledge of the system and also the more abstract sphere that incorporates knowledge of the appropriateness of language usage. This sphere, they add, encompasses paralinguistic (facial and gestural) and proxemic (spatial) knowledge, sensitivity to the level of language use in particular contexts and relationships (formal or informal), and sociocultural knowledge. Thus, they contend that for a person to be considered as being communicatively competent, s/he must have the capacity to make appropriate, natural and socially acceptable utterances in every communicative context (O'Grady et al., 1989:318-319).

Also, O'Grady et al. recount that during the early days of language teaching, a majority of the methodologies focused on grammar, pronunciation, and vocabulary. The contemporary communicative language teaching approach, however, focuses on learners and how they might learn to utilize language, a shift that has challenged many teachers to comprehend the specific components of the new teaching approach. This communicative language teaching approach is a cover term for many techniques which claim to have a communicative design that emphasizes content instead of form. The experiential feature of second language learning is foregrounded when learners are provided with genuine communicative activities that entail pertinent thrusts and subject matters. The new approach's axiom is that a language would be acquired subliminally when learners interrelate with second language speakers employing real-life subject matter (O'Grady et al., 1989:319).

In addition, O'Grady and his coauthors speak briefly about many linguists advocating for the making of sociocultural context an integral component of any communicative method, since language always functions within such a context. They note that second language teachers and researchers have also discovered that any second language teaching method must be grounded on psycholinguistic discernments of cognitive processes connected to mastering a second language. They suggest, however, that more research must be conducted to enhance our current knowledge of the language learning process in order to establish better teaching methods in the domain of learner strategies (O'Grady et al., 1989:319).

Communicative Power

The subject of communicative power has not received much attention and, thus, following Levinson, there is a need for the development of a theory or idea of communication to address this shortcoming (Levinson, 1983:113). His observation on the notion of communicative power is as follows:

> ...the communicative power of language can never be reduced to a set of conventions for the use of language. The reason is that wherever some convention or expectation about the use of language arises, there will also therewith arise the possibility of the non-conventional *exploitation* of that convention or expectation. It follows that a purely conventional or rule-based account of natural language usage can never be complete, and that what can be communicated always exceeds the communicative power provided by the convention of the language and its use (Levinson, 1983:112-113).

Also, given Leech's analysis, communicative power can be demonstrated by discerning illocutionary force (i.e., the communicative design or plan underpinning a speaker's remark) from the illocutionary act (i.e., the fulfillment of that communicative goal) (Leech, 1983:200).

Language Development

Like the subject of communicative power, that of language development as it relates to language as a mode of communication has also not received much attention. Nonetheless, Hakuta's relatively brief analysis of the work on what is called *motherese*—i.e. the language mothers employ when communicating with their children—is quite enlightening. He starts with the observation that although arguable, models of correlations between motherese and child language appear to indicate that the functional and interactional characteristics of the parent-child circumstance explicate much of language development (Hakuta, 1986:113).

Next, as mentioned earlier, Hakuta asserts that this interest in communicative meaning, which is a shift from the earlier emphasis on grammar, is not so much a shift in scientific progress but more so a shift in style. He adds: "For example, the shift from grammar to meaning did not occur because grammar failed as a theory of language acquisition but because researchers were interested in charting new territory" (Hakuta, 1986:113).

African Linguistic Evidence

To start with, H. Ekkehard Wolff tells us that African cultures have established particular designs of language utilization for a successful *Mode of Communication* just as other cultures across the globe have done. As such, a level of agreement at a particular instance of communicating exists in Africa. Wolff mentions an example from the Itsekiri and Okpe in the Delta region of Nigeria who employ contrasting words for such things as "blood," "fire," and "firewood" based on whether the wood is being used during the day or at night. He also states that words of avoidance, or taboo, and respect are among the best-known instances of culturally determined language utilization for successful communication in Africa. For this manner of communication, he notes the case of the Xhosa and Zulu in Southern Africa who call it *hlonipa* (Wolff, 2000:305).

Next, Øyvind Dahl broaches how language is employed in communication to reduce uncertainty, act effectively, and defend or strengthen the ego in Madagascar. He points out that in particular instances, words are utilized to ward off anxiety, and in other instances, they are used to establish more deeply satisfying ways of communication. Accordingly, as he also points out, the objective of communication is to increase the number and consistence of a person's "meanings within the limits set by patterns of evaluation that have proven successful in the past, the emerging needs and drives, and the demands of the physical and social setting of the moment" (Dahl, 1999:9).

Also, as Barbara Trudell informs us, multilingualism in multiple "home languages" compel people to communicate in multiple languages all the time within their circles of loved ones due to their extended family and intermarriage connections. As means of communication that mirror the close connections, stronger sentimental attachments underlie these languages

(Trudell, 2009:1). Trudell quotes a 35-year old Kenyan woman called Carol Ngugi of having articulated this type of sentiment in describing her own linguistic context as follows:

> I speak Kikuyu with my mother and my family upcountry. My dad always insisted that I speak Swahili with him, so I do. But my husband [also an ethnic Kikuyu] only speaks English, so we use that; with the rest of the immediate family and my close friends I speak English or Swahili, or Kikuyu if I want the story to come out a certain way (quoted by Trudell, 2009:1).

Trudell adds that because each of the three languages entails an emotional significance to Ngugi, she therefore connects each with her personal relationships (Trudell, 2009:1).

In addition, according to B. Rotimi Badejo, multilingualism, as "the ability of an individual to speak more than one language," is a prominent *Mode of Communication* in education and conveying information in societies across Africa. He supports this postulate by providing evidence from many African countries that are multilingual and the three that are monolingual (i.e. Botswana, Lesotho, and Swaziland). Since repeating everything he says about the many countries covered is unnecessary, it suffices to state that his investigation covered the language situation across Africa South of the Sahara at the time of his study and a categorization of the languages based on the countries' language policies, with the attendant statistical measurement for each language policy. Also, since looking at an example of a multilingual country will provide the expected result, let us look at what Badejo says about one of the monolingual countries—i.e. Botswana. As he puts it, in "Botswana 99 percent of its population speak Tswana which is the unchallenged lingua franca of the country. Nevertheless, Tswana has to be content with sharing official language status with English. Although…English is used as the language of instruction…Tswana is used in early and higher primary school years. The situation is the same in the mass media; both languages (Tswana and English) are used for national broadcasts" (Badejo, 1989:47).

Furthermore, it should be noted that across Africa, *Mode of Communication* manifests itself not only among the living but also between the living and the departed. An example of this phenomenon is *Ifa*, which I have defined elsewhere as "a geomantic divination system based on the ability to interpret 256 ancient verses" (Bangura, 2015a:65), with its attendant mathematics to be discussed in the next section.

Indeed, the essence of *Ifa* as a *Mode of Communication* between the living and the dead has been treated very well by Mwalimu Toyin Falola in seven of his more than 120 books: namely, *Yoruba Gurus: Indigenous Production of Knowledge in Africa* (1999), *A Mouth Sweeter than Salt: An African Memoir* (2005), *Yoruba Identity and Politics* (2006), *Ibadan: Foundation, Growth and Challenge 1830-1960* (2012), *The African Diaspora: Slavery, Modernity, and Globalization* (2013), *Esu: Yoruba God, Power, and the Imaginative Frontiers* (2013), and *Counting the Tiger's Teeth: An African Teenager's Story* (2014). The reader will gain a comprehensive knowledge about *Ifa* as an essential African *Mode of Communication* because the lessons Mwalimu Falola imparts in these books include the definition, Orisa, divination system, divination verse, festival, traditions, Will Mend Our Broken World belief, poetic song, *aya* (oracle of the heart), priest, oracle, corpus, chronicles and traditional healing of the religion. The abundant information on these themes in the preceding texts can inspire one to even write a monograph on the topic.

African Mathematical Evidence

Beginning with Claudia Zaslavsky, she provides a description of a mathematically communicative act with many examples in her examination of vocalized and gestural counting across the African continent. She points out that gestures are utilized either as substitutes for the vocalized word or to call attention to a particular aspect. She also mentions that gestures may be employed to circumvent the taboo of counting living things and to overcome language barriers between people of different backgrounds. In certain contexts, she adds, fingers are utilized by both Western-educated and uneducated individuals as an aid in cognitive arithmetic (Zaslavsky, 1999:238). What follows is one instance of vocalized and gestural counting from the Shambaa language spoken principally in northeast Tanzania (Zaslavsky, 1999:238-239):

6 = *mutandatu* = *ntatu na ntatu* = 3 + 3.

7 = *mufungate* = *funga ntatu* = bind 3 (fingers); from the ten fingers seven remain; therefore a subtractive formation.

8 = *munane* = *ne na ne* = 4 + 4.

9 = *kenda*, an alien word, used here, as in other languages without the class prefix.

Next, Scott Pakin provides LATEX letters that can be utilized to arrange or process the data for text that is to be printed for African languages. These letters allow one to write African mathematical equations very quickly with minimum inconvenience. The equations start and end with the symbol \.

According to Pakin, the symbols are to be employed "sporadically throughout a document (e.g., phrases such as "β-decay")...the characters all need the T4 font encoding (a text editor that helps one to style the text that s/he is writing), which is provided by the fc package" (Pakin, 2017:16). The fc package entails fonts developed by Jörg Knappen to support the T4 font encoding for African languages.

Also, Laurie Butgereit has developed the "Math on MXit" application as a *Mode of Communication* mathematics education in South Africa. She describes the application as an "instant messaging system" that operates on cellular phones to be used by learners to contact mathematics tutors at particular times. The system allows a tutor to use a computer and keyboard in order to respond to questions from as many learners as possible. The learners remain anonymous and will not know from the application the other learners employing the system (Butgereit, 2007:107). She states the following reasons for developing the application:

> It is common cause that the average mathematics mark among high school learners in South Africa has declined. Traditional "math clubs" and "math extra lessons" often do not work because of a number of reasons, including (1) it being "uncool" to be seen going to math club; (2) the stigma of going to extra lessons; (3) lack of suitable tutors; (4) suitable tutors are employed full time and are not available in the afternoon; (5) learners are not available to return to school in the evenings for meetings; (6) other scheduling conflicts between learner and tutor (assuming a tutor could be found); and (7) costs of extra lessons (Butgereit, 2007:107).

In addition, as I mention in the preceding section, there is a range of mathematical operations that undergird *Ifa* as a *Mode of Communication* between the living and the dead. In his January 31, 2016 E-mail to a small group of African intellectuals, Mwalimu Emanuel Babatunde challenged us "to build valid mathematics on Odu Ifa besides its binary permutations, etc." I have replicated his E-mail with its provocative questions verbatim as follows for the sake of precision:

> Brother Oluwatoyin Adepoju,
>
> Thank you for this pertinent and deep analysis of the possible uses of Ifa as a mathematical system of claims about human nature and a magico-religious matrix. As one interested in African Metaphysics, I am fascinated by these possibilities, while I also accept your caution that Ifa contains, like most magico-religions, internal contradictions. Since you

see it as a possible mathematical system, may I posit the following suggestions?

1. How does a religious matrix based on revelation and symbolic classification transform into an empirically validatable and refutable source of knowledge that will not depend on dogma and persecution to justify and corroborate its views as valid?

2. Since Mathematics is the language of science because it is logic in symbols, in order to make anything scientific, it must be mathematically replicable, refutable, and verifiable. When will Yoruba Renaissance occur that will move claims of Ifa from the realm of belief to the realm of fact as Lock-Facta non Verba noted there must be a move from words and to facts?

3. I am fascinated by your mapping. When and what can we do to move Yoruba metaphysics to the point when it can provide mathematical knowledge for the laws of gravity?

When Galileo provoked a movement to the nature of science as the defining matrix for understanding the relationship between heavenly bodies in the universe when he propounded the theory of heliocentrism—the sun, not the earth, as the center of the universe, he dealt a blow to magico-religious epistemologies of human experience and debunked the idea of heaven the earth hell or Gehenna—hell below the earth. All the powers of the Catholic Church, even the Medici Royal line, abandoned him. He came close to losing his life but for the support of Jesuits illuminati who confirmed the veracity of his views. Happily enough, the globe has moved away from that pernicious history of that imposed knowledge. What do we need to do to build valid mathematics on Odu Ifa besides its binary permutations, etc.?

Akinwumi Ogundiran (January 31, 2016), Nimi Wariboko (January 31, 2016) and I (Bangura, January 31, 2016) responded to Mwalimu Babatunde's E-mail. What stood out to me the most in our responses is Ogundiran's urging that "one must learn Ifa from the Babalawo and/or Iyanifa in order to know what Ifa is and is not" in order to respond to Mwalimu Babatunde's call. While Ogundiran's impetus may be correct, it nonetheless got me to think that if the lessons on *Ifa* that Mwalimu Toyin Falola and many other scholars have taught us are not enough to gain a rich knowledge of *Ifa,* then anyone seeking more knowledge from a "Babalawo and/or Iyanifa in order to know what *Ifa* is and is not" must want to be a Babalawo and/or Iyanifa himself, not just wanting to know about *Ifa*. Besides, as Moyo Okediji, a scholar of *Ifa* at the University of Texas at Austin, states when asked about whether he is on his way to becoming a Babalawo, "Scholars do not make good *Awo*, because we are sworn to disclose knowledge, conserve and expand knowledge—and the idea behind *Awo* implies classification, secrecy, cooperation, and concealment" (Okediji, February 8, 2014:1).

Nonetheless, as I stated in my response to Mwalimu Babatunde's call, to the best of my knowledge, there are two ideas that can be explored and subsumed into a mathematical essay on *Ifa*: (1) *Ifa's* Fractal Mathematics which Ron Eglash (1999) and I (Bangura, 2011a & 2015a) have discussed briefly in our works that needs elaboration; and (2) *Ifa's* Symbolic Numerical Series which somehow corresponds to the supertrinity of the Egyptian ennead (Trinitarian Mathematics). Hopefully, this will finally force European mathematicians to explain how Plato's series ended with *Timaeus* given the missing numbers and incoherent calculation. Another blessing is that it will continue to solidify the idea of a massive scale of the contact between the Yoruba and Pharaonic Egypt, as Mwalimu Cheikh Anta Diop (Diop, 1981/1991:324) had urged.

We must recall that Mwalimu Diop had posed the challenge to African intellectuals that until Africa is able to reclaim the historical and promethean consciousness that is embodied in the achievements of Ancient Egypt, the history of Black Africans and that of humanity in general will "remain suspended in air" (Diop, 1955 & 1967/1974:xiv-vi). According to Mwalimu Diop, such a history can never be written correctly "until African historians dare to connect it with the history of Egypt." In his view, even the study of languages, institutions, etc. cannot be treated properly until this is done: "in a word, it will be impossible to build an African Humanities, a body of African human sciences, so long as the relationship does not appear legitimate" (Diop, 1955 & 1967/1974:xiv).

It behooves me to end this section with at least one illustration of an *Ifa* mathematical calculation and mentions of other inroads into mathematical calculations dealing with *Ifa* that have been made by other scholars. Moyo Okediji illustrates the *Ifa* mathematical calculation of the visual allusions underlying *Oyeku Meji* as follows (Okediji, 2013:1):

2 + 2 + 1 = 5: 2215
2 + 2 + 2 = 6: 2226
2 + 2 + 3 = 7: 2237
2 + 2 + 4 = 8: 2248
Total = 8926

According to Awo Falokun, "*Oyeku* is an elision of the phrase *0 yeye iku*, meaning 'Spirit of the Mother of Death.' In simple terms, *Oyeku* is darkness—the complete contraction of matter into what physics calls a black hole. In human terms, this can mean physical death" (Falokun, 2018:1). *Meji* refers simply to a "pair." Other *Ifa* mathematical calculations that have been done include trigonometry (Abiodun, 2014), the 16 principal/the god *Fa's* eyes *Odu* calculations (Assata Shakur Forums, 2008), the 256 octagram *Odu* calculations, and scientific computer calculations (Folorunso et al., 2010; Alamu et al., 2013; Ajayi, 2016).

Conclusion

In the Ancient Kemetian language, the term for *Mode of Communication* is best represented as *Mode of Discourse, ret/sep/qi khen,* which also denotes "manner, habit, characteristic, behavior, or state of condition of talk, oration, comparing, speech, word, report, good report, a good thing" (see Budge, 1978 for this translation). In addition to verbal communication, of course, Ancient Kemetians also used several other media for *Mode of Discourse.* One medium comprises sacred signs or hieroglyphs, which were used as the ratified language of the gods and utilized in monumental art. Penelope Wilson describes hieroglyphs as follows:

> The hieroglyphs of ancient Egypt [in which] the word for pictorial writing was *medu-netjer*, which means "words of god," were an exalted mode of communication, used to speak with the gods or to guide rulers in the afterlife (for instance, hieroglyphs might describe a set of spells, rituals, and the route to the afterlife for use by the king or queen). Indeed, hieroglyphs differed from everyday writing and the average Egyptian was unable to read them—only the elite were taught the pictorial signs...hieroglyphs can be read either left-to-right or right-to-left and that animals, birds, people, serpents, and fish were drawn in profile, all facing the same way, so that readers would know which way to read the line. Also, individual signs derived from everyday objects in the Egyptian world, show that learning hieroglyphs is not only a linguistic exercise, but also an insightful introduction to Egyptian culture and daily life (Wilson, 2003:1).

Another medium of communication in Ancient Kemet was the Tarot, which is a language system in itself (coded and linked directly to alchemy, Astrology, gnosticism, ritual magic, and Qabala) written in a book for use by practitioners of occult. The following is the later fate of the Tarot after it was appropriated by Westerners as described by Debbie J. Challis: "As a system of occult meaning and esoteric guidance, Tarot was forced underground in Medieval Europe. Disguising the Tarot as a game was a way of enabling practitioners to continue its usage without persecution. It was called 'The

Devil's Picture Book' by the Christian Church and heretics using it were put to death. This is why records are patchy and the Tarot appears to only to resurface at certain times. Those in the know always used it, but secretly if they needed to" (Challis, 2011:1).

The other medium of communication in Ancient Kemet encompassed Emojis (i.e. small digital images or icons used to express ideas, emotions, etc.) which we still use to this day in digital channels. And, as Nick Hammond correctly points out, while "it is unlikely that our future written/typed communications will only include images, Emojis are another valuable tool which we can use to connect with each other" (Hammond, 2015:2).

Chapter 4

Semanticity

In my article titled "Domesticating Mathematics in the African Mother Tongue," I state that Anousschka Bergman, Kathleen Curie Hall and Sharon Miriam Ross (2007) point out that "*Semanticity* calls for all signals of communication to possess a meaning and function. It is vital that a speaker/writer and a hearer/reader share a similar idea about what is being talked or written in order to have successful linguistic communication. Even when a person hears a word s/he does not know, s/he nevertheless assumes that it must possess some meaning" (Bangura, 2014:18). Before delving into the topic, it makes sense to begin with general definitions of *Semantics* and *Semanticity.*

Beginning with *Semantics,* David Crystal provides the following general definition: "The study of meaning in language" (Crystal, 1992:347). Next, *Semanticity* is defined by William O'Grady, Michael Dobrovolsky and Mark Aronoff as a "system [that] conveys meaning through a set of fixed relationships among tokens, referents, and meanings" (O'Grady et al., 1989:407). They add that "Human language conveys meaning through arbitrary symbols. Bee dancing conveys meaning, but within a very limited range, as do bird calls and song, along with nonhuman primate vocalizations" (O'Grady et al., 1989:407).

To explore language semanticity, I segment the remainder of this chapter into the following three sections: (1) linguistic expositions on the subject; (2) African linguistic evidence, and (3) African mathematical evidence. In the end, conclusion is drawn.

Linguistic Expositions on Semanticity

The subject of semanticity, by belonging to one of the most studied subfields of linguistics—i.e. Semantics, has been examined in a plethora of books, articles, and video and audio lectures. Thus, it is impossible to cover in one section everything there is about the subject. While it would suffice to simply provide summaries of the various topics covered by John Lyons who has done the most extensive work on Semantics in his two-volume book similarly titled *Semantics* for the reader to get a good background on the subject, this would miss the new findings that emerged on the subject after his work was published in 1977. What are presented in this section, therefore, are brief synopses of the following 19 aspects of Semanticity I consider to be the most germane to the subject matter of this book: (1) semantic theory, (2) semantic field, (3) semantic rules, (4) semantic differentiation, (5) generative semantics, (6) semantic information, (7) semantic features, (8) semantic markings, (9) semantic representation, (10) semantic analysis, (11) semantic integration, (12) semantic triangle, (13) semantic presupposition, (14) semantic broadening and narrowing, (15) computational semantics, (16) semantic shift, (17) semantically defective square, (18) semasiography, and (19) semantic subreption. These aspects are discussed separately for the sake of simplicity.

Semantic Theory

Semantic theory, following Francis J. Dinneen, can be denoted as a framework that is utilized to provide an interpretation of the sentences produced by the syntactic component of a grammar. The need for this theory arises from the desire to link syntax and semantics. For instance, the oft-repeated example *Colorless green ideas sleep curiously* exhibits a sense in which grammar is independent of meaning. Thus, Dinneen maintains that "distinctions concerning degrees of grammaticalness point out the need for a formal specification of how these degrees are to be judged. There are sentences that we would call ungrammatical yet acceptable, and there are other sentences that are grammatical but unacceptable. It is where semantic and grammatical criteria overlap that this problem is raised" (Dinneen, 1967:379).

In addition, Dinneen points out that another feature has to do with the more explicit awareness of traditional grammatical concepts such as universal grammar as being quite correct. He does concede, however, that there were obvious gaps in the traditional rendering of syntax and that some inconsistencies existed in the distinction of substantial versus formal universals (Dinneen, 1967:379).

Semantic Field

Commencing with Paul A. Gaeng's brief definition, a *Semantic Field* simply means the "sphere of reference" of a concept (Gaeng, 1971:138). He provides the following example to explain the notion:

> The concept to which the French word *temps* refers is not the same as that which is reflected in English *time*, which is only a partial equivalent of the French term. The Semantic Field (an expression of for sphere of reference) of *temps* covers that of English *time, tense* (grammatical 'time'), and *weather* (for example, *Quel temps fait-il?* 'What is the weather like?') Conversely, English *time* covers part of the French concept of *temps* and also French *heure* 'time of the day' (for example, *Quelle heure est-il?* 'What time is it?') (Gaeng, 1971:138).

Also, Gaeng suggests that instead of describing abstract referents as individual referents, it would be better to describe them as spheres of reference. This is because, according to him, while the relationship among the three components of meaning—i.e. (1) the symbol (the word), (2) the referent (the object or concept symbolized), and (3) the thought or reference (the mental image of the referent)—is straightforward for concrete referents like *parent, child, lion,* etc., it may be restrictive for abstract nouns such as *love, belief, peace*; descriptive adjectives such as *pretty, bad, unfair,* etc.; and verbs. To support his point, Gaeng gives the example of the verb *know,* which "has a very large sphere of reference, especially if we compare it with the Romance languages and other Germanic languages which differentiate between knowing in the sense of 'being familiar with a fact' and 'being acquainted with someone or something,' for instance German *wissen* and *kennen,* Dutch *weten* and *kennen,* French *savoir* and *connaitre,* Spanish *saber* and *conocer,* Italian *sapere* and *conoscere,* or Russian *znat'* and *umet'*" (Gaeng, 1971:135-136).

Furthermore, Gaeng informs us that synonyms (words or phrases that mean exactly or nearly the same as other words or phrases in the same language) lack spheres of reference in that one can, for instance, talk about a *high mountain* and a *tall mountain,* but *he is high* and *he is tall* mean two completely different things. Despite this shortcoming of synonymy, Gaeng, however, believes that it remains an important concept for investigating meaning, This is due to the fact that synonymy can essentially be treated as the opposite of polysemy (the coexistence of many possible meanings for a word or phrase) given that in synonymy two or more lexemes (basic lexical units of a language, consisting of one word or several words, considered as abstract units, and applied to families of words related by forms or meanings) may connect to the same sememe (a semantic language unit of meaning, analogous to a morpheme—the meaningful morphological unit, or forms of words, of a language that cannot be further divided; e.g. *out, go, -ing,* forming *outgoing*), while in polysemy two or more sememes connect to only one and the same lexeme (Gaeng, 1971:141-142).

Next, according to John Lyons, the theory of semantic fields, also called Field-Theory, concerns the study of lexical structure, also referred to as the structure of a vocabulary. The focus of this theory can be on either (a) the lexemes of a language as a starting point; or (b) the objects, properties and relations external to a language. In the first case, the interest is on the meaning of a lexeme in relation to the meanings of other lexemes in a language system; in the second case, the interest is on designation—i.e. how a lexeme or a class of lexemes are assigned in a particular language. The distinction between these notions of meaning and designation has been imperative for the development of Field-Theory, which has been linked to the idea that the thought patterns and philosophies of societies have been shaped by languages (Lyons, 1977:251-252).

In addition, Lyons points out that another concern of Field-Theory is the examination of sense. Therefore, the vocabulary of a language is investigated in terms of an integrated system of lexemes interrelated in sense. This system is in continuous flow throughout the history of a language because as previous lexemes disappear, new ones emerge; also continuously changing throughout time are the relations between a particular lexeme and its neighboring lexemes. Hence, a broadening in the sense of a lexeme leads to the narrowing of one of more of its neighboring lexemes (more on semantic broadening and narrowing later). Consequently, traditional diachronic semantics seeks to document the historical evolutions in the meanings of individual lexemes singularly, as opposed to analyzing the historical evolutions of the entire structure of a vocabulary. Thus, Lyons calls for both diachronic and synchronic linguistic analysts to address systems of interrelated elements, since diachronic linguistics necessitates, and is contingent upon, synchronic linguistics. He adds that an analyst must, therefore, compare a series of successive synchronic language-systems, if s/he is to provide a historical description of the evolution of a language (Lyons, 1977:252).

Semantic Rules

On the one hand, based on what is referred to as the "Standard Theory," Lyons distinguishes three sets of rules for a language system. The first set comprises transformational rules (T-rules). These rules yield a set of surface structures to which the phonological rules (P-rules) apply and produce for each structure its phonological representation (PR). The T-rules are obligatory, a proposition that became a component of the standard version of transformational grammar. Also, as Lyons mentions, the proposition is accompanied by "an immediate corollary, by virtue of the semiotic principle that meaningfulness implies choice, the presupposition that transformations do not affect meaning" (Lyons, 1977:412). He further argues that "it is only in so far as this proposition is held to be true that one can maintain the principle that all the information that is relevant to the semantic interpretation of a sentence is present in its deep structure" (Lyons, 1977:412-413). The second set is composed of semantic rules (S-rules, which are the main focus of this section), and the third set is made up of phonological rules (P-rules). The S-rules apply to and yield a set of semantic interpretations (SIs) that are the base of the deep structures of a language system (Lyons, 1977:412).

On the other hand, Dinneen presents a slightly different perspective on semantic rules. Grounding his analysis on what is known as "Montague Grammar," Dinneen starts off with a definition of semantic rules as "unique translation rules, representing a phrase resulting from a syntactic rule in an intensional logic formula as a specified function of the translation of its parts. A second stage of 'interpretation' matches model-theoretic semantics with that of intensional logic" (Dinneen, 1995:548). Thereafter, Dinneen says that a preference for a particular analysis emerges when "syntactic-rule/semantic-rule-association" is hypothesized (Dinneen, 1995:548).

Semantic Differentiation

Starting with a general definition, *Semantic Differentiation* or *Semantic Differential*, according to Crystal, is "a technique devised…to find out the emotional reactions of speakers to lexical items" (Crystal, 1992:346). He describes how the technique is utilized as follows:

> Subjects are asked to rate words in terms of a seven-point scale with opposed adjectives at each end—for example, 'good-bad', 'active-passive', or 'tense-relaxed'. The word *bus*, for instance, might be rated strongly with respect to 'bad', 'active', and 'tense'—by someone who did not take much pleasure from bus-commuting. When the responses of subjects are obtained, it is possible to draw conclusions about the main affective dimensions in terms of which a language's concepts are organized, and to make comparisons between cultural groups (Crystal, 1992:346).

Next, Ralph Fasold edifies us about the semantic differential scales alluded to in the preceding excerpt. As he points out, semantic differential scales for listener responses are often utilized with the "matched-guise technique," an experimental method that has become the ideal for gauging language attitude. He characterizes the "pure matched-guised technique" as a method which seeks to completely control all of the variables, but language. To do so, a number of bilingual speakers fluent in the languages being studied are recruited. The speakers are tape-recorded reading precisely the same passage, once in each language. The taped passages are rearranged so that they appear that each speaker has only been recorded once. A group of bilingual speakers from the same community would then be asked to listen to the recordings and rate the speakers on different aspects such as intelligence, likeability, social class, etc. When the same speaker is rated differently in the different "guises," the difference in language is then designated to be the reason for the differences. Fasold, therefore, depicts the technique as being both (a) "direct in the sense that the listeners are explicitly asked to give their opinions of the speakers' characteristics" and (b) "indirect in the sense that listeners are asked to react to *speakers*, not languages, and they are not aware that they are hearing the same person in each guise" (Fasold, 1984:150).

Also, it behooves me to mention here a shortcoming of the *Semantic Differentiation* technique that has been identified by Kenji Hakuta. As he observes, the use of the technique to compare compound and coordinate bilinguals on tasks involving information-processing has yielded a set of confusing results. This is due to the fact that, as he argues, rating individual words as they relate to their connotations on various dimensions of meaning has been poorly specified in terms of the relationship between these different tasks and the original compound-coordinate distinction. He adds that while the distinction is designed to be at the level of *denotative meaning* (i.e. to what the words of the two languages refer), many researchers comparing compound and coordinate bilinguals have instead employed measures that delineate the *connotative* meanings (the abstract meanings or intensions) of words (Hakuta, 1986:98-99).

Generative Semantics

Touting the impact of *Generative Semantics* in linguistic research, Anthony Arlotto begins by describing the subject as "a new way" of examining and illustrating language by developing descriptive and theoretical models and investigating the phenomenon of language change. He then indicates that the models emerged from thorough and comprehensive work on small segments of living languages (Arlotto, 1972:213).

Thereupon, Arlotto delineates some of the commonly agreed-upon postulates dealing with the arrangement and framework of synchronic grammatical descriptions. He explains that from a generative perspective, a grammar, or language description, encompasses the following three major parts: (1) phonology, (2) syntax, and (3) semantics. A fourth element that could be added to these parts, he says, is the vocabulary or lexicon of a language. He then describes the structure of a generative grammar as follows:

> The central (= starting) component is the syntax. Here we begin with the sentence or complete utterance from which we derive all the acceptable sentences of a language...the syntactic component has two outputs: the one that enters the semantic component is called the *deep structure* and contains all the information necessary for the semantic interpretation or understanding of the sentence; the output that enters the phonological component is called the *surface structure* and contains all lexical and grammatical elements arranged in proper order...each component is essentially a series of rules which characterize or describe the knowledge a speaker has of his own language (Arlotto, 1972:213-214).

Arlotto thereafter broaches the transformational component of grammar that illustrates the difference between the deep and the surface structures. First, he notes that the phonological part of a generative grammar likewise comprises a series of rules. Operating on base forms of morphemes, these rules yield a particular phonetic output. The morpheme structure rules convey information about the possible phonetic configurations of the morphemes in a specific language. Therefore, in the English language, for instance, a word does not start with a stop consonant preceded by a fricative: i.e. English words do not begin with the clusters *ts*- and *pf*-. Morphemic rules inform us how particular phonemes are pronounced in certain environments: e.g., for a word like *logs*, the plural morpheme (*-s*) is pronounced [*z*]. One more rule instructs us to voice the final consonant in the plural stems of words like *knife* and *life*: plurals *kniv-es* and *liv-es*, respectively (Arlotto, 1972:216-217).

Second, according to Arlotto, a language's vocabulary is a compilation of all the morphemes in that language. For this compilation, he mentions the following three components that generative grammarians say comprise it: (1) *phonological information*—the base form and marks of the phonological rules, whereby the ideal morpheme of the language would be placed in the vocabulary with only one base form; (2) *semantic information*—the meaning of the word; and (3) *syntactic information*—the manner in which the word is employed for configuring words, thereby designating the part of speech of whether a word is a noun, a verb, and so on (Arlotto, 1972:217-218).

Third and finally, Arlotto states that generative grammarians are solely interested in the changes in grammar when they examine language. Hence, for these analysts, surface forms are not of interest to them (Arlotto, 1972:218).

Semantic Information

Lyons apprises us about several features of *Semantic Information*, an aspect he says is a theoretical approach used to explain the nontechnical (or everyday) concept of "information." To start off, he makes a distinction between *Signal Information* and *Semantic Information*—whereas the former deals with identifying a signal (i.e. acoustic information for spoken utterances and visual information for written utterances), the latter deals with interpreting the signal vis-à-vis the message that it encodes. He adds that a signal is said to be informative when it transmits some semantic information to the receiver. Thus, he argues that even though signal information and semantic information are distinct, they, nonetheless, are intertwined when decoding and interpreting utterances (Lyons, 1977:41). He adds that this aspect is evident in cases where an analyst must determine which facts are more informative or significant than others; it is in these cases that the content of semantic information and that of signal information are parallel (Lyons, 1977:50).

Next, Lyons proffers the view that as far as *Semantic Information* is concerned, the proposition that when the occurrence of a signal or a part of a signal is totally predictable, then it is characterized to convey no signal-information, is acceptable. He also maintains that this proposition is equally acceptable for semantic information. The proposition, he adds, is congruent with the information-theory postulate that what is totally shaped by its context conveys no information at all (Lyons, 1977:45-46).

In addition, Lyons makes the observation about the simple model of communication (described in the previous chapter) being incapable of accommodating the poetic utilization of language signal-information, and that it fuses semantic information. This is because, according to him, many of the elements of poetry such as alliteration, assonance, chiasmus, meter, rhyme, rhythm, etc. employ the elements of the medium. Therefore, he states, poetic function relates not only to poetry, but also to its orientation toward the message "encoded-as-a-signal" (Lyons, 1977:54).

Semantic Features

To begin with, the general denotation for *Semantic Features*, emulating Crystal, can be stated as minimal contrastive elements of the meanings of words in a language. In certain approaches, he informs us, a minimal contrastive element is called a *semantic component*, which would then be used to analyze a word like *boy*, for example, as "young," "male," and "human." He also notes that the *semantic features hypothesis* postulates that in child language acquisition, the sequence in which a child's lexical items appear reflects the kind and intricacy of the semantic attributes of the items (Crystal, 1992:346).

Also, taking the point of view of *componential analysis*, defined by Lyons as an approach employed to describe the meanings of words and phrases based on the axiom that the sense of a lexeme can be analyzed in terms of a series of more common sense-components or semantic features, he states that the approach facilitates the analysis of some or all semantic features that are common to several different lexemes in a vocabulary. Given this potential, he, therefore, argues that componential analysis is connected with conceptualism, since the sense-components could be perceived as atomic concepts and the senses of certain lexemes can be perceived as molecular concepts (Lyons, 1977:317).

In addition, O'Grady and his colleagues point out that *Semantic Features* is an approach used to establish associations between the intensions of words and the abstract concepts made up of smaller components. They also avow that componential analysis is particularly efficacious when utilized to delineate differences and similarities among words with associated meanings. They add that componential analysis yields the most impressive results when used to examine a series of words alluding to categories of entities with shared linguistic elements (O'Grady et al., 1989:170-171).

Semantic Markings

Lyons discusses a couple of aspects pertaining to *Semantic Markings*, which he denotes as lexemes that are more specific in sense compared to the related semantically unmarked ones. He provides the following example to illustrate the difference, a portion of which is worth repeating here:

> 'Lioness' is more specific in sense than 'lion', as 'bitch' is more specific in sense than the formally unrelated 'dog.' For 'lioness' and 'bitch' denote only females whereas 'lion' and 'dog' can be applied, in many contexts, to either males or females; and it is for this reason that the collocations 'male lion' and 'female lion', and 'male dog' and 'female dog', are equally acceptable. In such contexts, the semantic contrast between 'lion' and 'lioness' and between 'dog' and 'bitch' is neutralized. In other contexts, however, and most obviously when the opposites are employed in a disjunctive question (*Is it a dog or a bitch?*) or a statement in which one is predicated and the other negated (*It's a dog, not a bitch*), the unmarked lexeme assumes a more specific sense which is incompatible with the inherently specific sense of the marked lexeme (Lyons, 1977:307).

Correspondingly, one aspect Lyons talks about is that, due to their more specific sense, all semantically marked lexemes are distributionally marked (lexemes that appear in a total set of linguistic contexts in which they can occur), but not all distributionally marked lexemes are semantically marked. To support his contention, he tenders the following example: "*X has a dog* can be uttered to make a true statement whether the animal referred to is male or female. But the proposition expressed by uttering *X has bought a big house* would be generally regarded as false if the house was in fact small rather than large in relation to the relevant norm" (Lyons, 1977:306).

The other aspect treated by Lyons entails *hyponymy* (a word of more specific meaning than a general or superordinate term applicable to it; for example, *spoon* is a hyponym of *cutlery*) and *quasi-hyponymy* (e.g., a relationship between 'square' and 'shape' or 'delicious' and 'taste') as being ubiquitous in the lexicons of English and other languages. According to him, this situation is born by the fact that hyponymy and quasi-hyponymy are the result of semantic marking and, hence, must not be considered as instances of polysemy (Lyons, 1977:308).

Semantic Representation

While Lyons simply informs us that *Semantic Representation* is the same as "deep structure," which could be utilized to rescue the idea of co-reference in certain cases (Lyons, 1977:192), Leech provides us with a substantive treatise on the topic. First, Leech explores the distinction between semantic representation and pragmatic interpretation. He begins with the proposition that the distinction is due to the fact that contrary views had in that past produced accounts of language that are implausible, if not absurd. He provides an example of implausibility in the semanticist position when an effort was made to subsume pragmatic occurrences into semantics. As he puts it, "According to the *performative hypothesis* put forward by the generative semantics school in the early 1970s, every sentence *S* in a language is in its deep or semantic structure a performative sentence roughly in the form *I state/declare/ask/etc. S*. By this means, the illocutionary force of an utterance was given as place in its semantic representation, which, since this school did not distinguish semantic representations with syntactic ones, was also its deep syntactic representation" (Leech, 1983:19).

Leech adds that the most absurd of the preceding proposition is that the deep structure of every poem (e.g., all of Shakespeare's 154 *Sonnets*) starts with a deleted performative that reads as follows: "*I imagine myself in and invite you to conceive of a world in which...*" (Leech, 1983:19-20). He argues that the exotic nature of this notion notwithstanding, its implausibility can be easily detected in the following example: "In a piece of expository written prose (say, an encyclopedia entry), every single sentence will have, in its underlying structure, a performative preface such as *I state that...*which has undergone deletion; so that a 100-sentence article will presumably have the same prefixed clause (or a similar one) repeated 100 times. Moreover, to make it more implausible, the performative will include a reference to the writer (*I*), even in formal expository prose of the kind in which the first-person reference is avoided for stylistic reasons" (Leech, 1983:20). Thus, for Leech, the performative hypothesis is nothing more than an endeavor to "grammaticize" pragmatic aspects, or illocutionary forces, and acceptable to those analysts who consider the grammatical paradigm of generative grammar to be completely sufficient. This is the only reason, he says, explains why these analysts would be oblivious to the fact that language occurs in situations (Leech, 1983:20).

Also, Leech points out that another implausible endeavor to state everything relating to meaning is the analysis of speech acts and speech situations, albeit not about language in general. An extreme example of this endeavor is defining word meaning in terms of what a word contributes to illocutionary force as in the following attempt by William P. Alston: "A meaning of W_1 is W_2 = *df.*. In most sentences in which W_2 occurs W_1 can be substituted for it without changing the illocutionary-act potential of the sentence" (Alston quoted by Leech, 1983:20). To Alston's postulate, Leech responds as follows: "Alston defines synonymy of words (W_1, W_2, etc.) as a matter of the smaller illocutionary-act potential of the sentences in which they occur; but since words do not normally constitute illocutionary acts in isolation, he has no way of explaining how words have meanings in themselves" (Leech, 1983:20).

Second, Leech examines the semantic representation of *declaratives*, *interrogatives*, and *imperatives*. He starts by reminding us that the three elements are basic sentence-types generally used as syntactic categories. Thus, they are traditionally differentiated from their attendant semantic or speech-act classifications: "assertion," "question," and "command," respectively. Accordingly, an analyst must also differentiate the classifications on the semantic and pragmatic levels. Unfortunately, however, as Leech laments, the English language has not been favorable to this "complementarist" perspective because the language lacks the adequate terminology for all three levels of meaning (i.e. syntactic, semantic, and pragmatic). This situation has made it effortless for analysts to subsume semantic lexicon to pragmatic lexicon and vice versa ((Leech, 1983:114).

In addition, Leech points out that the absence of an acknowledged concept for the common category of logical units of which commands, questions, and propositions are subcategories contributes to the terminological shortcoming. As a result, the concepts "propositional content," "predication," and "sentence radical" have variously been employed to describe commands, questions, and propositions. Leech provides the examples "*You will sit down, Will you sit down,* and (*You*) *sit down!*" as all having a common content describing a sitting down to take place in the future. He adds that while the logical forms of all three terms differ, they are, nonetheless, propositional (Leech, 1983:115).

Semantic Analysis

For *Semantic Analysis,* I turn again to Leech. For starters, he states that on the logical/semantic level, the three syntactic oppositions (affirmative/negative, declarative/interrogative, assertive/non-assertive) are utilized to determine whether they correlate with three logical oppositions (predication/propositional content; factuality operators—fact and non-fact) (Leech, 1983:159).

Next, in his semantic analysis of some illocutionary verbs, Leech enlightens us that the technique used is akin to componential analysis in that some discrete variables are isolated in terms of which the meanings of the verbs differ strikingly. Selected assertive, commissive, directive and expressive types of verbs are therefore analyzed. The similarity among these variables makes it easy to place them within a single table for analytical purposes (Leech, 1983:218).

Semantic Integration

Based on his reflection of social processes in linguistic structures in the United States, *Semantic Integration,* says William Labov, appears to emerge via a core series of relations of equality and quality upon which all speakers of the English language agree. For example, on the notion of *common sense,* he finds that an overwhelming majority of English speakers agree that it falls within the purview of superordinate judgment (i.e. "good judgment"). He also finds that there is an equally high agreement on the collocation of *practical* (or *everyday*) with *common sense.* He further discovers that while there is no plain concept to illustrate the quality of "not being learned from books," there is, nevertheless, a high degree of agreement in this feature of *common sense* (Labov, 1984:120).

In addition, Labov argues that there must be a social procedure in which the uttermost alternatives are restrained in group interaction by sacrificing core values for semantic integration to occur. He, therefore, calls for more studies to be conducted in order to ascertain whether such a procedure actually works (Labov, 1984:120).

Semantic Triangle

Crystal tells us that the *Semantic Triangle* is a specific model of meaning which proposes that "meaning is essentially a threefold relationship between linguistic forms, concepts, and referents" (Crystal, 1992:347). Givi Amaglobeli provides the following definition using different, albeit related, concepts: "Semantic Triangle relates...three basic components to each other as they constitute one whole system. These are the Object, the Meaning (of the Object), and the Sign" (Amaglobeli, 2012:37).

Furthermore, Amaglobeli notes that an object gets its meaning and then expressed by a linguistic sign—i.e. word. He adds that the same system and the same parts make up the Stoic concept comprising "Meaning (Thought), Sounds (words) and the Object. The components of this [S]toic concept [are] the same as those of the modern Semantic Concept. Three elements link to form one whole system. According to the Stoics, two of these elements

(Sounds and Object) are of a material form and one of them is of a non-material (Meaning/Thought). This last element of a three-component system the Stoics gave the name 'Lekton'" (Amaglobeli, 2012:37).

Semantic Presupposition

From Stephen C. Levinson, we learn that *Semantic Presupposition* refers to a number of theories that seek to develop a new and well-articulated semantic relation that would become a part of formal semantic theories. The impetus for the approach was birthed by the possibility of ushering in a new and distinct semantic relation that could be added to the existing ones via the idea of presupposition. Unfortunately, as Levinson points out, the idea of semantic presupposition has been mostly abandoned because of the nature and properties of the idea when adequately explored (via, for example, the assumption of bivalence, truth-value gaps, etc.). The idea was substituted with an independent and distinct category of pragmatic inferences referred to as *Pragmatic Presuppositions*: i.e. presuppositions that "are best described as a relation between a speaker and the appropriateness of a sentence in a context" (Levinson, 1983:174, 177).

Despite the shortcomings of *Semantic Presupposition*, Levinson still goes on to explain a number of its features. First, he identifies two major categories within which semantic theories of presupposition have been analyzed, although he argues that the attempts have been incorrectly positioned. One of these categories entails truth-conditional theories which clearly prognosticate what cannot be encapsulated in semantics. The other category comprises theories which suggest that all semantic relations can be explained in terms of glossing sentences into atomic ideas or semantic attributes (Levinson, 1983:199).

Second, according to Levinson, presupposition has been defined in such a manner that logical consequence relation cannot be affected by negation. This was done in order to include presupposition into truth-conditional theories. Levinson, therefore, insists that the logical structure of semantic theory must be restructured completely in order for these theories to be accommodated into the structure. Such reorganization, he opines, could be warranted if the elements of presupposition are ensnared, although he is doubtful that such a theory can be successful in principle (Levinson, 1983:199).

Third, Levinson points to two cardinal elements of the behavior of presuppositions that lead to the grim fate of semantic presupposition theories. One of these elements is "defeasibility": i.e. the notion that presuppositions do not always persevere in particular discourse contexts since they are subject to deletion by attributes of the contexts. The other element is the oddity of the "projection problem": i.e. the idea that there are ways in which presuppositions are synthesized when complex sentences, whose components yield the inferences being considered, are constructed (Levinson, 1983:199, 13).

Fourth and finally, Levinson alludes to efforts undertaken by some linguists to subsume presupposition within a semantic theory by utilizing atomic concepts or semantic primes or attributes. The elements of these particular semantic models, he says, are not as well defined compared to those of logical models. He, therefore, contends that contrary to the claim by some analysts using these particular semantic models that "semantic presupposition is a perfectly viable concept, indeed the only viable one, when modelled within a feature-style semantics," the truth of the matter is that these models "simply cannot handle the projection problem" (Levinson, 1983:203). Hence, he concludes that since "presuppositions are not stable, context-independent aspects of meaning," they do not belong in semantics, but in pragmatics (Levinson, 1983:204).

Semantic Broadening and Narrowing

The twin concepts *Semantic Broadening* (or *Extension*) and *Semantic Narrowing* represent the opposite sides of the idea of semantic change. Before discussing the definitions and examples provided by O'Grady with his colleagues (1989) and Arlotto (2972) for the concepts, I will begin with the following quite salient point made by the former: "Although changes in word meaning take place continually in all languages, words rarely jump from one meaning to an unrelated one. Typically, the changes are step by step…" (O'Grady et al., 1989:206).

O'Grady and his fellow scholars go on to define *Semantic Broadening* as "the process in which the meaning of a word becomes more general or more inclusive than its historically earlier meaning" (O'Grady et al., 1989:206). Among the four examples they provide to explain their definition is the word "bird," whose old meaning was "small fowl" and its new meaning is "any winged creature" (O'Grady et al., 1989:206). Calling the phenomenon *Semantic Extension*, Arlotto defines it as the increase in the number of things to which a word refers; or put differently, the meaning of the word widens. He cites "the English word *salary*, which means 'wages of any kind.' The Latin ancestor of [the] word, *salārium*, was of more limited usage, meaning 'the wages of a soldier'; even further back, usage of *salārium* was limited to the allotment that a soldier received for salt (*salārium* ← *sal* 'salt')" (Arlotto, 1972:177).

Thereafter, O'Grady and his colleagues define *Semantic Narrowing* as "the process in which the meaning of a word becomes less general or inclusive than its historically earlier meaning" (O'Grady et al., 1989:206). Among the four examples, they also tender to explain this definition is the English word "hound," whose old meaning is "any dog" and its new meanings is "a hunting breed" (O'Grady et al., 1989:206). Using the same name, *Semantic Narrowing*, Arlotto defines it as "the number of things a word refers to is reduced, or we might say that a word becomes more specialized" (Arlotto, 1972:178). He adduces the following example and explication to underpin his denotation: "the French word *saoul*...in modern French means 'drunk, intoxicated by means of liquor'. At a previous stage of French, the word meant simply 'seized, possessed' by anything. Perhaps because within the given culture, the primary agent for seizing one's soul or personality was liquor, the word *saoul* narrowed its field of reference and came to mean exclusively 'seized by the power of alcoholic beverages' or simply 'drunk'" (Arlotto, 1972:178).

Computational Semantics

O'Grady and his fellow scholars present a brief review of the semantic procedures and representations that have been suggested by scholars working in the area of computational linguistics. They begin with the observation that lexical items encompass fields for semantic information that include the type of semantic attributes that are necessary for a verb's subject or which thematic functions are needed or allowed by a verb. They go on to suggest the following two senses of the English word *flounce* as semantic fields (O'Grady et al., 1989:436):

Word: flounce-1
Semantic Restrictions: female human subject

Word: flounce-2
Semantic Restrictions: clothing subject

O'Grady and his colleagues add that while the semantics of words is a vital part of any language system, there still remains to be addressed the larger issue of the semantics of sentences and paragraphs (O'Grady et al., 1989:436).

Next, O'Grady and his coauthors proffer that two approaches have been proposed for semantics and language analysis. One of these approaches pertains to syntactically based systems. In this approach, a sentence is first syntactically analyzed and then a semantic representation is constructed thereafter. The problems that emerge here, according to O'Grady and company, have to do with what is referred to as the "mapping problem," which involves how to get from one representation to the other. The other approach involves semantically based systems. In this approach, a semantic representation is first developed and, occasionally, with no syntactic analysis. After that, the semantic representation is determined before a sentence is mapped into the various syntactic forms that occur in the basic sentence. O'Grady and his colleagues also note that semantically based systems have been utilized by linguists working in the area of discourse and text analysis such as for comprehending stories. But, they further point out that it is imperative for both approaches to meet the goal of assigning word, sentence and text meanings if they are to remain relevant. Also important for the two approaches to thrive, they add, is to be capable of determining matters like references and scope (O'Grady et al., 1989:437-438).

Semantic Shift

On this topic, I again cull from the work of O'Grady and his colleagues. To begin with, as part and parcel of two other aspects of syntactic change (semantic broadening/extension and semantic narrowing), discussed earlier, these scholars first define *Semantic Shift* as "a process in which a word loses its former meaning and comes to refer to a new, but often related, set of things" (O'Grady et al., 1989:206). Among the three examples they furnish is the English word "immoral," whose old meaning was "not customary" and new meaning became "unethical" (O'Grady et al., 1989:207).

Next, as O'Grady and his fellow scholars indicate, "Sometimes, a series of semantic shifts occurs over an extended period of time, resulting in meaning that is completely unrelated to the original sense of a word" (O'Grady et al., 1989:206-207). To lend credence to their observation, they recount the following evolution of the English word *hearse* as follows: The word "originally referred to a triangular plow. Later, it denoted a triangular frame for church candles and later still was used to refer to the device that held candles over a coffin. In a subsequent shift, it came to refer to the framework on which curtains were hung over a certain tomb. Still later, *hearse* was used to refer to the coffin itself, before finally taking on its current sense of the vehicle used to transport a coffin" (O'Grady et al., 1989:207).

Semantically Defective Square

One can glean from Joseph H. Greenberg's work at least two features of the *Semantically Defective Square.* One feature is his characterization of the concept as corresponding to "formally defective squares," whereby analysis is allowed when there are formally perfect non-phonologic variations, even though definite meanings cannot be ascribed to the relevant morphs—i.e. actual linguistic forms. He gives this example: "the sets *deceive*: *receive*: :*decep-tion*: *recep-tion*: :*decei-t*: *recei(p)-t* justify the division *de* + *ceive* and *re* + *ceive*" (Greenberg, 1957:22).

The other feature discussed by Greenberg is that the *Semantically Defective Square* follows the same precepts as those for the perfect squares. These precepts, he asserts, can be applied when the technique is exclusively formal and the semantic differences of the morphs do not play a part in the analysis (Greenberg, 1957:24).

Semasiography

Semasiography, we are told by Gaeng, comprises "descriptive-representational" tools in the development of the writing stage. He also informs us that the concept originated from the Greek words *sēmasia* "meaning" and *graphē* "writing," through which the common meanings a writer intends to convey are represented in pictures (Gaeng, 1971:156).

In addition, according to Geoffrey Sampson, *Semasiography* refers to the depictions of what words, texts, concepts, or actions mean with visible marks "without reference to any specific spoken language" (Sampson, 1994:118). He supports his claim by restating it and providing an example as follows:

> It is indisputable that there exist systems of communication by visible marks which are independent of any particular spoken language. One example is the road sign system, which, for instance, uses the contrast between circular and triangular shape to distinguish command from warning, and displays a red and black car side by side to signal 'No overtaking'. Some of these signs (such as the one just cited) are partly iconic, others (e.g. the white horizontal bar on a red disc for 'No entry') are wholly arbitrary, but almost all of them are entirely independent of spoken language. It makes no sense to ask whether the first sign cited should be read as 'No overtaking' or as 'Overtaking is forbidden', or whether it should be read as an English or as a German phrase (Sampson, 1994:119).

Nonetheless, Sampson concedes that "no semasiographic script ever used in practice has approached the degree of generality and flexibility possessed by all spoken languages. Each such system has been limited to expressing messages relating to some narrow, limited domain, such as traffic discipline" (Sampson, 1994:119).

Semantic Subreption

Arlotto elucidates that *Semantic Subreption* is another form of semantic change referring to "the continued existence of a signifier, while the signified changes either in some real or some intellectual way (i.e. in terms of its understanding) without a disruption or sharp break in the history of the word" (Arlotto, 1972:181). Among the many examples provided by Arlotto to explicate *Semantic Subreption* is this one concerning the evolution of military hardware from the medieval to the modern era: "we still use the word *artillery* (← Middle English *artillerie*) to refer to the physical means of ground warfare. The word has continued on, but with new weapons of destruction replacing older ones, the reference for *artillery* has shifted slightly, at one point meaning catapults and crossbows, and at present meaning mortars, howitzers, and cannons" (Arlotto, 1972:181). In sum, according to Arlotto, the psychologically significant aspect of *Semantic Subreption* is that while speakers of a language are aware of the similarities of a word when used to mean different things, they still ignore those differences (Arlotto, 1972:183).

Also, Andrew Meehan expresses the fact that as a tool, *Semantic Subreption* "may be intentional and malicious, or attributable solely to ignorance" (Meehan, 1912:1). He adds that the phenomenon "may affect the primary, substantial reason or motive of the grant, or constitute merely a secondary or impellent cause of the concession" (Meehan, 1912:1). Citing canon law as an example in which "the concealment or suppression of statements or facts according to law or usage should be expressed in an application or petition," he argues that the effects of *Subreption* are "equivalent to obreption, which consists in a positive allegation of what is false" (Meehan, 1912:1).

African Linguistic Evidence

Beginning with Funwi F. Ayuninjam, we learn from him that in Mbili, a language spoken in Cameroon, there is abundant intersecting between the prefixes and the semantic classes; in certain instances, there is intersecting even within a noun class—a hybridized-classification that is somewhat frequent in Bantu languages. He also informs us that nouns and verbs in Mbili have independent group semantic features; as such, they are classified as two distinct categories, although they are connected in the sense that they share syntagmatic relations by which they are linked by forms. He further states that nouns and adjectives derive from verbs, but not vice versa, based on valency of range of usage, morphological dependence, semantic dependence, and semantic range (Ayuninjam, 1998:201-202). Moreover, he explains that as it pertains to semantic features, there exist verbs which express aspectual meanings but do not have overt grammatical markers, and that selectional restrictions in these features are analogous to collocational restrictions, albeit some are semantic and others are formal (Ayuninjam, 1998:334-335).

In his analysis of the *Semanticity* of the verb in Mbili, Ayuninjam first states that the semantic perspective of the language underscores its use and function, but which can escape a simple syntactic explanation. He adds that from a semantic perspective, the logical analysis of the constituents is more vital than their linear order (Ayuninjam, 1998:303). Thereafter, he segments his discussion into seven sections.

In the first section, Ayuninjam discusses the agentive, affected, benefactive, and attributive. He says that a subject generally performs the function of agentive—i.e. the organic element that facilitates the action signaled by the verb. The direct object generally performs the function of affected—i.e. the element directly affected by the action signaled by the verb and, thus, the term "affected." The affected element can be either organic or nonorganic. An indirect object generally performs the function of benefactive, or dative agent—i.e., of organic element being passively signaled by the occurrence or condition. And, the subject complement generally performs the function of attribute of the subject. As Ayininjam points out, there is a single principal kind of attribute in the Mbili language—the current or existing attribute that materializes with stative verbs. He speculates, however, about the possibility of the existence of a resulting attribute that could materialize with dynamic or process verbs (Ayuninjam, 1998:306-307).

Ayuninjam in the second section broaches the agentive and instrumental subject. He states that in addition to serving as an agent, the subject also performs the instrumental function in a clause, albeit this function is much less limited compared to that of agent. There is a subcategory of instrumental referred to as "natural forces," which possess the ability to trigger an action—examples include fires, floods, and winds. Other instrumental agents, referred to as "immediate causes," mostly act as instruments which organic agents employ to trigger actions. Nonetheless, for most case grammarians, the instrumental is regarded as a distinct case; no verb, because of its propositional content, needs it. Ayuninjam, however, provides many examples of instrumentals appearing as immediate causes and those appearing as dynamic natural agencies (Ayuninjam, 1998:307-308).

In the third section, Ayuninjam explores the benefactive subject. He points out that from a case grammar perspective, the benefactive acts as the holder of a non-agentive member or an object in the conveyance of possession. The benefactive can be either a gainer or loser, or either positive or negative. Thus, according to Ayuninkam, a subject can perform a benefactive function with Mbili verbs such as *bɛ* ("lose"), perceptual verbs like *yu'* ("hear"), which differ from non-perceptual verbs such as *kɛ* ("look") (Ayuninjam, 1998:308).

The fourth section entails Ayuninjam's examination of the locative, temporal, and eventive types of subject. Here, he indicates that clauses do possess a subcategory of subjects that act locatively to identify an action's spatiality, its temporality, and its representation by the verb. He supplies many examples to demonstrate the existence of the three types of subject that are operative in the Mbili language for the reader who is interested in more details (Ayuninjam, 1998:309-310).

Ayuninjam in the fifth section investigates the locative and eventive object. In this section, he talks about the fact that even though the object is generally linked to nominals acting either as affected parties or as benefactives, there are also a small number of nominals that are utilized in the position of the object in order to indicate places and occurrences in the Mbili language. He illustrates this *Semanticity* by providing many examples of a place where people trade as a locative and the act of trading as an eventive (Ayuninjam, 1998:310-311).

In the sixth section, Ayuninjam looks at the locative and affected object. He does not say much about the characteristic of this aspect here. He simply asserts that the Mbili language is pregnant with nominals that are employed to function as locatives and affected entities. He then provides many examples to support the assertion (Ayuninjam, 1998:311-312).

The seventh and final section encompasses Ayuninjam's treatment of the affected object. Here, he informs us that an affected object is one that is discerned only because of the action demonstrated by the verb. This type of verbs, he says, generally specify "once-and-for all," new actions like creating, doing, and inventing a novel entity. He adds that these actions are not usually replicable once they are effectuated. He provides a few examples of this phenomenon and illustrates their non-repeatability (Ayuninkam, 1998:312).

Next, Denis Creissels from the perspective of *Semanticity* explicates that "African languages that have gender generally have a morphologically complex form of plural marking, characterized by a fusion of gender markers and number markers, and variations in gender and number manifest themselves through morphemes affixed to the head noun and to (some of) its modifiers, in an agreement relationship. In these languages, plural marking tends to function on a 'semantic' basis, which means that plural markers tend to be present in every noun phrase referring to a plurality of individuals, irrespective of their communicative relevance" (Creissels, 2000:246-247). To illustrate this phenomenon, Creissels mentions the case of Hausa whose plurals separate into six fundamental kinds, most of them having two or more subtypes, and a few that are entirely irregular plurals. He adds that plural formation in Hausa involves both suffixes and vowel insertion (e.g., *gàrema:* pl. *garèmani:* 'large hoe'), consonant reduplication (e.g., *ko:fa:* pl. *ko:fo:fi:* 'doorway'), and change in tone (Creissels, 2000:247).

Also, from John R. Watters' exposition on *Semanticity*, we notice that in Central Sudanic languages such as Ma'di and Western Nilotic, the perfective demands subject-verb-object (SVO) and the imperfective demands a subject-object-verb (SOV) word order. He also shows us that most of the time in these languages, the first object has a benefactive semantic role (i.e. the one who benefits from the action), which is often called the "indirect object" (Watters, 2000:199).

In addition, Felix Banda explicates how he used the Semanticity and Grammatical Judgment Test (an adapted variety of the Grammaticality Judgment Test within the tradition of Transformational Grammar) to show the differences and similarities among English as a Second Language (ESL) learners in Zambia's primary, secondary and tertiary educational institutions vis-à-vis the probability of an "accessibility hierarchy" that spans from semantic features to formal features of language. From the perspectives of declarative and procedural knowledge, he argues that the two types of features reflect self-determining positions of linguistic depiction. Thus, if the features are not naturally hierarchically connected to each other, then it is not farfetched to state that a feature need not wait for the result of the other in order for it to be operative. Accordingly, Banda particularized the two kinds of linguistic representations along the formal and the semantic categories (Banda, 2000:1-2). The following is his rationale for using this technique:

> Knowledge represented within the semantic dimension can be said to be aligned to cognition as it is more closely related to the process of conceptualisation, than the intricate forms and modalities of syntax and morphology found in the formal dimension which tend to be sentential or language-specific. It has to be noted from the onset that semantic development is not synonymous with cognitive development, although the two could be more closely aligned, than morphosyntactic development and cognitive development. Therefore, the more sentential and language-specific a form is, the more difficult it is to process and access it (Banda, 2000:2).

Banda goes on to argue that in the end, because of their close link to the conceptualization procedure and, thus, to cognition and conception of the world, lexical-semantic occurrences are more effortlessly processed and accessible compared to syntactic-grammatical occurrences. Nonetheless, he cautions that an analyst must not ignore the fact that semantics and morphosyntax are connected when it comes to collocation regularities that deal with the denotations of lexemes and the fact that the end goal of grammar is to convey denotation. He also, however, points out that it has been observed by some linguists that it is not implausible to envisages semantics as free of morphosyntactic particularities of language so that its rational constituent can be straightforwardly identified without utilizing the formal particularities of grammar (Banda, 2000:2-3).

The findings from Banda's investigation lead him to propose that if classroom instruction is to parallel a learner's innate disposition, then it should be based on denotation and augmented by "operating conditions" assignments to generate meaningful grammatical forms that are usually unreachable and abstruse to learners. As such, teachers in a multilingual classroom must facilitate a context whereby learners will be provided techniques that will allow them to expand on the denotations and forms of English, thereby preventing them from fossilizing their native languages into English "vernaculars." Banda says that this technique calls for instruction that is geared toward creativity and reflection on the types of interaction the learner experiences in the real world (Banda, 2000:10).

Furthermore, Robert Nicolaï in his investigation of the typological resemblance that exists between the Mande and Songhay languages spoken in West Africa seeks to determine whether the resemblance reflects a case of linguistic convergence as proffered by some scholars. But, as he points out, the premise on which this proposition hinges has yielded a very different elucidation of the isomorphisms in a study of languages in contact. He, therefore, suggests that the typological resemblance itself should be analyzed more critically in order to increase our knowledge of the languages and the procedures of language change in West Africa. Correspondingly, for his study, he (a) presents a precise geographical and sociolinguistic context of the Mande and Songhay languages and the suggested genealogical connections between them; (b) provides an account of the basic resemblances that make up the isomorphisms; and (c) broadens the coverage of the procedures of change, the general history of the languages in the region, and how the linguistic structures are arranged and justified. The results that emerged from Nicolaï's systematic analysis propel him to conclude that linguistic and historical hypotheses can be fused together to yield substantive findings on typological resemblance, albeit prudence must be exercised given the newness of the endeavor (Nicolaï, 2007:1).

African Mathematical Evidence

Starting with what Cheikh Anta Diop educates us about Problem 64 of the *Rhind Papyrus* it is evident that the mathematical calculation invokes the notion of *Semanticity,* since dealing with a distribution of differences is an arithmetic progression that encompasses a meaning and function. Correspondingly, as Diop states, the problem "consists of dividing ten loaves of bread among ten persons, in such a manner that the difference between two consecutive persons is one-eighth of a *hekat.* One obtains the same result as the scribe applying the classical formula of an arithmetic progression" (Diop, 1981/1991:270):

$$l = a + (n - 1)d$$

where l = the last term, a = the first term, and d = the common difference: $^1/_8$.

Next, we glean from Dirk Huylebrouck's article on mathematics in central Africa before the advent of colonialism in the region that the following number words in the Logo language spoken principally in North East Congo is accompanied by the semantic representation that comes after it:

> *1 = alo; 2 = iri; 3 = na; 4 = su; 5 = nzi; 6 = kazya; 7 = nzi drya iri (= 5 + 2) ; 8 = nzi drya na (= 5 + 3); 9 = nzi drya su (= 5 + 4);*
> *10 = mudri; 11 = mudri drya alo; 12 = mudri drya iri; 16 = mudri drya kazya (= 10 + 6); 17 = mudri drya nzi drya iri (= 10 + 5 + 2); 18 = mudri drya nzi drya na (= 10 + 5 + 3); 19 = mudri drya nzi drya su (= 10 + 5 + 4);*
> *20 = nyaba alo (= 20 × 1); 30 = nyaba alo drya mudri (= 20 × 1 + 10); 40 = nyaba iri (= 20 × 2); 50 = nyaba iri drya mudri (= 20 × 2 + 10); 60 = nyaba alo (= 20 × 3); 70 = nyaba na drya mudri (= 20 × 3 + 10); ... ; 100 = nyaba nzi (= 20 × 5);*
> *200 = nyaba nzi iri (= (20 × 5) × 2); 300 = nyaba nzi na (= (20 × 5) × 3); ... 600 = nyaba nzi kazya (= (20 × 5) × 6); 700 = nyaba nzi nzi drya nayba nzi iri (= (20 × 5) × 5 + (20 × 5) × 2); 800 = nyaba nzi nzi drya nayba nzi na (= (20 × 5) × 5 + (20 × 5) × 3); 700 = nyaba nzi nzi drya nayba nzi su (= (20 × 5) × 5 + (20 × 5) × 4);*
> *1000 = nyaba nzi mudri (= (20 × 5) × 10)* (italics as in the original; Huylebrouck, 2006:142-143).

The *Semanticity* for the preceding number words is described as follows:

> *The symbol / is the semantic representation for the word "one" or a "unity". For instance, the group of symbols ////////// corresponds in the same way with the word ten, but also with the related morphemes "-ty" and "-teen". [...] M can in the case of the English language for example be "-ty", "hundred", "thousand" etcetera* (italics as in the original; Huylebrouck, 2006:143).

Huylebrouck posits that the foregoing evidence underscores the fact that the power of Mathematics hinges upon its general capacity to devise logical linguistic rules for all languages across the globe despite the significant differences (Huylebrouck, 2001:142).

Next, from Jonathan M. O. Chimakonam, we learn about several characteristics pertaining to the nexus between *Semanticity* and quantification in African logic. First, he asserts that meaning is disguised for Africans. Many utterances made by Africans, whether the utterances are encoded in formal or in metalanguage, do not lead a receiver straight to their semantic thrust. He illustrates this proposition with the example of an Igbo saying that "Okonkwo is a tortoise" and explains that the expression does not denote that Okonkwo is an animal, but that he is "crafty." Similarly, as Chimakonam notes, natural African languages possess signatures that are distinctive from what they appear to encompass (Chimakonam, 2013:417). As he also explicates this phenomenon,

> ...an assignment of semantic meaning to a logical signature is called interpretation. Now, an interpretation of say a first-order predicate language assigns a denotation to all non-logical constants in that language. In addition, it determines a domain of discourse, i.e., subject matter and scope which, specifies the range of the quantifiers. In other words, an interpretation also tells the African logician which variables are free and which are bound by which quantifiers. Normally, under a given interpretation, each term is assigned an object that it represents and each sentence is assigned a truth value but unlike in the western logic, this is not done arbitrarily in African logic (Chimakonam, 2013:417).

Second, Chimakonam points out that the semantic of African logic is engendered from the subject matter of what is referred to as "logical custom" instead of "logical form." This means that a non-African analyst must be schooled in the signature of African utterances so that s/he is not misdirected by the literary acclimatization of those utterances. In essence, it is imperative for a non-African analyst to know what an African denotes when s/he expresses logical statements which in most cases differ from the literal references of those expressions. In essence, a sentence's truth value hinges upon the semantics of the discourse context. To explain this point, Chimakonam states that Africans do not, for instance, say that "snow falls in Abakaliki during the wet season" and arbitrarily ascribe the value true to the statement when they know that it is really untrue. This is a case of the nexus between African logic and "relevance logic," whereby "the claims of the premises must be relevant to the conclusion and the negation of the conclusion in necessarily non-complementary with the premises. So in African logic, an interpretation provides actual semantic meaning to the terms and formulae of the language" (Chimakonam, 2013:417-418).

Third, as Chimakonam tells us, while the exploration of the elucidations of traditional languages would be referred to as "formal semantics" in Western logic, it is called "customary semantics" in African logic. Concomitantly, African predicate logic is emerging as a systematically profitable domain for assessing truth values. In this regard, an explanation is provided for true, true-false, or false based on a formulaic assessment and a variable designation y which connects a constituent of the discourse context with a variable. From an African logician's perspective, this phenomenon cannot be done arbitrarily; instead, it could be done only in terms of the subject matter of the discourse context. Put differently, reality dictates how the formulae and variables should be assessed as true, true-false, or false (Chimakonam, 2013:418). Thus, Chimakonam provides the following rules for assigning truth value:

> Variables: each variable u with an assignment y evaluates to $y(u)\longmapsto {}_{M}T(u)vF(u)$
>
> Functions: given terms $t1,\ldots t_n$ that have been evaluated 1o elements $g1,\ldots g_n$ of the domain of discourse, and a n-ary function symbol f, the term $f(t,\ldots t_n)$ evaluates to $(j(f)(g,\ldots g_n)$ (Chimakonam, 2013:418).

Given the preceding rules, Chimakonam proposes that a formula is hypothecated a truth value based on the real value of the subject matter it constitutes, as if logicians are empowered to do so; what is really done, he maintains, is the designation of the subject matter represented by a formula or variable. It is here, according to him, that "the power and discretion of the African logician ends, the values for such formulae or variables naturally reveal themselves to the logician. Hence, the values to be assigned to any given formula and variable in African logic are determined by the subject-matter of the domain of discourse" (Chimakonam, 2013:418-419).

Conclusion

Semanticity for Ancient Kemetians/Egyptians can be regarded as *ā*, one of the words for "mean" which denotes what an entity "means," "states," or a "condition" of the entity (for this translation, see Budge, 1978). Thus, as Ellyn Lucas Arwood tells us, in terms of developing formal *Semanticity* of time, semantic meaning in Kemetian culture was not about the here-and-now (Arwood, 2011:348). In addition, as Jan Assmann also informs us, "world reference of Egyptian hieroglyphic writing lies in the *materiality* of the sign and not in what we call *semanticity* [and] Egyptian cursive scripts took this path and developed within the independent laws and paths of graphic systems" (Assmann, 1994:17-18). Moreover, as Assmann further points out, while Hieroglyphics continued its pictorial authenticity, it demonstrated that it is "not a 'visible language' in the complete sense, but is more than just a script, involving more than just language reference. This 'more' is based on its pictorialness; it is therefore 'world reference'" (Assmann, 1994:18).

Chapter 5

Pragmatic Function

Elsewhere (Bangura, 2014) based on the work of Anouschka Bergman, Kathleen Curie Hall and Sharon Miriam Ross (2007), I state that "a communication system is said to have a *Pragmatic Function* when it serves some useful purpose. Such functions include language used to help individuals to stay alive, to influence others' behaviors, to find out more about the universe, to ask for food, to yell for help, and to influence voting behavior. Some communicative acts such as gossip are usually questioned, but they do have a pragmatic function in helping us to understand our social environment" (Bangura, 2014:20). Given this backdrop, I must begin with general definitions of *Pragmatics*, from where the idea of *Pragmatic Function* emerged, and the idea itself, before getting into the discussion of its nexus with African languages and mathematics.

To start with, David Crystal provides the following general definition for *Pragmatics* as "the study of language from the point of view of the users—especially of the choices they make, the constraints they encounter in using language in social interaction, and the effects their use of language has on the other participants in an act of communication" (Crystal, 1992:310). He goes on to describe the two subareas of *Pragmatics*. One subarea is *General Pragmatics* in which "the principles governing the communicative use of language, especially as encountered in conversation" are investigated (Crystal, 1992:310). The other subarea is *Applied Pragmatics* in which "verbal interaction in such domains as counselling, medical interviews, language teaching, and judicial sessions, where critical problems of communication are of crucial importance," are studied (Crystal, 1992:310).

Also, William O'Grady, Michael Dobrovolsky and Mark Aronoff characterize *Pragmatics* as a corpus of knowledge involving the interpretation of a sentence in terms of "the speaker's and addressee's background attitudes and beliefs, their understanding of the context in which a sentence is uttered, and their knowledge of the way in which language is used to communicate information" (O'Grady et al., 1989:181). They add that the preferences privileged in pragmatic analysis are independent of grammatical rules. Instead, these pragmatic preferences are a reflection of the different perspectives people have in a society (O'Grady et al., 1989:181). These scholars further mention that the word *meaning* (which itself has many denotations), logic, paragraph meaning, etc. have also been analyzed in the area of *Pragmatics*, which leads them to define the area generally as "the study of how language is used in communication" (O'Grady et al., 1989:438).

Next, Francis P. Dinneen, while not explicitly providing a definition of *Pragmatic Function*, does so implicitly by restating the following extract from Bronislaw Malinowski's *Coral Gardens*, II:62:

> In all [of] the child's experience, words when seriously uttered *mean* in so far as they act. The intellectual function of words probably develops later, and develops as a byproduct of the pragmatic function...Thus the source of the magical attitude towards words is, if the theory developed here is correct, to be found in the use of words by infants and children. Thus start also those profoundly pragmatic ways of learning how to use the word by learning how to use its counterpart in the reality of behavior (Malinowski quoted by Dinneen, 1967:300).

Based on the preceding excerpt, Dinneen, therefore, proffers that language is a mode of social undertaking and collaboration, and that the denotation of a remark in a specific set of situations is to be envisioned in terms of its impacts on the environment, which speech attempts to conserve or change (Dinneen, 1967:300-301).

In order to examine the notion of pragmatic function of language, I divide the discussion of the rest of this chapter into three parts. The first part deals with the linguistic expositions on the subject. The second part entails African linguistic evidence. And, the third part encompasses African mathematical evidence. The chapter ends with a conclusion.

Linguistic Expositions on Pragmatic Function

A variety of pragmatic functions of language have been delineated by a number of linguists. The following are discussed here: (a) deictic functions; (b) conversational implicatural functions; (c) presuppositional functions; (d) speech act functions; (e) conversational structural functions; (f) ideational, interpersonal, and textual functions; (g) tact maxim functions; (h) communicative grammatical functions; and (i) anaphora controlled functions. For the sake of coherence, these pragmatic functions are presented in this section separately. Also, it is proper for me to mention here that most of the discussion in this section is based on the works on Pragmatics by Geoffrey N. Leech (1983) and Stephen C. Levinson (1983), which are the most consulted on the subject, and the work on Discourse Analysis by Gillian Brown and George Yule (1983), which is equally the most consulted on the latter subject; coincidentally, all of these works were published in the same year—i.e. 1983.

Deictic Functions

To begin with, *Deictic Functions* can be defined as the concern about the manner in which the link between language and context is manifested in language systems. Levinson defines *deixis,* the root word of these pragmatic functions, as a Greek borrowing for "pointing or indicating, and has as prototypical or focal exemplars the use of demonstratives, first and second-person pronouns, tense, specific time and place adverbs like *now* and *here,* and a variety of other grammatical features tied directly to the circumstances of utterance" (Levinson, 1983:54).

Next, according to Levinson, the fundamental concern of deixis is about the methods in which languages grammaticalize or encode the characteristics of a speech event or context of utterance. Deixis therefore also deals with the methods in which the interpretation of utterances hinges upon the analysis of that particular context of utterance. Levinson cites the pronoun *this* as not naming or referring to any specific entity each time it is employed; instead, it is a place-holder or a variable for a certain entity based on the context—e.g., a gesture (Levinson, 1983:54).

Also, Levinson argues that despite the fact that many elements of deixis are prevalent in natural languages and are deeply grammatricalized, deixis does not belong in the area of semantics, but in that of pragmatics. This is because (a) since semantics does not incorporate all conventional aspects of meanings, then most deictic occurrences cannot be regarded as semantic; (b) deixis directly deals with the connection between the structure of languages and the context in which they are employed; and (c) deixis deals with the encoding of many various attributes of the circumstances surrounding an utterance and within that utterance itself. Nonetheless, Levinson cautions that "all such categorizations are theory-dependent, and on the view that...pragmatics concerns those aspects of meaning and language-structure that cannot be captured in a truth-conditional semantics, the grammatical category of deixis will probably be found to straddle the semantics/pragmatics border" (Levinson, 1983:55).

In addition, Levinson identifies two groups of approaches that have been utilized to analyze deixis. The first group comprises philosophical approaches. Levinson says that philosophers normally refer to deixis as indexical expressions, or simply indexicals, and analyze them in terms of truth-conditional semantics that addresses particular natural language expressions. He also points out that expressions with the context-dependent characteristic such as morphemes designating tense, first and second-person pronouns, and demonstratives have attracted the attention of philosophers. It was these philosophers that named deixis indexical signs and proffered that they impelled a referent by an existential relation between sign and referent. Levinson adds that the emergence of these philosophers' interest on the subject was based on the following two aspects: (1) "whether all indexical expressions can be reduced to a single primary one" and (2) "whether this final pragmatic residue can be translated into some external context-free artificial language" (Levinson, 1983:57). He further notes that these philosophers argued that the first aspect was possible by translating all indexicals, or as they called them "egocentric particulars," into expressions that have *this*, where the second aspect is noted as a subjective episode. These philosophers also contended that all indexicals necessitate a facet of token-reflexivity (meaning that they refer to themselves); thus, the pronoun *I* would be considered as "the person who is uttering the token of the word *I*." Levinson argues that while this analytical approach may seem appealing at first glance, it does, however, present many challenges: for example, even though token-reflexivity or self-referring expressions are plentiful in natural languages, they pose serious challenges for logical analysis, especially when deictic expressions cannot be rendered into context-independent terms without losing meaning (Levinson, 1983:57-58).

Thereafter, Levinson suggests that a way to permit context-relativity would be to treat a proposition stated by a sentence in a context as a basis of prospective spheres "*and that context* to truth values." Here, he denotes context as "a set of pragmatic indices, co-ordinates, or reference points (as they are variedly called) for speakers, addressees, times of utterance, places of utterance, indirect objects, and whatever else is needed. Sentences can, therefore, express different propositions on different occasions of use" (Levinson, 1983:58). He further suggests that indexicals could also be addressed in terms of the specification of the content of an utterance as a two-pronged process: (1) "the 'meaning' of utterance is a function from contexts (sets of indices) to propositions" and (2) the propositions "are in turn functions from possible worlds to truth values" (Levinson, 1983:58-59).

Thereupon, Levinson states that a subject of deep theoretical importance to the philosophers is the nexus between indexical reference and the elements of reference in general. As he retells, in the early stages, many philosophers who wanted to investigate reference focused very little on indexicals; they later started to analyze them as particular types of expression needing indices or context co-ordinates. They then started to inquire about whether many types of referring expressions could be secretly indexical in certain usages, including (a) *referential usage*, whereby "the description might, in fact, be wrong but the reference succeeds in any case"; and (b) *attributive usage*, whereby "the speaker would not have any particular individual in mind" (Levinson, 1983:59-60). Levinson, however, finds these usages problematic for the following reason: "It is the speaker's intention and the addressee's successful location of the intended referent that matter in the first usage, not the exact aptness of the description, so that we could call this usage *speaker reference* (as opposed to *semantic reference*)" (Levinson, 1983:60). In essence, Levinson is not convinced that these philosophical approaches sufficiently address the many complex and different deictic expressions that eventuate in natural languages.

The second group of approaches that have been used to analyze deixis is composed of the descriptive variety. Expressing disappointment with the very little amount of descriptive work that has been done on *Deictic Functions* and, consequently, the lack of suitable theories and analytical frameworks, Levinson proceeds to discuss a spate of exploratory categories that have been suggested, with some examples of their implementation.

For starters, Levinson discerns three traditional classes of deixis. The first traditional class is the *person deixis* which deals with the encoding of the *roles* played by participants in a speech occurrence in which the particular utterance is made. The grammaticalization of the speaker's reference to himself/herself is encoded as *first person*, the speaker's reference to one or multiple addressees is encoded as second person, and the reference to persons or things that are not the speakers or addressees are encoded as *third person*. Levinson mentions that pronouns and their attendant predicate agreements are the ways participant-roles are commonly encoded in language (Levinson, 1983:62).

The second traditional class is the *place deixis* which pertains to the encoding of spatial locations *relative* to the location of those participating in a speech occurrence. Levinson states that most languages grammaticalize at least a difference between close to a speaker in terms of *proximal* and close to an addressee in terms of *distal* or *non-proximal*. He adds that in the English language, place deixis are usually encoded in demonstratives such as *this* versus *that* and in deictic adverbs such as *here* versus *there* (Levinson, 1983:62).

The third traditional class is the *time deixis* which relates to the encoding of temporality *relative* to the time an utterance is made or a text is written. Levinson notes that time deixis are either *coding time* or *receiving time*. He also points out that in the English language, time deixis are normally grammaticalized in deictic adverbs of temporality such as *now* and *then, today* and *this week*, and primarily in tense (Levinson, 1983:62).

Next, in addition to the preceding traditional classes, Levinson broaches two additional classes of deixis that have emerged. The first is *discourse* or *text deixis* which encodes the reference to parts of the evolving discourse in which the utterance, with the text referring expression, is situated. Examples of discourse deixis provided by Levinson are the utilizations of *that* and *this* in (a) "Puff puff puff: *that* is what it sounded like" and (b) "*This* is what phoneticians call a creaky voice" (Levinson, 1983:62-63). The second additional class is the *social deixis* which encodes the social differences between "participant-roles," especially features of the social relationship that exist between speaker and receiver(s) or speaker and a particular referent (Levinson, 1983:63). Levinson also asserts:

> In many languages, distinctions of fine gradation between the relative ranks of speaker and addressee are systematically encoded throughout, for example, the morphological system, in which case we talk of *honorifics*; but such distinctions are also regularly encoded in choices between pronouns, summons forms or vocatives, and titles of address in familiar languages. Deictic systems in natural languages are not arbitrarily organized around the features of just any of the different kinds of medium and context in which languages are used. Rather there is an essential assumption of that basic face-to-face conversational context in which all humans acquire language...Further, it is generally (but not invariably) true that deixis is organized in an egocentric way. That is if (for the purposes of semantic or pragmatic interpretation) we think of deictic expressions as anchored to specific points in the communicative event... (Levinson, 1983:63).

Since their work is on Discourse Analysis, Brown and Yule approach the idea of *Deictic Function* slightly differently. First, they indicate that deictic forms such as *I, you, this, that, here, there,* and *now* are among the most evident linguistic features for which contextual information is required in order to explain them. They say that the interpretation of these deictic features in a stretch of discourse calls for an analyst to gain knowledge about at least who the speaker and hearer are, and the place and time the discourse was constructed (Brown and Yule, 1983:27).

Second, Brown and Yule address the question about whether deictic expressions like the temporal *now*, the spatial *here*, the first person *I*, which relate directly to *deictic context*, can be interpreted based on the typical processes for deducing what information is pertinent for explaining these expressions. They begin by asserting that there indeed exist common systems for identifying points temporally and spatially. The time of an utterance can be said to have covered between 9:00 a.m. to 12:00 p.m. on April 10, 2019, referring to the good common systems of clock and calendar. The place of the utterance could then be stated in terms of latitude and longitude should there exist good relevant instrumentation. It is less likely, however, that a speaker can recall the exact date and time of a message the further away it was temporally established (Brown and Yule, 1983:51).

For deictic expressions such as *here*, Brown and Yule suggest that the standard processes can allow an analyst to establish the space spans for their interpretation. Thus, expressions like *here* and *now* can be employed in "displaced contexts." In essence, deictic expressions may keep a normal deictic core but must be explained in terms of (a) the context of the utterance in which they are manifested; (b) the appropriate normal temporal explanation of an utterance; (c) the knowledge and intention of the speaker (or analyst) in referring to the utterance as spatially located (Brown and Yule, 1983:52-53).

In terms of the *deictic role* of an individual, Brown and Yule maintain that what is of interest is the designation of first, second, and third-person pronouns and his/her status or *social role.* These include the terms of address used by a social superior to a social inferior and vice versa, between peers, etc. (Brown and Yule, 1983:54).

Conversational Implicatural Functions

Beginning with a general definition for *Conversational Implicatural Functions*, they can be denoted as the bases of how utterances can convey more than what is actually said, written, or communicated through gestural language. Thus, it is only fitting that as Levinson asserts, *conversational implicature* (or *implicatore* for short) has emerged as one of the most important notions in the subject of Pragmatics and its prominence in the subject hinges on five sources. The first source, according to him, is implicature's standing as a paradigmatic instance of the nature and power of pragmatic explications of linguistic occurrences. The elements of this source exist outside the structure of language, in certain common principles for cooperative interaction; nevertheless, these principles have a great impact on the structure of language. In essence, the idea of implicature appears to provide important functional accounts of linguistic details (Levinson, 1983:97).

The second source of implicature's prominence in Pragmatics identified by Levinson has to do with how the subject contributes to the provision of an explicit explanation of how what is actually said can mean generally more than what is uttered: "i.e. more than what is literally expressed by the conventional sense of the linguistic expression uttered" (Levinson, 1983:97). He, therefore, argues that semantic theory can only account for a small portion of how language is used to communicate because the theory accounts for the substantial gap between what is directly expressed and what is conveyed. Thus, he also asserts that implicature is better situated to address that substantial gap (Levinson, 1983:98).

The third source of the prominence of implicature in Pragmatics, according to Levinson, is that the idea of implicature appears to cause significant simplifications in the content and organization of semantic explanations. He suggests that this notion is beyond the semantic approach which normally takes one of the following two perspectives: (1) there are different meanings of words, thereby rendering them ambiguous; or (2) the denotations of words are normally indistinct and ever-changing and are affected by collocational milieus. Thus, he also suggests that implicature provides a solution in that it permits an analyst to aver that "natural language expressions do tend to have simple, stable and unitary senses (in many cases anyway), but that this stable semantic core often has unstable, context-specific pragmatic overlay—namely a set of implicatures" (Levinson, 1983:99).

The fourth source of the prominence of implicature in Pragmatics Levinson edifies us about is that implicature appears to be indispensable when different fundamental details of language are to be adequately considered. He cites the case of particles such as *well, anyway, by the way* as needing certain meaning specification in a theory of meaning as any other word in the English language; but in determining their denotation, we end up drawing upon pragmatic procedures that yield implicatures. Consequently, some syntactic rules appear to be sensitive to implicature, which in turns places limitations on what can be a potential lexical element in natural languages (Levinson, 1983:100).

The fifth and final source of implicature's prominence n Pragmatics Levinson apprises us about concerns the extensive explanatory power of the postulates that engender implicatures. This makes it possible for a wide assemblage of seemingly unconnected particulars to be explicated by utilizing a small number of fundamental propositions. This is the reason, according to Levinson, "English has no lexical item *nall* meaning 'not all', for why Aristotle got his logics wrong, for 'Moore's Paradox' [i.e. the evident absurdity involved in asserting a first-person present-tense sentence such as *It is snowing, but I don't believe it is snowing* or *It is snowing but I believe that it is not snowing*], for why obvious tautologies like *War is war* can convey any conceptual import, for how metaphors work and many other phenomena besides" (Levinson, 1983:100).

Next, Levinson enlightens us about Herbert Paul Gracie's theory of implicature, which he (Levison) says is a concept about how language is used by people. Levinson starts off by stating that Grice's proposal is that undergirding the execution of conversation is a series of comprehensive assumptions based on fundamental rational considerations for the effective and efficient utilization of language to achieve cooperative communication outcomes (Levinson, 1983:101).

For Brown and Yule, Grice's use of the concept *implicature* is geared toward considering what a speaker implies, means, or suggests as sui generis from what s/he literally utters. They add that Grice believes that the conversational denotations of words employed in a conversation are shaped by *conventional implicatures.* These *conventional implicatures,* they also say as Levinson does, are the results of a basic principle of conversation with its four attendant maxims to which speakers generally adhere during a conversation (Brown and Yule, 1983:31).

Levinson (1983) and Brown and Yule (1983) go on to provide a summary of what Grice says are the *maxims of conversation* or common propositions that underlie the cooperative employment of language that together express a basic *cooperative principle* as follows (Levinson, 1983:101-102; Brown and Yule, 1983:31-32):

> *The cooperative principle*
> make your contribution such as is required, at the stage at which it occurs, by the accepted purpose or direction of the talk exchange in which you are engaged
>
> *The maxim of Quality*
> try to make your contribution one that is true, specifically:
> (i) do not say what you believe to be false
> (ii) do not say that for which you lack adequate evidence
>
> *The maxim of Quantity*
> (i) make your contribution as informative as is required for the current purposes of the exchange
> (ii) do not make your contribution more informative than is required
>
> *The maxim of Relevance*
> make your contributions relevant
>
> *The maxim of Manner*
> (i) be perspicuous, and specifically:
> (ii) avoid obscurity
> (iii) avoid ambiguity
> (iv) be brief
> (v) be orderly

The preceding maxims, suggests Levinson, indicate what participants in a conversation must do so that they can communicate in a highly cooperative, efficient and rational manner. He adds that the maxims enjoin participants to provide enough information that is clear, relevant and sincere when they engage in a conversation (Levinson, 1983:102). Brown and Yule, however, point out that Grice's list is not exhaustive and that the maxims do not possess equal weights (Brown and Yule, 1983:32).

Presuppositional Functions

A basic definition for *Presuppositional Functions* is specific raison d'être of how background assumptions against which the main import of an utterance or statement can be assessed. Levinson deliberates five major aspects of these functions. First, by way of introduction, he describes *Presupposition* as a pragmatic inference that appears to be situated in the actual linguistic system of sentences, although such inferences cannot be regarded as semantic in the restricted sense since they are quite sentient of contextual elements. He recounts that presupposition as a subject emerged as a focal area of linguistic theory between 1969 and 1976 because it raised very important questions about all sorts of generative linguistic theories at the time. Despite the widespread interest in presuppositional analysis, however, as he suggests, the technical perspective of presupposition remains on specific pragmatic inferences or assumptions that appear to hinge upon linguistic elements and which can be identified with the use of certain linguistic tests (Levinson, 1983:167-168).

Second, from a historical perspective, Levinson chronicles the roots and evolution of the subject of presupposition, which he says was birthed by the debate in philosophy dealing with the inherent features of referring expressions and reference. At the core of the debate on logical theory was the cogitation of how referring expressions in natural language should be interpreted vis-à-vis confined logical languages. Levinson recalls that one group of early philosophers working on the subject proffered the perspective that (a) uses of sentences, particularly assertions, have presuppositions; (b) sentences themselves have presuppositions; and (c) speakers hold presuppositions. The other group dismissed this point of view, arguing that it yields inconsistencies. This latter group tendered the *theory of description*, which posits that "definite descriptions like *The so & so* have nothing like the simple logical translation that one might imagine. Whereas they occur in natural languages as subjects...in logical form they are not logical subjects at all but correspond instead to conjunctions of propositions" (Levinson, 1983:170-171).

Third, based on the initial observations on presuppositional phenomena, Levinson notes that the earlier proposition that presuppositions are maintained in negative statements or sentences (i.e. the constancy under negation rule) is the preliminary operational measurement for determining presuppositions. This straightforward measurement entails taking a statement or sentence, negating it, and seeing what inferences remain and that are shared by the negative and positive statements or sentences. In essence, while negation makes changes to the entailments of a statement or sentence, the procedure leaves the presupposition unchanged. Consequently, according to Levinson, presuppositions appear to be bind to certain words or general features of surface structure. He calls these presupposition-generating linguistic elements *presuppositional triggers* and lists them (with attendant descriptions and examples which are omitted here) as follows: (a) definite descriptions, (b) factive verbs, (c) implicative verbs, (d) change of state verbs, (e) iteratives, (f) verbs of judging, (g) temporal clauses, (h) cleft sentences, (i) implicit clefts with stressed constituents, (j) comparisons and contrasts, (j) non-restrictive relative clauses, (k) counterfactual conditionals, (l) questions, and (m) other phenomena that could be better addressed under various features of pragmatic theory (for details, see Levinson 1983:177-185).

Fourth, focusing on the problematic properties of presupposition, Levinson begins by asserting that constancy under negation is not a copious approach for teasing out a collection of logical and correlative inferences. He mentions that presuppositions have the following two characteristics: (1) "they are *defeasible* in (a) certain discourse contexts, (b) certain intra-sentential contexts"; and (2) "they are apparently tied to particular aspects of surface structure" (Levinson, 1983:186). He also states that the first characteristic eliminates any possibility for a semantic theory of presupposition while the second characteristic could be used to differentiate conversational implicatures from presuppositions (Levinson, 1983:186).

Next, Levinson points out that defeasibility is among the critical features of presuppositional behavior and against which all theories of presupposition must be evaluated. He adds that another difficult feature of presuppositions is what is called *the projection problem*: i.e. how presuppositions behave in complex sentences (for more details, see Levinson, 1983:186-198).

Fifth, Levinson illustrates the two kinds of presuppositional explanation. One is semantic presupposition. He identifies two major groups of semantic theories employed to explain presupposition. One is the truth-conditional group of theories which clearly states what cannot be encapsulated in semantics. The other is the group which states that every semantic relation can be defined apropos translations of sentences into semantic attributes or atomic ideas. Levinson indicates that for analysts to include presupposition into truth-conditional theories, they have to treat presupposition as a special kind of entailment in which a logical consequence relation can be denoted in a manner that exempts it from negation. He adds that, as mentioned earlier, defeasibility and the projection problem assure the ill-fate of semantic theories of presupposition (Levinson, 1983:199).

The other kind of presuppositional explanation encompasses pragmatic theories of presupposition. According to Levinson, the earlier variety was programmatic, using pragmatic concepts to offer workable definitions of presupposition. Despite using varying jargon, these theories employed two fundamental concepts: (1) *appropriateness* or *felicity* and (2) *mutual knowledge* or *common ground* or joint assumptions (Levinson, 1983:204-205). Concomitantly, Levinson makes the following observation: "The idea, then, was to suggest that there are pragmatic constraints on the use of sentences such that they can only be appropriately used if it is assumed in the context that the presuppositions indicated by the presupposition-triggers are true" (Levinson, 1983:205). Thus, he further argues that "to utter a sentence whose presuppositions are, and are known to be, false, would merely be to produce an inappropriate utterance, rather than (on the semantic view) to have asserted a sentence that was neither true nor false" (Levinson, 1983:205).

Based on Levinson's discussion, four pragmatic theories on presupposition can be noted to have emerged after the earlier ones. The first of these latter theories is the *Montague grammar* framework which constructs clause from the bottom up instead or from top to bottom as is done in transformational generative grammar. Therefore, for this framework, associated with any word, clause or syntactic functioning is a semantic representation or extension expression. The second theory which seeks to address the projection problem also deals with contextual defeasibility problems. The theory likewise treats presuppositions as non-truth-conditional characteristics of the meaning of linguistic expressions. This theory begins with the generation of all *potential presuppositions* of a sentence as a complete set. Thereafter, through a cancellation process, it eliminates those propositions that do not survive and leaves those that do as the *actual presuppositions* of a sentence uttered in a specific context. The third theory postulates that all of a sentence's entailments do not have the same face value. Thus, following this theory, a viable semantic representation would have an ordered assortment of these entailments categorized into (a) *background* and (b) *foreground* entailments, which are determined by grammatical form (including stress) instead of logical scrutiny. The fourth theory focuses mostly on logical form, or the structure of a semantic representation, in the generation of pragmatic inferences. Two complex logical devises used in this theory are (1) *lambda extraction* employed to build complex characteristics and (2) *group-* or *gamma-operator* utilized to establish shared terms (Levinson, 1983:207-221).

Speech Act Functions

Speech Acts Functions can be generally defined as how to use words to do things. To assert that among all of the topics regarding *Semantic Function*, *Speech Act Functions* have enjoyed the most exploration is hardly a matter of dispute. It is, therefore, no coincidence that Levinson begins his examination of *Speech Acts* by indicating that (a) psychologists have looked to concepts underlying speech acts to suggest that they may be the necessary condition for acquiring language in general; (b) literary critics have turned to speech act theory to analyze subtleties in texts and the inherent features of literary genres; (c) anthropologists have utilized speech act theory to explain spells and ritual in general; (d) philosophers have found speech act theory to be efficacious in investigating ethical statements; and (e) linguists have employed various notions of speech act theory to probe issues in syntax, semantics, second language learning, etc. He also mentions that for linguists working in the area of pragmatics, speech acts, in addition to implicature and presupposition, are core features which any general pragmatic theory must consider (Levinson, 1983:226).

But, before going any further, I will first briefly discuss the origin of the theory of speech acts, what they are, and the basic types. As Brown and Yule apprise us, Speech Act Theory can be traced to John Langshaw Austin's observation (in a series of his lectures published by his students as *How to Do Things with Words* in 1962) that sentences can be used not only to report states of affairs, but also to perform certain acts in some circumstances. Such utterances, which Austin called "performatives" and the particular circumstances needed for their success he named "felicity conditions." According to Brown and Yule, Austin's performatives "are not just a group of ritual sentence forms, but are also a subset of the utterances in the language which can be used to perform acts" (Brown and Yule, 1983:231). Leech (1983), Levinson (1983), Brown and Yule (1983), and O'Grady et al. (1989) all mention three basic speech acts: (1) the *locutionary act*, which correlates with an utterance of a sentence with its specific meaning; (2) the *illocutionary act*, which indicates the intention of the addresser when s/he utters a sentence; and (3) the *perlocutionary act*, which is the effect of the addresser's utterance on the addressee.

Levinson dissects the notion of *Speech Acts* into seven sections. In the first section, he discusses the philosophical background of speech acts. He points out that the 1930s witnessed the philosophical surfeit of *logical positivism*, whose core precept was that a sentence can fundamentally be *verified* (i.e. checked for its falsity or truth), become fatuous. This is because most aesthetic, ethical and literary discourses, in addition to most everyday utterances, simply lacked meaning. But, instead of accepting this verity as a *reductio ad absurdum*, proponents of logical positivism celebrated it as being positively delightful and made it pervasive in the philosophical circles existing at the time. They propagated the slogan "meaning is use" and insisted that utterances are only relevant in terms of the activities, or *language-games*, in which they play a part (Levinson, 1983:227).

Levinson in the second section examines the thesis that "speech acts are irreducible to matters of truth and falsity." He summarizes the formulation of the thesis as follows: (a) every utterance expresses a proposition(s) and performs an action(s); (b) there is an advantaged level of action referred to as an *illocutionary act*—in this case, the complete speech act made in a typical utterance that consists of the delivery of the propositional content of the utterance—when a speaker utters a linguistic expression, or *doing* something with words; this action is traditionally related to the form of the particular utterance, distinguishing it from *perlocutionary acts* (or *effects*)—in this respect, speech acts as perceived at the level of their consequences, or actions that go with the main illocutionary act and carried out through the main action; and (c) while a certain *illocutionary force* (a speaker's intention in producing an utterance) may be conveyed efficiently in different ways, the explicit performative is at least one type of utterance that is capable of directly and traditionally expressing it (i.e. the *illocutionary force*) (Levinson, 1983:223-224).

In the third section, Levinson investigates the antithesis of "the reduction of illocutionary force to ordinary syntax and semantics." He explicates the antithesis as a call for abandoning the effort to develop a particular theory for illocutionary force since the elements that challenged the tenet of doing things with words can be subsumed under the conventional theories of truth-conditional semantics and syntax. He argues that this position can be simply attacked by pointing out that explicit performatives will be no different from other sentences which can be checked or demonstrated to be true, accurate, or justified by their use (Levinson, 1983:2546-247). I must add here that Leech (1983) opposes both the thesis and the antithesis, dubbing them "the performative and illocutionary-verb fallacies."

Levinson in the fourth section analyzes the "collapse of the antithesis." He notes that the disintegration of the antithesis in its complete form hinges upon its lack of tenability, not only on the semantic front but also on the syntactic front. The semantic problems of the antithesis include (a) the inability of truth-conditional semantics to deal with "non-assertoric" utterances; (b) the impossibility to support a coherent and consistent implementation of the semantic idea of truth conditions to sentences; (c) the failure to provide a semantic distinction for the performative clause, whether overt or covert; (d) the inability to show how the meanings of some performative adverbs (e.g., *honestly*) are parallel in the explicit performatives, especially when certain adverbs such as *hereby* (e.g., *I hereby urge you to go*) can only modify explicit performatives; and (e) the inability of performative verbs to handle general problems pertaining to the truth-conditional evaluation of declaratives (Levinson, 1983:247-257). The syntactic problems of the antithesis include (a) many cases in which explicit performatives do not make reference to the speaker; (b) sentences appear to entail more than a single illocutionary force—e.g., tag questions; (c) the need for an unusual and uninspiring rule of performative deletion for most cases of every implicit performative; and (d) the syntactic mechanisms required to address *indirect speech acts* (i.e. the acts of conducting illocutionary acts indirectly; for instance, one might say "Could you go away?", thereby asking the hearer if s/he could go away or requesting that s/he goes away) can duplicate the outcomes of a performative analysis, or performative hypothesis, without turning to performative clauses (Levinson, 1983:251-263).

In the fifth section, Levinson broaches "indirect speech acts as a problem for the thesis and the antithesis." He tells us that the idea of *indirect speech acts* makes sense only if an analyst accepts the concept of *literal force* or *literal force hypothesis*: i.e. the postulate that illocutionary force is embedded into sentence form. He also points out that both the *idiom theory* (i.e., the proposition that the indirectness of many putative cases of indirect speech acts is only actually evident) and *inference theory* (i.e. the proposition that indirect speech acts have the literal force that is linked with the surface form of a pertinent sentence) which have been suggested to rescue the hypothesis have not fared well. This is because, for the idiom theory, (a) idioms are not compositionally analyzed but recorded whole in a vocabulary with their semantic equivalence; (b) whenever there is a grammatical reflex of indirect force, an analyst must embarrassingly claim that an idiom exists; (c) there is a significant comprehension challenge given the many ways ambiguity can manifest itself in a stretch of discourse; and (d) the non-compositional nature of idioms makes them idiosyncratic to speech communities just as the arbitrary sound-meaning equivalences of lexical items do. As for the inference theory, it fails to address the motivations of indirect speech acts—for instance, some speakers prefer contortions over simple and direct statements (Levinson, 1983:263-273; see also Leech, 1983).

Levinson in the sixth section treats "the context-change theory of speech acts." He asserts that the "set-theoretic sense" of Context-change Theory ("the analysis of speech acts as operations on context—i.e. as functions from contexts into contexts") is a prime prospect for a pragmatic theory of speech acts (Levinson, 1983:276). He describes the theory as follows:

> A context must be understood here to be a set of propositions, describing the beliefs, knowledge, commitments and so on of the participants in a discourse. The basic intuition is very simple: when a sentence is uttered more has taken place than merely the expression of its meaning; in addition, the set of background assumptions has been altered. The contribution that an utterance makes to this change in the context is its speech act force or potential (Levinson, 1983:276-277).

In the seventh and final section, Levinson talks about "beyond theories of speech acts." Here, he deliberates on three important aspects. First, he mentions that any speech act theory must place utterances into speech act classes, no matter how they are delineated. The challenge then becomes how to make such placements either arbitrarily or an effort is made to determine correctly the functions of sentences in context. Second, Levinson points out that external disciplines engaged in the investigation of natural language use are subverting speech act theory. One of these disciplines is the *ethnography of speaking*, which engages in the cross-cultural examination of language usage. A major concept of this scholarly endeavor is the idea of a *speech event*: i.e. a "culturally recognized social activity in which language plays a specific, and often rather specialized, role like teaching in the classroom, participating in a church service, etc." (Levinson, 1983:278-279). Third and finally, Levinson informs us that the study of language acquisition takes speech acts way beyond their narrow conception. This is because instead of focusing on the grammatical systems that undergird a child's early utterances, analysts of language acquisition focus on the functions the utterances perform and the interactional context to which they contribute (Levinson, 1983:281; see also Leech, 1983).

Conversational Structural Functions

Generally, *Conversational Structural Functions* can be denoted as activities concerning the well-known primary sort of talk. Thus, as Levinson indicates, *conversation* involves the act of two or more participants freely speaking in turns and commonly taking place outside certain institutional surroundings such as congressional hearings, judicial hearings, city council meetings, school board meetings, etc. He adds that it is only fitting that some analysts turn to conversation to delineate pragmatic elements, as conversation is evidently the first type of language usage, the genre through which we first encounter language, and the material for acquiring language. Therefore, according to him, different attributes of pragmatic organization can be demonstrated to be at the core of conversation (Levinson, 1983:284).

Thereafter, Levinson segments his analysis of *Conversational Structure* into four major sections. Section one compares discourse analysis and conversation analysis. He begins by stating that while the two approaches are principally concerned with how coherence and sequential ordering in a stretch of discourse are developed and viewed, they, nonetheless, differ in terms of how they conduct their analyses. He characterizes discourse analysis as the utilization of the methodology, theoretical propositions and traditional terms such as "rule" and "well-formed formula" that are the staples of linguistics to analyze stretches of discourse. The approach seeks to advance the basic techniques employed in the field of Linguistics beyond the sentence as a unit of analysis. To accomplish this, the following two techniques are utilized: (1) "the isolation of a set of basic categories or units of discourse"; and (2) "the formulation of a set of concatenation rules stated over those categories, delimiting well-formed sequences of categories (coherent discourses) from ill-formed sequences (inconsistent discourses)" (Levinson, 1983:286).

Next, Levinson describes conversation analysis as a simple empirical technique that eschews precipitous theory building. The essentially inductive methods used by conversation analysts go from the specifics to the general. In this case, a researcher begins with a search for recrudescing tendencies in conversations that naturally take place, instead of beginning with the development of categories to straightjacket the data. The researcher would thereafter focus his/her analysis on the inferential and interactional outcomes of discourse participants' choices between different utterances. Thus, for conversation analysts, the more and disparate data they can get, the happier they are with their analyses (Levinson, 1983:286-287).

Section two focuses exclusively on conversation analysis. This approach, as Levinson points out, was birthed by ethnomethodologists, a seceding group of sociologists. These analysts named their subject *Ethnomethodology*: the investigation of "ethnic"—i.e. participants' own—techniques for developing and explaining social interaction. The impetus for their subject emerged from their conviction that the genuine entity of sociological analysis is the series of techniques employed by members of society themselves to elucidate and act within their social milieu. This development was a response to the quantitative methods and random application of data to what are generally believed to be unbiased categories that had become the mainstay of sociology in the United States (Levinson, 1983:294-295).

Section three addresses preference organization. According to Levinson, the concept *preference* is not geared toward accounting for the desires of a speaker or hearer; instead, it is a term for a structural occurrence similar to the linguistic term *markedness* (i.e. the way words are added or changed to yield a particular meaning), particularly as employed in morphology (i.e. the study of the forms of words). As he further explains, preferred or unmarked (i.e. the normal meaning) classes seem to contain less morphological information than the “dispreferreds” or marked ones, and there is a greater chance of morphological asymmetry in the unmarked types (Levinson, 1983:332-333).

And, section four is about pre-sequences. As Levinson mentions, the concept *pre-sequence* is systematically-ambiguously employed to indicate a particular type of turn and a particular type of sequence embodying that type of turn. One group of pre-sequence is that of *pre-announcements* which are attempts for story space that function to get approved entry to a continued turn at talk (Levinson, 1983:345-349).

Brown and Yule explore *Conversational Structure* slightly differently by focusing on two specific aspects of the topic: (1) conversational implicature and (2) conversational interaction. Starting with conversational implicature, they define it as a general proposition (called the *Cooperative Principle*) with four maxims to which participants in a conversation will normally adhere (this principle and the attendant maxims were discussed earlier). They add that in order to delineate conversational implicatures, an analyst must be knowledgeable about particular facts of the world. They also point out that conversational implicatures hinge upon the literal or conventional meaning of an utterance made in a particular context that is shared by the participants in a conversation (Borwn and Yule, 1983:31-33).

Next, as it relates to conversational interaction, Brown and Yule state that it is an approach used to determine whether the coherence of a sentence partially hinges upon our expectation that, based on the adjacency pair formula, what comes after a question should be considered a response to the question. They also mention that while this point may appear obvious, “it is exactly the type of point that, because of its obviousness, is rarely made explicit in the analysis of language” (Brown and Yule, 1983:231). This obvious point, they add, “captures one important aspect of how we assume that two formally unconnected utterances placed together form a coherent piece of discourse. They do so because there is an assumed coherent structure of discourse over and above the more frequently described structure of sentential form” (Brown and Yule, 1983:231).

Ideational, Interpersonal, and Textual Functions

I must begin by stating that the three functions are discussed together here because that is how Leech, who has examined the subject in-depth and from whose work the discussion in this section relies, presents them. Based on Michael Alexander Kirkwood Halliday’s work, Leech denotes the three functions as follows:

1) "*The ideational function*: language functioning as a means of conveying and interpreting experience of the world. This function is subdivided into two sub-functions, the *Experiential* and the *Logical* sub-functions" (Leech, 1983:56).

2) "*The interpersonal function*: language functioning as an expression of one's attitudes and an influence upon the attitudes and behavior of the hearer" (Leech, 1983:56).

3) "*The textual function*: language finctioning as a means of constructing a text, i.e. a spoken or writing instantiation of language" (Leech, 1983:56).

Leech adds that his disagreement with Halliday's conceptualizations is how he (i.e. Halliday) amalgamates all three functions into grammar. For Leech, while ideational function can be subsumed into grammar because it channels ideas to a hearer via a sense-sound association, interpersonal and textual functions are decidedly pragmatic because from a speaker's perspective they are "input constraints" and "output constraints," respectively, on grammar (Leech, 1983:57).

Tact Maxim Functions

Tact Maxim Functions can be generally defined as the uses of a variety of illocutionary acts with an emphasis on the heuristic approaches of explication whereby politeness is examined from the addressee's end instead of that of the addresser. According to Leech, different situations require different types and degrees of politeness strategies. He identifies the following four general types of these functions that are imperative for initiating and sustaining courtesy and considerate behavior toward others:

1) "COMPETITIVE: The illocutionary goal competes with the social goal; e.g., ordering, asking, demanding, begging" (Leech, 1983:104).

2) "CONVIVIAL: The illocutionary goals coincides with the social goal; e.g., offering, inviting, greeting, thanking, congratulating" (Leech, 1983:104).

3) "COLLABORATIVE: The illocutionary goal is indifferent to the social goal; e.g., asserting, reporting, announcing, instructing" (Leech, 1983:104).

4) "CONFLICTIVE: The illocutionary goal conflicts with he social goal; e.g., threatening, accusing, cursing, reprimanding" (Leech, 1983:104).

Leech designates the first two types as mostly having to do with politeness, since the first seeks to reduce negative politeness between an addresser and an addressee and the second seeks to promote positive politeness between them. For the third type, politeness is irrelevant; and, for the fourth type, politeness is impossible. Leech therefore calls for analyses to be focused on competitive and convivial *Tact Maxim Functions* with their accompanying negative and positive politeness strategies, respectively (Leech, 1983:105).

It behooves me to end this section by mentioning that in addition to the *Tact Maxim,* Leech also presents five other maxims of politeness. He briefly describes all six of them as follows:

1) "TACT MAXIM (in impositives and commisives)

2) Minimize cost to *other* [(b) Maximize benefit to *other*]" (Leech, 1983:132).

3) "GENEROSITY MAXIM (in impositives and comissives)

4) Minimize benefit to *self* [(b) Maximize cost to *self*]" (Leech, 1983:132).

5) "APPROBATION MAXIM (in expressives and assertives)

6) Minimize dispraise of *other* [(b) Maximize praise of *other*]" (Leech, 1983:132).

7) "MODESTY MAXIM (in expressives and assertives)

8) Minimize praise of *self* [(b) Maximize dispraise of *self*]" (Leech, 1983:132).

9) "AGREEMENT MAXIM (in assertives)

10) Minimize disagreement between *self* and *other*

11) Maximize agreement between *self* and *other*" (Leech, 1983:132).

12) "SYMPATHY MAXIM (in assertives)

13) Minimize antipathy between *self* and *other*

14) Maximize sympathy between *self* and *other*" (Leech, 1983:132).

Communicative Grammatical Functions

In general terms, *Communicative Grammatical Functions* can be defined as the various ways grammar is applied to the pragmatics of sentences. To start with, as Leech points out, any grammatical category—be it interrogation, modality, or negation—can be analyzed on three clear-cut levels. One level is the syntactic, which is used, for instance, to form clauses or negative sentences in the English language. Another level is the semantic, which is employed to form the sentence instead of the force by taking into account, for instance, the negative operator vis-à-vis propositional logic. And, the other level is the pragmatic, which is utilized, for example, to form negative "uninformativeness": i.e. the sub-maxim that a negative sentence is less informative and, when combined with the maxim of quantity, yields that a negative sentence would be avoided if a positive one can be substituted for it. In sum, according to Leech, these three levels are what communicative grammar seeks to interrelate (Leech, 1983:152, 101).

Next, Leech considers how one could analyze his/her data given the three levels: i.e. whether to begin with an analysis of the syntax, analyze the semantics next, and then end with the pragmatics, or the other way around. While he concedes that there is no scientific precedent for how to order these levels of analysis, he, nonetheless, believes that it is natural and practical to begin with syntax (which is distinct), then semantics (which is relatively context-variable and opaque), and end with pragmatics (which deals with the force of the sentence). He also admits that implicit in the selection of this approach is the privileging of the addressee, "who arrives at a force of an utterance by decoding its sense and then, by heuristic problem-solving, works out its force, or pragmatic interpretation" (Leech, 1983:152-153).

Anaphora Controlled Functions

A general denotation for *Anaphora Controlled Functions* is they are the uses of words referring to or replacing words used earlier in a sentence in order to avoid repetition. I must mention here that *anaphora* is one of the three subclasses of pronouns or elements that constitute *endophora* (the set of relationships among pronouns or elements having the same reference within a text, contributing to textual cohesion), with the other two being *cataphora* (the use of a pronoun or element that refers to or stands for a later word or phrase) and *self-reliance* (when a pronoun or element refers to itself). There is also *exophora* (a relation where a pronoun or element points outside the text to the context in which it exists). According to Brown and Yule, from the perspective of a style of writing or public speaking characterized by more elaborate grammatical structures and more conservative and technical vocabulary, pronouns constitute the prototype of the anaphoric expressions speakers and writers utilize to designate "given" units. They note that pronouns are commonly uttered with low pitch in spoken discourse, thereby operating as referring expressions that are lexically and phonologically "attenuated." They also mention that pronouns are employed as elements for testing theories of reference because they do not have "content" (Brown and Yule, 1983:214).

In addition, Brown and Yule point out that while anaphora as a concept is sometimes used to refer to other linguistic elements, it has, nevertheless, been historically employed to refer mostly to "pronominal referring expressions." Consequently, an "antecedent-anaphora relation" is one in which a relationship exists between a full nominal expression and a pronominal expression. They also mention that since most linguistic analyses are done from the perspective of the hearer, the experimental and theoretical investigations of pronouns, or anaphors, have been done mostly on the interpretation instead of the production of pronouns (Brown and Yule, 1983:214-215).

African Linguistic Evidence

Pragmatic Function is one of the most widely-explored subjects on African languages. The rest of this section entails synopses of some of the findings presented by various scholars on this subject. First, in my book on the African National Anthem titled *Nkosi Sikelel' iAfrika* ["Lord Bless Africa"] (2011), I delineate the *Pragmatic Function* of the song by delineating its deixis, presuppositions, implicatures, and speech acts, respectively. Composed by the South African Methodist Clergyman Enoch Mankayi Sontaga what later became the African National Anthem in 1897, I demonstrate that the song was geared toward addressing the aspirations of Black South Africans living in a period of high cultural, economic, political and social oppression. I also show that the song is an outcome of the politico-religious movement of the time, which took the form of the religion of the oppressed and, in turn, emerged as the ideological expression of the progressive tendencies of the anti-colonial resistance (Bangura, 2011:1; see also Meli, 1988:32).

Also, in my essay, "An Analysis and Description of the Temne Language: Preliminary Notes" (1990), I allude to the fact that the Temne language (its speakers constituting the largest linguistic group in Sierra Leone, or 35.5% of the country's estimated 6,312,212 in 2018—see the *CIA World Factbook* [2019] the name Temne is an English loan word; its speakers call it Ka Themnê) in Sierra Leone has two varieties: (1) the Sanda and (2) the Konike. The former variety is spoken on the coast and the latter in the interior. I further mention that according to the speakers of other Sierra Leonean languages, the most prestigious variety in terms of *Pragmatic Function* is the Sanda variety. This variety is employed as a High by speakers of Loko (2.9% of the Sierra Leonean population) in the northcentral part of Sierra Leone, speakers of Susu (about 1%) on the northwest border between Sierra Leone and Guinea, speakers of Limba (6.4%) in the central part, and speakers of Fula (3.4%) in the northern part (Bangura, 1990:1).

In addition, I discuss in my *Multilingualism and Diglossia in Sierra Leone* (1991) that in the case of Sierra Leone, in terms of *Pragmatic Function*, English (a non-indigenous language) has been the standard language for over 100 years, albeit its use is restricted to formal administrative, business and technical communication and, therefore, serves as the High variety. Krio, which is lexified largely by English, but with some vocabulary from Portuguese and several African languages, serves as the Low variety for English and as a High variety for speakers of the rest of the other languages in the country. As mentioned earlier, Temne, the language with the largest number of speakers, is used in normal conversations by its members and as a High variety by the Loko, Susu, Limba, and Fula speakers. Mende, the language with the second largest number of speakers (33.2%), is used in normal conversations by its members and as a High variety by the Sherbro, Krim, Vai, Gola, Kissi and Kono speakers. Thus, as I postulate, diglossia (i.e. a situation in which two languages, or two varieties of the same language, are used under different conditions within a community, often by the same speakers; the term is usually applied to languages with distinct "high" and "low" or colloquial varieties) may simply remain stable for a long time in Sierra Leone. This is mainly because, as I point out, the relatively low literacy (approximately 48.8%), the lack of wider familiarity with English among the citizens, and the absence of serious development of *nationalism* (i.e. feelings that develop from and in support of nationalities) with the desire for a national language as its symbol make English predominant for *nationist* (i.e. the more pragmatic problems of governing) purposes in Sierra Leone. I also postulate that a few languages in a multilingual society are used more or less naturally by their members and as High varieties by others; these languages are therefore diglossic within their own communities. But, as I further indicate, it is highly probable that there are occasions whereby some speakers of indigenous languages in Sierra Leone use their tongues at home with friends and relatives, and only employ English in public functions and with speakers of other languages—a situation which approximates diglossic distributions. There are also moments whereby these speakers can reverse these uses, especially when they become quite familiar with the people with whom they interact, or simply code-mix (Bangura, 1991:45-47; for more on this issue as it pertains to Africa-wide and its Diaspora, see Bangura, 2017).

Second, Øyvind Dahl in his examination of the Malagasy language offers numerous examples of the *Pragmatic Function* of the communication features in the language, including communicative style, cultural values, etc. He discusses the language's pragmatic operative conditions upon which the society widely relies: i.e. "truth," "honor," "non-confrontation," "self-assertion," and "directness" (Dahl, 1999:125-125). He also explains some of the features of the *Pragmatic Function* and structural attributes that are embedded in the use of the language by focusing on a special case of communicative decision-making in the public domain. His conclusion is that "In a society that values commonality as much as the Malagasy society, the procedures for argumentation, decision making, and choice of speech mode have great influence on how the partners of communication decode each other's message" (Dahl, 1999:143).

Third, Denis Creissels by investigating the *Pragmatic Function* of plural markers in African languages that have significant relationships with their morphological nature and the existence versus non-existence of a gender system discovers that African "languages devoid of a gender system generally have a single plural marker with the morphological status of a plural affix, and such plural markers tend to be used on a 'pragmatic' basis, that is to be employed only when plurality is both communicatively relevant and not implied by the context, at least in the case of nouns that do not refer to persons" (Creissels, 2000:246). He mentions the Eastern Sudanic branch of Nilo-Saharan and all branches of the Afroasiatic phylum as African languages with exceptional cases of morphologically complex plural marking (Creissels, 2000:246).

Fourth, during excavations in the 1970s near what is now called Border Cave in the Lebombo Mountains situated between Swaziland and South Africa, a small piece of the fibula of a baboon, which has been dated to approximately 35000 BC, was discovered. The bone is marked with 29 distinctly delineated notches. At more than 37,000 years old, the bone, which resembles the calendar sticks still in use to this day by the San people in Namibia, is one of the oldest mathematical objects in existence (www.math.buffalo.edu/mad/Ancient-Africa). It has been argued that since the *Pragmatic Function* of the Lebombo Bone was for lunar phase counting, then women must have been the first mathematicians, since keeping track of a menstrual cycle calls for a lunar calendar (*The Internet Encyclopedia of Science*, 2007:1).

Fifth, the Association for the Development of Education in Africa (ADEA) proffers the view that a major *Pragmatic Function* for African languages is to serve as a major engine for the continent's educational and sustainable development. Thus, the organization's members, who include academics and government ministers from across Africa, state as their objective the reform of the educational system. Their objective is based on the premise that "language is without a doubt the most important factor in the learning process for the transfer of knowledge or skills…is mediated through the spoken or written word. The paradox is that educational plans and programs are often designed to pay more attention to the structure of the educational system and curricula than to language policy" (ADEA, 2001:1).

Sixth, Ogone John Obiero recounts the efforts to revitalize the Olusuba language to serve a *Pragmatic Function*, although perceived by him as a burdensome one for learners in Kenyan schools. As he relays the story, the contact between the Luo and the Suba ethnic groups which began approximately two centuries ago advantaged the Luo, thereby causing a shift toward the Dholuo, which is the language of the Luo. This was due to the numerical superiority of the Luo and the fact that the Olusuba language of the Suba was unwritten and lacked official status at all levels. In 1995, the Kenyan government via its Ministry of Education launched a language development initiative for the Olusuba language. Among the energetic efforts was a campaign to teach the language in schools and used as a medium in radio broadcasts. In the 2000s, the Olusuba language is once more in contact with the Dholuo language, albeit this time it is in the classroom. Along with English and Kiswahili as official languages, the Suba school children have also been taught Dholuo as a Mother tongue. Hence, due to Kenya's language policy that supports trilingualism in early primary school, Obiero argues that the "teaching of Olusuba in school amounts to some sort of quadrilingualism, that, apart from being a deviation from the policy, is an obvious burden on the individual learner" (Obiero, 2010:267).

Seventh, as Dagnachew Teklu informs us, the African Union (AU) acknowledges the *Pragmatic Function* of African languages to include fostering communication, sharing of information, solidifying African identity, and establishing a position for Africa on the global stage. Correspondingly, the organization is quite worried about the large number of language deaths and the many languages that are on the verge of dying, and has called on African states to launch policies that would preserve their languages. The factors cited for this phenomenon are the use of foreign languages for official communication which discourages the development of indigenous languages, the dwindling numbers of speakers of some languages, the unfavorable external factors that force certain communities to abandon their languages for neighboring ones, and the vestiges of colonialism and their continued impact during the postcolonial era (Dagnachew, 2006:1-2). One can add to these factors the forces of neocolonialism (i.e. the use of economic, political, cultural, or other pressures to control or influence other countries, especially former dependencies) and globalization (i.e. the process by which businesses or other organizations develop international influence or start operating on an international scale).

Eighth, Sabelo J. Ndlovu-Gatsheni analyzes how the Zimbabwe National Union-Patriotic Front (ZANU-PF) attempted to superimpose a *Pragmatic Function* for *Chimurenga* (i.e. situating the birth of the nation within a series of nationalist revolutions going back to the 1890s) and *Gukurahundi* (i.e. the elimination of enemies and opponents through physical violence) that mirrors a "nationalist monologic history" of Zimbabwe that buttresses the political party's assertion as the forebear and protector of the postcolonial state. Concomitantly, since it was launched in 1963, the party has worked pertinaciously to maintain that it is the sole genuine organized body with an inalienable vocation to deliver the colonized people of Zimbabwe from the shackles of settler colonial rule (Ndlovu-Gatsheni, 2012:1). ZANU-PF's claim, however, has not gone without challenges, which Ndlovu-Gatsheni describes as follows:

> ...this hegemonic drive has always encountered an array of problems, including lack of internal unity in ZANU-PF itself, counternarratives deriving from political formations like the Zimbabwe African People's Union (ZAPU); labor movements; and critical voices from the Matebeleland region, which fell victim to *gukurahundi* strategy in the 1980s. With the formation of the Movement for Democratic Change (MDC) in 1999, which soon deployed democracy and human rights discourse to critique the ideology of *chimurenga* and the strategy of *gukurahundi*, ZANU-PF hegemony became extremely shaky and it eventually agreed to share power with the MDC in February 2009 (Ndlovu-Gatsheni, 2012:1).

Ninth, Jacobus Alwyn Kruger Olivier, writing about multilingualism in South Africa, begins by describing how the phenomenon's *Pragmatic Function* is for South Africans to employ and perpetuate more than a single language in a given context. In essence, he adds, the phenomenon may allude to the actuality that many languages are spoken in the country and that many South Africans know two or more languages (Olivier, 2009:1).

Next, Olivier tells us that the English language has enjoyed a significant privilege in South Africa ever since the British arrived there in 1795. The significance of the language skyrocketed after the British occupation of the territory in the 1800s that saw a power shift from the Dutch to the British settlers. English has enjoyed two conflicting sources of support. The first source is that as a language of political power, the English language enjoyed the support of the European colonists and the Dutch/Afrikaans movement that operated at the time. The second source is that English was perceived as a language of apartheid alongside the Afrikaans language during the resistance struggle of Blacks against the Afrikaans, as apartheid laws forced the teaching of Afrikaans on them—a policy that led to the Soweto uprising (in June of 1967). As a very important medium of communication across the globe, English is predominant in the areas of the Internet, media and science and, thus, the language is linked to economic prosperity and political power in South Africa. While attempts are being made to advance multilingualism in South Africa, English still remains the most used language in the media and by politicians and other government officials. Correspondingly, the levels of understanding of the English language vis-à-vis other languages are the following: Afrikaans, 50%; Northern Sotho, 19%; Sesotho, 28%; Setswana, 14%; siSwati, 27%; isiNdebele, 3%; IsiXhosa, 24%; isiZulu, 32%; Tshivenda, <1%; and Xitsonga, 24% (Olivier, 2009:1-2).

In addition, the picture painted by Olivier for a *Pragmatic Function* of the African languages in South Africa is not as bright as that of English, since there is no transparent lingua franca among the African ones. The following instances support this assertion: (a) related languages such as the Nguni languages comprising siSwati, isiNdebele, isiXhosa and isiZulu, and Sotho languages comprising Northern Sotho, Sesotho and Setswana or often only fairly understandable; (b) isiZulu enjoys geographical prominence in the KwaZulu-Natal Province, isiXhosa enjoys the same status in the Eastern Cape, and Sesotho enjoys a similar status in the Free State; (c) due to their usage, higher status is accorded to isiZulu and Sesotho among speakers of Tshivenda and Xitsonga in for instance the Gauteng Province; (d) the levels of understanding of African languages, other than a language used at home, by the various South African linguistic groups are as follows—Afrikaans, 11%; English, 14%; Sesotho, 30%; Setswana, 20%; siSwati, 42%; isiNdebele, 19%; isiXhosa, 13%; isiZulu, 16%; Tshivenda, 13%; and Xitsonga, 21% (Olivier, 2009:2-3).

Tenth, following Kwesi Kwaa Prah, the *Pragmatic Function* of multilingualism in the urban centers of Africa can be considered as either a "bane" or a "blessing." He points out that over the past several decades, multilingualism has had a major effect on the lives of Africans, especially those living in urban areas. Coming from various parts of their countries to live in these urban areas, these Africans are compelled to learn one another's languages for daily interactions, although not in an organized and punctilious way but as a mutually beneficial process. This phenomenon can be observed in every urban area across the African continent. Prah denotes the process as almost as easy as the cross-border nature of African languages is yielding a type of multilingualism that manifests a regional feature to most of these languages. Thus, he explores the usefulness of the phenomenon, its potential to be put into full service for the continent's development, and the facilitating and hindering challenges it presents to the development (Prah, 2010:169-181). While he finds both positive and negative aspects of the phenomenon, he is, nonetheless, convinced that the following will occur to reconcile the two aspects:

> The harmonization of African orthographic conventions will help bring order into the identification of "separate" speech forms. This is because cognate and proximate languages will be written with the same rules of spelling. The result of this will be that for such proximate varieties readers will be able to read across dialectal variants. This implies that readerships will become potentially larger than is currently the case. On the economies of scale this will help make literature production more economically lucrative (Prah, 2010:181).

Eleventh and finally, Barbara Trudell purports that just like in other parts of the globe the *Pragmatic Function* of multilingualism in Africa South of the Sahara is frequently a practical answer to the connections among linguistic communities, the necessity to collaborate and exchange goods and services, and the imperative for having amicable relationships with one another. She notes that the nature of multilingualism among the people in the region is a reflection of African civilization's long history of participating in large-scale trade relationships, enjoying technological innovations with others, and establishing political ententes. She adds that this type of multilingualism is evident not only among linguistically-related language communities, but also in language communities that are geographically contiguous. She further mentions that the level of articulacy in each of the languages may differ, ranging from basic proficiency in the languages to native-speaker type proficiency (Trudell, 2009:1).

African Mathematical Evidence

To commence with, we cull from the work on the Lebombo Bone by Richard Mankiewicz (2001) the *Pragmatic Function* of this most ancient human mathematical tool. As he points out, the bone is a testimony to the existence of a very sophisticated accounting system which enabled humans to master time, and it is the first visible hint of the emergence of calculation in human history. The Lebombo Bone, with its six phases, indicates that it constitutes a binary calendar. Also, according to Franz Gnaedinger,

> The pattern consists of 14 by 14 dots, linked by two more dots. The 29 dots of the top line can be read as a lunar calendar. There are 30 spaces

> between and next to the 29 dots. Read the spaces and dots as follows: 30 spaces plus 29 dots plus 30 spaces plus 29 dots plus 30 spaces ..., yielding 30 29 30 29 30 29 30 29 30 29 30 29 30 29 30 ... nights or 30 59 89 118 148 177 207 236 266 295 325 354 384 413 443 472 502 ... nights for 1 2 3 4 5 6 7 8 9 10 11 12 13 14 15 16 17 ... lunations. There are 2 x 13 x 14 dots below the top line, yielding 364 days; add the dot in the middle of the bottom line and you obtain 365 days for a year (Gnaedinger, 2005:1).

In light of the purpose for which it was invented, the Lebombo bone is an exemplar of early Africans' endeavor to develop a system on communication with a *Pragmatic Function.*

Next, as Mwalimu Rev. Alfred Munda SamForay teaches us, the Mende language spoken primarily in Sierra Leone and Liberia has a numerical system based on the decimal structure as the Western system. Unlike the Western system, however, for linguistic *Pragmatic Function,* the Mende count only to 19. The number 20 is called *nugboyogo*r (i.e., *numugboyogor*), meaning that one person finished. The reason the Mende call the number 20 as *nugboyogor* is that they count 10 fingers and 10 toes and, thus, one person finished. The Mende counting system simply adds one until 40 (*nufelegboyogor*), meaning two people finished, repeats the same procedure all the way to 99, and then there is no Mende word for 100 and beyond, or for 1,000. Mwalimu SamForay adds that he was told by a Congolese professor friend of his a long time ago that the reason there is no word for a thousand or million in many African languages is that no one had a thousand or a million anything when the languages were formulated (SamForay, E-mail correspondence, May 03, 2019).

Mwalimu SamForay also edifies us that although the Ancient Kemetians/ Egyptians invented the number zero (0), without which the binary digit system on which modern computing is based would not have been possible, many other African societies may not have found a need for such a number. Nonetheless, he adds, the Mende counting system does actually have a word for zero quantity: *gbewogbeh* (SamForay, E-mail correspondence, May 03, 2019).

Also, we are apprised by Mwalimu Ndi Chick Loveline Ayoh about the *Pragmatic Function* of the counting system of Moghamo, a language spoken mainly in the North West region of Cameroon, specifically in Mbu I, Baba II, Mbei, Mefebe, Batibo, Bajai, widikum communities. These communities used to be under a large clan called the Widikum and broke up into various sections during an ethnic war that occurred among them for personal interests. Over time, there have emerged slight variations in the tongue as various communities seek to differentiate themselves from the others. Nevertheless, since these people were great traders as they used to sell their goods to foreign traders who invaded their area to buy their main cash crop that was Kola nut, in order not to be cheated or to make a mistake in counting the Kola nuts, they sold them in large quantities. In counting, they count one to ten as follows: 1=*Moh*, 2=*Beh*, 3=*Tat*, 4=*Quay*, 5=*Thay*, 6=*Refohe*, 7=*Sambe*, 8=*Fame*, 9=*Bog*, 10=*Wom*; and 100=*Quefebe* (Ayoh, E-mail communication, May 12, 2019),

Malimu Ayoh further informs us that the main mathematical idea that this culture had is the following: counting is done in tens, and takes into consideration a person's ten fingers. For example, five kola nuts are collected in one hand and five in the other hand. As a person counts from one to ten, one Kola nut is borrowed from the lot and kept aside to indicate the first ten of tens, making it one hundred. That is, 10 times 10 = 100, indicated by 1 kola nut; instead of $10^2 = 100$, it becomes $10^1 = 100$. Therefore, 100 times 10 = 10 Kola nuts = 1,000; instead of $10^3 = 1000$, it becomes $10^2 = 1,000$, and 100 times 100 = 100 Kola nuts = 10,000, instead of $10^4 = 10,000$, it become $10^3 = 10,000$. This mathematical idea regroups counting of large numbers by those who are not formally educated by summarizing them into 100s by using 1, 10s, 100s, and 1,000s in the same way (Ayoh, E-mail communication, May 12, 2019).

In addition, Fareeaa Abdoola, Penelope S. Flack, and Saira B. Karrim after stating the general denotation of *Pragmatics* as how we employ language in social intercourse go on to discuss the impact of facilitating pragmatic skills through role-play among learners with language learning disability at a school for such learners in eThekwini located in the KwaZulu-Natal province of South Africa. The tests used assessed the effectiveness of role-play in terms of stylistic variation and requesting clarification pragmatic skills and students learning mathematics, reading, and written expression. The researchers find that role-play was indeed quite effective for the students in the two pragmatic skills. The researchers therefore suggest that role-play be employed as a therapeutic method to facilitate learning among students with a language learning disability (Abdoola et al., 2017:1-2).

Furthermore, from Merryl Ford and Adele Botha (2010), we garner the *Pragmatic Function* of E-education (electronic education) in the fields of Mathematics and Science across South Africa. The rationale that underlies the E-education initiative, according to these researchers, is as follows:

> The goal of South Africa's e-Education White paper is for every school learner in the country to be ICT capable by 2013 and for teachers to use information and communication technology (ICT) to enhance teaching and learning. Many initiatives have been undertaken to provide schools with computer labs and infrastructure in order to reach these goals, with varying degrees of success. In the meanwhile, the ICT landscape in South Africa and worldwide has changed dramatically with the advent of the cell phone and its almost

> ubiquitous penetration into all levels of society, especially amongst the youth. Indigenous services that utilise this new ICT platform have emerged, that specifically target the youth and address very South-African issues, such as affordability (Ford and Botha, 2010:1).

Ford and Botha also mention the MXit application (discussed in an earlier chapter) as a cellular phone application being employed to facilitate "instant messaging" and "real time text-based chatting" between mathematics learners and tutors at a very small cost compared to that of a Short Message Service SMS). MXit as a teaching and learning platform, they add, constitutes a pragmatic framework for integrating cell phones as ICT tools into the education system across South Africa (Ford and Botha, 2010:1).

Conclusion

The notion of *Pragmatic Function* is evident in one of the Ancient Kemetian terms for "mean," *āq*, which denotes how a person uses language to "keep the true meaning of an entity, be right, behave rightly, be exact, be correct, be proper" (see Budge, 1978, for this translation). Thus, as we glean from Eitan Grossman and Stéphane Polis' exploration of forms and function in the Ancient Kemetian language, for example, the interrogative construction, especially questions that are introduced by *ist* or *is*, possesses a specific *Pragmatic Function* in that "the proposition directly under question (P) is presented by the speaker as the closed option for elimination, being contrary to the speaker's expectations, and a polar inverse inference option is invited (I) for the hearer to access the speaker's intended point of view" (Grossman and Polis, 2014:2).

Other examples cited by Grossman and Polis include (a) registers of the various types of form-function couplings in the historical inscriptions of Ramses III at Medinet Habu; (b) cases of formal identity or one-form many functions such as the polyfunctional and functional identity, and many forms-one function such as the negation of the imperfectives; and (c) grammaticalization of a future construction referred to as the "First Future" and later the "Allative Future." Also, according to these scholars, all of these grammatical constructions that have experienced grammaticalization can be appropriately elucidated by pragmatic procedures (Grossman and Polis, 2014:2-3).

Chapter 6

Interchangeability

In "Domesticating Mathematics in the African Mother Tongue," drawing from the work of Anouschka Bergman, Kathleen Curie Hall and Sharon Miriam Ross (2007), I mention that "when individuals are able to both transmit and receive messages, they can engage in *interchangeability.* This is because an individual can both produce messages by gesturing, speaking, or writing and understand the messages of others by looking, listening, or reading" (Bangura, 2014:21). This chapter broaches the subject of *Interchangeability* by exploring the linguistic expositions on it and the African linguistic and mathematical evidences related to it. Before doing so, however, a bit more needs to be said about the conceptualization of the concept. Beginning with David Crystal's more general definition, *Interchangeability* refers to a proposed defining attribute of the capacity of the human language system to be reciprocally conveyed and received by constituents of an identical species. He notes, however, that this is not the case for every animal species—an example being in a species where females and males employ different kinds of calls (Crystal, 1992:189).

Also, quoting Charles Francis Hockett, John Lyons states that *Interchangeability* means that "any organism equipped for the transmission of messages in the system is also equipped to receive messages in the same system" (Hockett, 1958:589; quoted by Lyons, 1977:81). Lyons, however, finds Hockett's further postulate somewhat idealized when he (i.e. Hockett) asserts that "any speaker of a language…is theoretically capable of saying anything he is able to understand when someone else says it" (Hockett, 1958:589; quoted by Lyons, 1977:81). This claim is idealized because, according to Lyons, we sometimes understand statements comprising expressions we never heard or read before and which we ourselves had not uttered or written. Nevertheless, he adds, Hockett's theoretical idealization pertaining to identifying speaker's competence with hearer's competence, if it is accepted as an idealization of the facts, is quite valid; and, it is a vital fact about language that we can practically use its system to send and also receive messages. Levinson, like Crystal, also mentions that this is not the case for many types of animal signaling behavior in which only one sex can make mating sounds while the other sex can only respond to them (Lyons, 1977:81).

In sum, *Interchangeability* allows us to understand why humans can both send and receive similar linguistic signals and are not limited in the kinds of messages they can utter and hear or read. A person can say "I am a Jamaican" even if he is a Trinidadian. This should not be conflated with being mendacious. What is important is that a speaker can somatically compose any and all messages no matter their veracity or connection to the speaker. Put differently, what a person hears, s/he can also say. In addition, this attribute is not available to every species.

Linguistic Expositions on Interchangeability

Relatively little theoretical exploration has been done on *Interchangeability*. As such, one continuously runs into the following four themes broached in works dealing with the topic: (1) synonymy, (2) complete feedback, (3) zoösemiotics or design features of animal communication, and (4) design feature of communication systems. These themes are briefly presented separately in the ensuing subsections for clarity. The chapter ends with a conclusion.

Synonymy

Synonymy is generally defined as the state of being synonymous: i.e. a word or phrase having the same or nearly the same meaning as another word or phrase in the same language. Thus, the immediate question that arises here is the following: What is the nexus between synonymy and *Interchangeability*? To provide answers to this question, I start with Lyons' exploration of "synonymy and 'normal' interchangeability. He begins with the observation that traditional semanticists have mostly treated synonymy as a link between lexical elements. He then considers the use of the concept synonymy to include classes of lexical elements that are assembled in a certain syntagmatic framework and also as individual lexical elements (Lyons, 1971:451). To support his proposition, he proffers the following relatively lengthy example which is worth ingeminating here verbatim:

> One might well say, for example, that the phrases *female fox* and *male duck* are synonymous with the lexical items *vixen* and *drake*, respectively. But it is important to notice that, in making this statement, one is assuming that the phrases and the lexical items are indeed interchangeable in the normal use of the language. By contrast **male cow* and *bull*, and **female bull* and *cow*, are not normally interchangeable (it may be assumed), even though one can easily imagine a situation in which the simplest way of explaining the meaning of *bull* (to someone who knew the meaning of *cow* and *male*) was by means of the normally unacceptable sentence *A bull is a male cow*. The reason why **female bull* and *male cow* are semantically unacceptable is that neither *bull* nor *cow*, unlike *fox* and *duck*, is

'unmarked' for the distinction of sex. So much would be undisputed by all semanticists (Lyons, 1971:451).

Lyons adds:

> But the condition of 'normal' interchangeability is here intended to exclude many semantically-compatible (significant) groupings of lexical items, as well as such semantically–incompatible phrases as **male cow*. The phrase *mature female bovine animal* (which might be given as a dictionary definition of *cow*) is undoubtedly well-formed, both grammatically and semantically. But it is probably far less 'normal' a phrase than even the semantically ill-formed **male cow*. The native speaker of English would not 'normally' construct such a phrase as *mature female bovine animal* and use it interchangeably with *cow* in his everyday use of the language. The question of synonymy therefore does not arise in the case of the lexical item *com* and the phrase *mature female bovine animal* (Lyons, 1971:351).

Lyons concludes with the observation that the most interesting question that emerges in these cases is about why a lexical item such as *cow* and a phrase such as *mature female bovine animal* cannot be employed interchangeably, not so much about whether the relation of synonymy is supported. He adds that the essence of this question has not been grasped by many semanticists (Lyons, 1971:351).

Next, in his analysis of "synonymy and interchangeability *salva veritate*" (the Latin phrase meaning in English "with unharmed truth," denoting the idea that two expressions are said to be synonymous if the substitution of one for the other does not change the truth value or meaning of any context in which either expression appears), Jeff Speaks makes a number of arguments related to *Interchangeability*. First, he avers that in order to know why two expressions may be synonymous an analyst must examine the attributes of those expressions in terms of the judgements that undergird them, instead of what is stated about them by lexicographers. This type of substitutability, he argues, however, does not provide an analyst with enough material to explain synonymy sufficiently. As he puts it, no explication of synonymy is satisfactory to delineate the "analytic/synthetic distinction" (Speaks, 2005:2). Analytic statements tend not to alter the forms of the words of a language and to use word order rather than inflection or agglutination to express grammatical structure. Synthetic statements are characterized by the use of inflections rather than word order to express grammatical structure.

Second, Speaks maintains that a way to address "the problem of quotational contexts" is to utilize the Tarskian perspectives of quote-names ("which are treated like single words of the language," albeit Speaks argues that "quote-names cannot be regarded as semantically simple, since that would lead to the language's containing infinitely many semantically simple expressions"). Thus, Speaks declares that while quote-names are perceived to be semantically simple, they, nonetheless, require a rule that limits the germane types of replacement to the replacement of entire words (Speaks, 2005:3).

Third, Speaks argues that if we are to accept the proposition that necessary truths are analytic, we must also accept the postulate that the verity of any sentence of the form "Necessarily, all and only *F's* are *G's* will imply that 'All and only *F's* and *G's* is analytic" (Speaks, 2005:3). He then proclaims that "...since every expression of the form 'Necessarily, all and only *F's* are *F's* is true, it is trivial to use substitutability *salve veritate* to define analyticity" (Speaks, 2005:3).

Fourth, Speaks advances the view that another attempt to denote synonymy that provides an answer to the difficulty presented by quotational and non-existential contexts is not to presuppose facts about analyticity. He, however, finds this solution quite weak and provides the following example to support his position:

> Consider, for example, the following two pairs of sentences: 'is a creature with a heart'/'is a creature with a kidney'; 'the first Prime Minister of Canada'/'John MacDonald.' These do not seem to be synonymous, as is seen by the fact that by replacing one with the other we can move from analytic and a priori sentences like 'Every creature with a heart is a creature with a heart'; 'The first Prime Minister of Canada is the first Prime Minister of Canada' to sentences which seem neither analytical nor a priori like 'Every creature with a heart is a creature with a kidney'; 'The first Prime Minister of Canada is John MacDonald' (Speaks, 2005:4-5).

Nonetheless, Speaks concludes that the foregoing expressions appear to be "substitutable *salva veritate* in extensional contexts" (Speaks, 2005:5).

In addition, János Tanács' exploration of analyticity and cognitive synonymy in terms of their connection to *Interchangeability* is quite informative. Tanács begins his inquiry with Willard Van Orman Quine's treatise of *Interchangeability* in his essay, "Two Dogmas of Empiricism," which appears in the collection of his logico-philosophical essays titled *From a Logical Point of View* (1961). Tanács says that Quine firmly rejects the postulate that there exists a difference between analytic truths (i.e. truths based on meanings free of matters of fact) and synthetic truths (i.e. truths based on fact). He adds that while Quine accepts the proposition that truth generally hinges upon both language and extra-linguistic facts, he rejects the proposition that the truth of a statement can be analyzed in terms of both a linguistic and factual component. Consequently, according to Tanács, Quine rejects the idea that statements can be classified into analytic versus synthetic categories, leading him (i.e. Quine) to also reject the notion that the synonymy of two linguistic items hinges solely upon the interchangeability in every context without the change of truth values. For Tanács, though, interchangeability only makes sense when it is relativized to a language whose extent is specific and relevant (Tanács, 1999:79).

Next, Tanács points out that Quine advances the notion that interchangeability *salva veritate* cannot ensure the desired form of cognitive synonymy in an extensional language. This is due to the fact that heteronym expressions can be interchanged. All that is required to provide the extensions of terms in advance, according to Tanács, is to consider extra-linguistic facts or analyze the link between a language and a required extra-linguistic fact. This analysis, he says, represents the cognition of observation itself. When extensionality is analyzed this way, he adds, it once again makes the meaning and the analyticity of terms unintelligible (Tanács, 1999:80).

Also, in examining the complication of analyticity and extensional coincidence in terms of interchangeability and cognitive synonymy, Tanács utilizes a simple approach by dividing a set of men into two clear-cut disjunctive and exhaustive subsets: (1) bachelors and (2) married men. This simple model allows him to bypass issues such as "whether a widow, divorced or young man is a bachelor or not" (Tanács, 1999:82-83). He goes on to state the utility of his model as follows:

> This model helps us to see that problems occurring without vagueness show another sort of inadequacy of the traditional idea of language. I will argue that there are some background assumptions of facts that are fixed or built into the language. One of these assumptions will be a social one in our case. The changing presupposed extralinguistic facts are to show what this assumption is and how the language really works. The problem and the connections amongst analyticity, meaning, cognitive synonymy and definition can be seen in a dynamic process not in a static state of a language (Tanács, 1999:82-83).

Furthermore, Tanács examines the married-unmarried conceptual framework vis-à-vis "husband-wife conceptual counterpart and bachelor." He suggests that in the European context (of the late 1990s when he wrote his article), the following terms are "cognitive synonyms and interchangeable without changing the truth values: husband = married man = a man having a wife = not bachelor"

(Tanács, 1999:82-83). He concedes that some of the expressions sound strange because they are seldom employed this way or denote the same meaning. This is because, according to him, it is tacitly and socially determined that marriage occurs between a man and a woman (Tanács, 1999:82-83).

Complete Feedback

Lyons asserts that *Complete Feedback* (i.e. the notion that speakers are able to hear what they are saying because through their auditory channels they are able to receive feedback on what they are vocalizing) is contingent on interchangeability because a speaker hears and has the capacity to observe and check his own performance. He posits that this phenomenon is both (a) a matter of observing and checking the signal for audibility in the specific physical states in which it is constructed and (b) requires the examining of one's own utterances for intelligibility and accuracy of configuration as they are constructed and making adaptations when necessary. He proposes that while it is unknown whether other non-human species have the capacity to control their signals by such feedback, it is quite vital that humans can, and do, utilize feedback to control language-utterances when producing them (Lyons, 1977:81-82).

In addition, Lyons indicates that due to the attributes of complete feedback and interchangeability, it is safe to suggest that the acquisition of a language-system by children in the early stages of their development is relatively in advance of their development of the systems of speech-production. Believing this to be the case, he also proposes that language acquisition goes through a sequence of identifiable stages (Lyons, 1977:90).

The question that emerges here then is the following: What are these stages of child language acquisition according to Lyons? He delineates three stages as follows:

> *Stage One*: this begins three months after birth when the child babbles (Lyons, 1977:89).
>
> *Stage Two*: this begins toward the end of the first year, referred to as the stage of holophrastic speech, which consists of grammatically unstructured one-word utterances (Lyons, 1977:89-90).
>
> *Stage Three*: this begins in the second year during which simple two-word and three-word utterances of what is referred to as telegraphic nature (i.e. missing overt markings of grammatical distinctions in, for instance the English, present versus the past tense, singular versus plural, etc.) are produced (Lyons, 1977:90).

Lyons goes on to state that going through the second and third stages of language evolution, children slowly enhance their command of the structure of segmental phonological differences employed in a language. Therefore, it is after children reach the third stage that an analyst can say they have acquired some type of language in terms of production. Lyons also suggests the possibility that children are capable of identifying sounds and interpreting some grammatical differences in the speech of adults even before they—i.e. children—show the capacity of making these grammatical and phonological differences in their own speech-production (Lyons, 1977:90).

Zoösemiotics or Design Features of Animal Communication

For starters, a general definition of *Zoösemjotics or Design Features of Animal Communication* is the following: the systematic inquiry into the features of animal communication. According to Crystal, this area of inquiry emerged as a component of a zoological method of communication, with the main objective being to pinpoint the major areas of relatedness between language and other communicative structures. He points out that the design features of communication initially investigated by analysts working in this area included displacement, duality of structure, and productivity. He adds that this area of inquiry emerged as a subject (within the linguistics subfield of semiotics) as a result of the efforts of analysts working on the topic. These analysts, he also states, employed different methods such as color, movement, smell, sound, and touch to investigate the forms of communication utilized within diverse species such as "the 'dancing' of honey bees, the courtship rituals of herring gulls, and the whistle patterns of dolphins" (Crystals, 1992:427).

Next, Slawomir Wacewicz and Przemyslaw Zywiczyński make two points about *Interchangeability* in their examination of Hockett's design features (discussed earlier) as they relate to language evolution. The first point is that the attribute of *Interchangeability* is comprised of rotating the sender-receiver positions in a manner that is archetypal of conversational exchange. Wacewicz and Zywiczyński add that Hockett cites bee dancing and gibbon calls as having the feature of *Interchangeability*, but that he does not believe it exists in other non-human communication structures (Wacewicz and Zywiczyński, 2015:32).

The second point Wacewicz and Zywiczyński make is that Hockett's framework is undergirded by a bias toward oral over gestural and other forms of linguistic communication as reflected in his treatments of "Vocal-Auditory, Channel, Broadcast Transmission and Directional Reception, Rapid Fading, Interchangeability and Total Feedback." In their comparative analysis of the way Hockett treats these features, they find that *Interchangeability* is the only feature that lacks a direct relationship with or at least concerns the elements of vocal/auditory modality. They therefore assert that this situation is a result of a deep-rooted quagmire stemming from the emphasis on the form or structure instead of the content or function of linguistic communication (Wacewicz and Zywiczyński, 2015:37). They go on to elucidate the conundrum inherent in Hockett's bias as follows:

> Favouring speech over its alternative(s) is unsubstantiated in three related ways. Firstly, as is now well known, sign languages are fully equivalent to spoken ones in practically every relevant respect, including morphosyntax, dialectisation, historical change, rough cerebral localisation, and acquisition by children...Secondly, language is largely modality independent, in that even if speech or sign [is] granted a certain special, primary, status, the actual communication acts can be carried out exploiting other channels...Thirdly, under normal circumstances, natural conversation is never unimodal, but rather multimodal,. With co-speech gesture, and even full body movement and facial expression being parts of the complex message and complementing it with nonredundant and communicatively important semantic information (Wacewicz and Zywiczyński, 2015:37-38).

Wacewicz and Zywiczyński add that the preceding aspect "can be further extended by abandoning the code-model of language altogether in favour of its alternatives, e.g. the distributed approach, on which language does not constitutively depend on any specific modality or means of transfer, but rather results from a network of social practices as a means of their coordination" (Wacewicz and Zywiczyński, 2015:38).

It is also worth mentioning here for the sake of elaboration that Wacewicz and Zywiczyński note that Hockett labels *Interchangeability*, like the other six features he propounds (i.e. *Arbitrariness, Cultural Transmission, Displacement, Duality, Productivity,* and *Specialization*), as a "key property of language," instead of a "design feature of language." This observation contradicts the authors' opening statement in the section of their paper titled "The Original Proposal" in which the issue is discussed when they say that Hockett "discussed the design features of language." They also state that Hockett treats neither *Interchangeability* nor the other six features as more or less important, but instead treats it as being basic to the designation of language (Wacewicz and Zywiczyński, 2015:31-32).

Design Feature of Communication Systems

To begin with, in terms of a *Design Feature of Communication Systems*, William O'Grady, Michael Dobrovolsky and Mark Aronoff define *Interchangeability* as the ability of every member of a species to both send and receive communication. They assert that while this is true for human language, it is not so for bee dancing done only by foragers or birdsong done only by male birds. They also mention that nonhuman primate vocalizations seem to exhibit *Interchangeability* (O'Grady et al., 1989:406).

Thereafter, O'Grady and his colleagues discuss three reasons the design feature is important for comparing communication systems. The first reason they proffer is that conflation often results when analysts do not differentiate between communication and language when comparing animal and human behavior. They also argue that while animals do communicate among themselves and even with humans and other species, the unanswered question is about whether the communication system used by animals is equivalent to human language (O'Grady et al., 1989:406).

The second reason tendered by O'Grady et al. is that by comparing the basic elements of human language and natural animal communication systems, the differences and similarities between the two systems can be demonstrated. These features which are referred to as *design features* are structured by privileging human language. Since as O'Grady and his partners do, and many other analysts also do, analyses of these features tend to emphasize the mental character of linguistic ability at the expense of vocal-auditory transmission. As a result, emphasis is also placed on the essence of the semantic and organizational structure of a system (O'Grady et al., 1989:406).

The third reason O'Grady and his coauthors allude to concerns the fact that by investigating this design feature, one can learn that while human language shares some design attributes with other communication systems, it does not have many other attributes those communication systems have. They provide the following examples to buttress their averment:

> Humans...are hopelessly inadequate at following scent trails, a feat the prosimians accomplish with ease; cannot change color for communicative purposes with the facility of an octopus; and are not as gifted as horses and many other mammals at assessing and interpreting subtle body gestures. Humans do possess an ability to symbolize that far exceeds that of bees and even (allowing for the most generous interpretation possible of recent experiments) our nearest relatives, chimpanzees and gorillas. Human language is also more flexible and productive in manipulating these symbols than any known animal communication system. Language is as suited for—and indeed as much a part of—human life patterns as the communication systems of our fellow creatures are for their modes of existence (O'Grady et al., 1989:409-410).

O'Grady et al. do point out, however, that investigations dealing with whether nonhuman primates display symbolic behavior and are capable of linguistic behavior has led to a spirited debate among certain researchers. According to O'Grady and his colleagues, some researchers have debunked the claim and have also characterized it as an example of "dressage or the Clever Hans Phenomenon" (O'Grady et al., 1989:409). *Dressage* refers to the art of riding and training a horse in a manner that develops obedience, flexibility, and balance. The *Clever Hans Phenomenon* refers to "a form of involuntary and unconscious cuing that sometimes happens in communication and the story itself is about a horse who responded to questions requiring mathematical calculations by tapping his hoof" (Rutledge, 2011:1).

African Linguistic Evidence

This section encompasses synopses of a number of findings limned by scholars who have examined *Interchangeability* as it pertains to African languages. They are presented in the chronological order in which they were published for the sake of lucidity.

For starters, Cheikh Anta Diop teaches us about the existence of numerous indigenous African writing systems, including "Adinkra, Afan-Oromo, Akan, Amharic, Bassa, Egyptian, Ethiopic, Mende, Meroitic, Nsibidi, Rock Arts, Tifingh, Vai, and Wabuti," and that foreign writing systems have been employed to develop orthographies for most and could be used to develop the same for the remaining languages, is proof that *Interchangeability* is a significant feature in African languages. In addition, Diop informs us that "it is a typically Negro African language that has been the oldest written language in the history of humanity. It began 5,300 years ago, in Egypt; whereas the most ancient testimony to an Indo-European language (Hittite) goes back to the XVIIIth Egyptian Dynasty (1470 BC) and this, probably under the influence of the political and cultural domination of Asia Minor by Egypt" (Diop, 1981/1991:1). Diop also states that Cameroon has a hieroglyphic script, the systematic development of which by the Ndyuya may be of contemporary date, but its origins are quite old. He further says that the syllabic script of the Vai in Sierra Leone and the cursive script of the Bassa have been well investigated, and in Sierra Leone, these scripts have been used to write modern texts (Diop, 1987:185).

Next, Isaac S. George Madugu shows the *Interchangeability* that operates in the structural behavior of different types of complex verbs in the Nupe (a Volta-Niger or Niger-Congo language native to Nigeria) and Yoruba (a Niger-Congo language native to Benin, Nigeria, and Togo) languages as they relate to causative constructions. These complex verbs freely allow sentence embedding that yield biclausal causative structures when they occur in simplex non-agentive causative and agentive non-causative sentences. Nonetheless, as it pertains to the non-agentive causatives, it is only the Yoruba language that permits "the verbs to be irregularly embedded into causative matrix sentences in such a way that the biclausal causative structure constitutes input to causative clause union, which compresses it into a uniclausal agentive causative sentence" (Madugu, 1985:295). Thus, on the one hand, *Interchangeability* holds in numerous cases of clausal and biclausal agentive causatives in the Yoruba language. For Nupe, on the other hand, the disappearance of the causative clause union from its grammars means that only the uniclausal agentive causatives operate in the language. In light of this evidence, Madugu asserts that "causative constructions in these languages demonstrate clearly that the CCU (i.e. causative clause union) rule is motivated by a diachronic process of moving from a pragmatic mode of expression to a syntacticized one, and where a particular target has been hit, the rule ceases to function" (Madugu, 1985:295).

Madugu provides two examples to illustrate his assertion. One example is that the apparent characteristic of the non-agentive causative complex verbs is that for some of them, the syntactic positions are interchangeable. Therefore, it is the case that the subject position operates as the object and conversely, thereby leading to no difference in semantic explanation. What explains the alternation is that a non-agentive causative "involves one and the same human experiencer both in the subject and in the object cause. That is someone learning through his senses or perceiving something inevitably evokes in him a certain emotional reaction expressed by a variety of emotive adjectives such as *happy, sad, surprised, amazed, shocked,* etc." (Madugu, 1985:307). The other example has to do with the fact that Yoruba sentences reveal three ways of expressing agentive causative: two are biclausal and the third is uniclausal, albeit all three structures entail the same semantic explanation. While *Interchangeability* operates for all of three sentential structures, the uniclausal one has a non-causative additional denotation such as "the mere sight of Olu annoys me" (Madugu, 1985:309).

Thereafter, we learn from Clyde-Ahmed Winters that the ancient Manding script dates back to about 7,000 years Before the Present (BP) and was used by the Mande to write or engrave inscriptions throughout Western Sahara. The script was a syllabic script similar to the Vai one. These scripts were developed by the demands of long-distance trade, as merchants needed them to keep record of their business transactions, and they were later used to preserve religious doctrines and writing obituaries. To the Mande, the *Kuma,* which means the word, is considered sacred. Therefore, written amulets with *Kuma* have long been recognized as containing magical power (Winters, 1986:208-209).

Winters, via his investigations, found that the Mande script is comprised of between 200 and 350 signs and around approximately different forms. The inscriptions are read from top to bottom or from right to left. Because of the frequent appearances of *pf* disyllabic roots, akin to the CVCV (consonant / vowel / consonant / vowel), CVN (consonant / vowel / nasal), or CVV (consonant / vowel / vowel), Mande was written in syllabary. The reduction of disyllabic roots evolved into the monosyllabic roots of CV. As a result, the same sign can be utilized to indicant contrasting phonetic sounds; hence, several characters depict disparate phonetic values. Six vowels are employed in the syllabary: a, e, é, è, o, and ò. The frequently appearing syllabic form is the monosyllabic. Even though compound nouns are not much in use in the script, derivative nouns and adjectives are formed by suffixes. The major suffixes employed are those of possession, or nationality, which are joined to names or serve to form names and verbs. Personal pronouns are all written in the second and third person singular. The most common verbs are the attributive, and the key nouns deal with agriculture, burial customs, death, and talismans (Winters, 1986:211-212).

Subsequently, Christiane Meierkord explains how *Interchangeability* functions in terms of the past tense verbs in the productions of speakers of the Black Englishes in South Africa. She starts off by pointing out that the English language, as does Afrikaans, functions as a tongue that has been adopted as a common language between speakers whose native languages are different in South Africa. Consequently, English is a mother tongue, a variety of language shift, a second language, or a foreign language for various South Africans. Although it is the mother tongue of primarily the descendants of the British settlers, it is now the first language of the Blacks and the Coloreds in the country. Nonetheless, a majority of South Africans speak English as a second language, having acquired it either in an institutionalized setting or informally, or in their day-to-day interactions with various English speakers. Meierkord adds that the 2001 census put the proportion of the speakers of English in South Africa at 8.2%. But as she is also quick to caution, this proportion "must not be interpreted as reflecting the number of mother-tongue speakers. Rather, the figures need to be read as including those Blacks and Coloreds whose families have undergone language shift from their indigenous African languages or from Afrikaans" (Meierkord, 2005:1).

Apropos to how this English language development relates to the past tense verbs in the productions of speakers of Black South African Englishes, Meierkord finds from her empirical data that as a speaker's exposure to English increases, his/her amount of utilization of the unmarked past tense phrases decreases—a sort of negative correlation. She also discovers that there are sequences which entail many occurrences of present tense verb forms that refer to the past and employed interchangeably—e.g., "the forms *took* and *take* and *said* and *say* in reporting clauses introducing direct speech reports" (Meierkord, 2005:14-15). She adds that this phenomenon is common in narratives in English and Bantu languages and, thus, should not be denoted as wrong. She further states that a speaker's life history and the makeup of his/her productions appear to dictate his/her utilization of past tense forms to communicate about past events. Thus, as she postulates, "The high frequency of present tense verb forms indicating the past in narratives may reflect a speaker's competence to adapt her speech to the communicative situation" (Meierkord, 2005:15-16).

Thereupon, Linda Hall (nee Kruyshaar) informs us about a notion of *Interchangeability* in *-be* relative tenses in the isiZulu language of South Africa. According to her, in the Nguni languages (spoken in Southern Africa, mostly in South Africa, Swaziland, and Zimbabwe, and include isiXhosa, isiZulu, siSwati, Hlubi, Phuthi and Ndebele—both Southern Ndebele and Northern Ndebele), the near and remote past tenses and the near and remote future tenses are frequently respectively interchangeable. The overlapping of these tense forms, she adds, illustrates the tense prototype of the absolute tenses of isiZulu (Hall, 2005:27).

A comme il faut question that emerges here is the following: What are these absolute tenses in isiZulu according to Hall? She states the following as the five absolute tenses in isiZulu: (1) remote past, (2) near past, (3) present, (4) near future, and (5) remote future. She also indicates that while there is no symbiotic connection between a category and the morphemes that designate it, the five absolute tense forms in isiZulu are designated in the common morphological system of the verb by an amalgamation of the subject agreement, categorical verb final, future tense and negative morphemes. She further posits that "These tense forms, wherein the interpretation of event time is grammatically marked (within the verbal morphology) in relation to coding time (the deictic centre), are called 'absolute tenses'. An absolute tense can thus be defined as the verb form which locates the event time relative to a reference point which coincides with coding time (the deictic centre)" (Hall, 2005:26). She goes on to illustrate these absolute tense forms via many sentences in isiZulu.

Also, E. Taljard, G. Faaß, U. Heid, and D. J. Prinsloo in their exploration of the development of a tagset for the Northern Sotho language (or *Sesotho sa Leboa*, also known by the name of its standardized variety Sepedi, or Pedi, is a Bantu language spoken primarily in South Africa) instruct us about why the level of *Interchangeability* would be intrinsically highest between languages that are genetically closely related. In order to do this, they first propose that the delineation of a tagset cannot be separated from its aim, or from its position and its delineation within a larger portrait of the construction of corpus annotation. Second, they use a multilevel annotation technique to gauge to a certain degree the morphological multiplex nature of Northern Sotho at the tagging level. Third and finally, they take into account particularities of standardization in regards to the reuse and sharing of attributes, and how they could be employed by other disjunctively written Bantu languages in South Africa (Taljard et al., 2008:111-112).

Therefore, according to Taljard and her colleagues, contrasting techniques are called for when tagging word class for isiZulu and Northern Sotho due to the different writing styles used by the two languages—while isiZulu uses a "disjunctive" writing system, Northern Sotho uses a "conjunctive" one. Despite the fact that the sequencing of techniques for part-of-speech tagging (also written as POS-tagging or PoS-tagging or POST) might vary, the instruments and the processes themselves are interchangeable. Thus, Taljard and her fellow scholars assert the following:

> The degree of interchangeability would naturally be highest between languages which are closely related genetically, and which share the same writing system. This implies that a tagset functioning on the morpheme level of Northern Sotho should with minor adaptations also be useful for languages such as Tswana and Southern Sotho, and possibly with a larger degree of adaptation to Venda and Tsonga, which are also written disjunctively. Furthermore, provided that conjunctively written languages such as Zulu, Southern Ndebele, et cetera are pre-processed using a morphological analyser, at least the principles underlying the tagset will also be reusable for these languages (Taljard et al., 2008:114).

Taljard and her colleagues continue by elaborating their earlier argument that annotation praxis, which includes the generation of a tagset cannot be separated from its aim, its position, its composition within the larger portrait of the framework of corpus annotation, and the measurement application or corpora that ensue (Taljard et al., 2008:115). A profitable question that arises here then is the following quite straightforward one: Why does this tagset design obtain in the case of isiZulu and Northern Sotho?

As Taljard and her companions also note, in spite of the ardent advocacy for standardization, caution needs to be exercised because an invariable commitment to the general scientific theorems of standardization may foist a "straightjacket" on systematic and theoretical analyses. They therefore recommend that instead of imposing stringent scientific standardization theorems on the process, interim principles should be organized around which a standard would inherently emerge. In essence, they say, a "default specification" can be arrogated for which there would be no rationality to circumvent it. They add that while standardization is a model procedure for which to aspire, linguistic actuality frequently makes it quite challenging to achieve (Taljard et al., 2008:114).

Taljard and her fellow scholars additionally point out that as it pertains to corpus annotation, part-of-speech tagging is a quite profitable technique because it contributes value to the existing corpus and the yielded corpus can be employed by other analysts. In order to reuse such corpus, standardization is required so that analysts can exchange the data and other attendant tools like annotated corpora. The employment of an annotated corpus for purposes other than the initial one would be quite beneficial, if a standard approach for annotating the data "content-wise" and of constituting the annotated text "formally" is established (Taljard et al., 2008:113). Taljard and company go on to state the following advantages for doing this:

> [First, it] would prevent…a "free-for-all" or "re-invention of the wheel" every time a new project is started. Secondly, standardisation should be aimed at integrating annotation with NLP (i.e. neuro-linguistic programming) components such as grammars and lexicons, and with tasks such as parsing, i.e. with a view to later use of the corpus in the NLP pipeline. In the third instance, standardisation of annotation practices will facilitate cross-language usability—an aspect that is of particular importance in the multilingual South African setup. In a situation where NLP development needs to be done for eleven languages, it is only reasonable to expect that tools developed for one language should be usable for other languages as well, especially in the case of the nine South African Bantu languages, since all these languages are genetically related and thus share common linguistic features. However, sharing of electronic resources and procedures for software development is only feasible if at least some measure of standadisation is adhered to (Taljard et al., 2008:113-114).

In addition, Karabo Molema in responding to a question on whether "African languages in South Africa are mutually intelligible to each other" posted on *Quora* (a website dedicated to users who ask and answer questions, develop question-and-answer blogs, and analyze statistics about other users; launched in June of 2010, and since then its user base has grown to over half a million active participants, its slogan is "The best answer to any question.") enlightens us about how *Interchangeability* manifests among some of the African languages in South Africa. She indicates that in general, her mother tongue Setswana is interchangeable with Southern Sotho and Northern Sotho.

The strategy employed by the speakers of these languages in order to understand one another is to speak what are referred to as "light versions" of the languages. Thus, in certain places, they must use slangs for some words and speak slower (Molema, 2016:4).

Molema adds that if a speaker of Setswana is to move to an area where Northern Sotho or Southern Sotho is spoken, the person would adjust to the language dynamics after a few weeks. She cites the cases of Bloemfontein and Pretoria to illustrate the situation as follows: in Bloemfontein, most of the residents being Tswana and Sotho speak a mixture of Sotho and Setswana; in Pretoria, the majority of the residents speak a mixture of Setswana and Northern Sotho (Sepedi). She notes that the closest approximation to these South African cases in Europe would be a city composed of Portuguese and Spanish speakers or a city comprising Dutch and German speakers who over time develop unique hybrid dialects. She adds that as it pertains to the other African languages in South Africa, *Interchangeability* is evident among isiZulu, isiXhosa, isiNdebele and siSwati by virtue of being members of the Nguni language family (Molema, 2016:4-5).

After that, we glean a notion of *Interchangeability* from Shola Sunday Olanipekun, Ebenezer Omoniyi Onabanjo and Clara Modupe Olayemi's illumination on linguistic innovation in African writing during the 21st Century. As these scholars observe, the African continent has the most interchangeable or intersecting languages across the globe in terms of contemporary fiction writing due to the "cross-linguistic collective consciousness" of the continent's third-generation writers. As Olanipekun and his colleagues colorfully put it, this outcome is due to the fact that "the world is of mixed linguistics perambulation with a form of language that is clearly reshaping language consciousness based on interdependency among nation-states of the world. In other words, these generations of African writers pose a literary poise of imprecise language bazaar to the world fictional scene" (Olanipekun et al., 2016:30).

Here, we must ask the following question: What is the reason for this innovative linguistic phenomenon among contemporary African writers according to Olanipekun and his fellow scholars? The answer to this question can be delineated from these scholars' brief review of the debates over the politics of language in African fiction writing. First, they point out that while some African writers proffer the viewpoint that European colonial languages are part and parcel of the linguistic literary milieu of African writing, others render this aspect as a useless endeavor because of those languages' lack of capability to capture many of the deep meanings of African thoughts. Nonetheless, for Olanipekun and his colleagues, there are two difficulties with the pro-African languages position. One difficulty is that most Africans cannot read texts in their own languages; thus, the small number of texts written in

African languages must still be translated into the European languages in other to reach a wider reading audience. The other difficulty is that alphabets have not been developed for many of these African languages (Olanipekun et al., 2016:29).

Second, Olanipekun and his colleagues mention that the "liberalists" argue that there is something basically wrong about expressing African thoughts in European languages. This school of thought, nonetheless, maintains that given the realities of the African literary landscape, the European languages can still serve a very useful purpose if they are "tamed," "nativized," and finessed in order to be assimilated into the African context. The European languages, proponents of this school also insist, must be made to mix with the milieus of African writers and readers in order to yield artistic products that will give pleasure through beauty. In essence, this new way of expression will enlarge the "trans-language" understanding of Africans represented in fiction writing. "In this sense," assert Olanipekun and his fellow scholars, "the future vision of the twenty-first century African writer's peculiarity with language is that of a linguistics 'contamination' and perforation with different languages at work" (Olanipekun et al., 2016:29).

Furthermore, Pieter Muysken in his 2011 essay, whose content is mostly repeated in his 2017 essay, apprises us about an idea of *Interchangeability* in a novel written by Ingrid Winterbach in the Afrikaans language of South Africa and titled *Die benedenryk* (2010). He discovers solitary manifestations of English adjectives, adverbs, nouns, verbs, and complete components and utterances from English in the following extract of Afrikaans-English code-switching by a young artist who was fulminating against the contemporary art vista in South Africa:

> ...Dis nog nie *multimedia* genoeg nie. Dis nog nie *take no prisoners* nie. Dis nog nie *confrontational* genoeg nie. Dit *challenge* nog nie sy eie aannames *stringently* genoeg nie. Dis nog niet genoeg van 'n *assault* op enige *established high culture* nie. "This is not yet mediamedia enough. This is not take no prisoners. This is not yet confrontational enough. This does not yet challenge. Its own assumptions stringently enough. This is not yet enough. Of an assault on any established high culture (quoted in Muysken, 2011:30-31).

Thereafter, Muysken considers how the preceding components are completely interchangeable with Afrikaans constituents of the same kind, thereby suggesting full categorical uniformity. He then goes on to discuss the following four likelihoods demonstrated with the various kinds of switched verbs: (1) "simple *harmonization*: match existing features"; (2) "*neutralization*:

add masking morphological material"; (3) "*null*: omit offending category in one language, generally the inserted one; this is of course related to the null strategy, of omitting (in this case morphologically) offending elements"; and (4) "*compromise* between the feature systems of the two languages" (Muysken, 2011:31-34). All this leads Myusken to suggest that the multiplex nature of the occurrences, the differences among the discoveries, and the reciprocal influences among the various arbitrating elements all necessitate an experimental technique with systematized elicitation and corpus-based investigation (Myusken, 2011:34).

Finally, H. Ekkehard Wolff tells us that *Interchangeability* is manifest in the fact that many African children acquire two or more languages before they even reach school age and frequently use the acquired languages interchangeably. This verity, he argues, obliterates the notion proffered by some that it is difficult to achieve multilingualism in putting African languages first in an education system. The ability of Africans to use two or more languages interchangeably, he mentions, even has "sociolinguists intrigued by the ways in which Africans communicate mainly in urban contexts—in what appears to be talking in two or more languages at the same time. The new academic terminology for this is *translanguaging* or polylanguaging" (Wollf, 2018:3). He therefore recommends that this great linguistic asset be utilized to teach via both African and European languages across an education system, since people already employ this approach in other arenas of society. It makes no sense to him why decision makers in the field of education continue to insist that only European languages, instead of "translanguaging," be used when teaching content subjects (Wolff, 2018:3).

Given this linguistic asset of *Interchangeability* possessed by Africans, Wolff also repudiates the argument by some that African languages are simply inadequate for teaching and learning at the university level. He characterizes the assumption as one based on racism and ignorance which is not supported by any empirical evidence. He alludes to the following instances to buttress his assertion:

> A recent PhD thesis currently being submitted at Rhodes University in South Africa, where I am a visiting Fellow, found that students with a background in languages other than English profit immensely from being assisted with teaching materials, terminology and translation aids in their mother tongues. At Rhodes, isiXhosa features as more than a language subject. It is used as a medium of instruction in support courses for Journalism and Media Studies. Pharmacy students are taught vocation-specific isiXhosa skills. Bilingual teachers in Politics, Commerce, Sociology and Economics are recognising the linguistic

> diversity of their classes by using students' lived experience as an important aspect of teaching and learning. There's more. The University of Limpopo offers multilingual studies, including a BA in Contemporary English language studies in both English and Sesotho sa Leboa. Masters and PhD students write their theses in any official language of their choice—recent examples have included theses in Sepedi, Tshivenda and Xitsonga. Both Stellenbosch University and the Cape Peninsula University of Technology offer multilingual glossaries in English, isiXhosa and Afrikaans for various faculties. These are also accessible online (Wolff, 2018:3).

Another argument against introducing African languages into an education system pertains to cost. He declares this reasoning as a myth. To support his declaration, he cites an empirical finding which reveals that "investment in such programmes in Africa…is usually less than 2% of a country's education budget—and is recovered within five years" (Wolff, 2018:2).

African Mathematical Evidence

First, the following excerpt culled from Claudia Zaslavsky's work on African Mathematics is a significant indication of *Interchangeability* in African mathematical practices through gesture:

> David Zarembka, an American volunteer teacher, lived and worked among the refugees from Rwanda in western Tanzania. At the market one day he asked the price of an item. "One shilling," replied the merchant, with a wave of his hand. David offered the shilling, but the merchant refused to sell the article. Again he waved his hand, with two fingers outstretched. At last David understood that the price was one shilling, twenty cents. The people of Rwanda accustomed to the Belgian franc, convert ten cents in shilling and two francs, indicated by waving two fingers (Zaslavsky, 1999:238).

The preceding extract clearly indicates how Africans utilize gestures as substitutes for the spoken word when exchanging messages and how they are also employed to highlight a point during communication.

Second, J. A. Andrea alluded to the *Interchangeability* that transpired in the field of Mathematics and a number of other intellectual disciplines among the empires along the Silk Road from 100 BC to around 1350 AD. The linguistic feature manifested itself due to the nature of the Silk Road, which Andrea describes as follows:

> The Silk Road, a complex network of caravan routes across the heart of Central Asia that connected and cross-fertilized the peoples and cultures of the Afro-Eurasian World, flourished from about 100 BCE to circa 1350 CE, with five periods of particular vitality. Long before the advent of a global "world system," the Silk Road served as a major medium for economic and cultural exchange, and constituted a true Afro-Eurasian System. Regardless of global shifts that witnessed the rise of new systems of transoceanic exchange after about 1450, the old roads of the Silk Road continued to bear traffic. Despite political and environmental shifts over the past 500 years, some of which have been quite disastrous, even today these ancient roads are important arteries for the lands and peoples that they connect (Andrea, 2002:2).

Andrea adds that the Silk Road span a huge territory covering East Asia, Europe, the Levant, and North and East Africa. This Afro-Eurasian world prospered for thousands of years due to the exchanges that took place via ideas, goods, and other items exchanged via the network. To this day, commercial activities take place on some of the routes of the network (Andrea, 2002:2).

Third, from the mathematical model of a packed bed designed by Gerald T. Marewo and D. J. Henwood for a cooling system in an international school building in Harare, Zimbabwe, one can notice the notion of *Interchangeability* in the framework. These researchers' mathematical model, which uses two terms (i.e. "heat dispersion in the fluid" and "heat loss to the environment"), is a two-phase process that allows the packed bed system to store coldness during the night to be used by the school's air conditioners during the day. Unlike previous studies that employed a simplified mathematical model in which one of the terms is left out, Marewo and Henwood say that their model, which uses both terms, makes it possible for them to derive and apply a solution. They also report that their utilization of gauged inlet temperatures resulted in a correlation between the measured and predicted outlet temperatures of the bed. By thoroughly analyzing the variations employing sensitivity analyses for both the heat convection transfer and air velocity, they discover that adjusting these parameters can increase the correlation between the predicted and measured data. This approach also allows them to conduct a parametric investigation "for heat storage with various materials and bed sizes, which indicates how the code may be used as a tool for improving design and operational parameters" (Marewo and Henwood, 2006:11).

Fourth, Palesa Rebecca Mochesela evokes the notion of *Interchangeability* when she talks about how the problem-based approach used to teach Grade 9 learners in South Africa in solving word problems led to an improvement in the learners' problem-solving performance and attitudes toward mathematics.

According to her, the impetus for the study was based on the evidence that traditional approaches that had been previously used to instruct these learners were not leading to successful results in their performance in mathematics. She also states that the success of the problem-based approach is due to its capacity to give learners lifelong skills and knowledge. This is because the approach is anchored on constructivism, which postulates that knowledge cannot be transferred from the teacher to the learner, but by the active construction of knowledge by the learner. She, nonetheless, points out that the constructivist perspective does not imply that the teacher is not vital in the learning process; it simply proffers that the teacher's role is to establish the opportunities and stimuli for learners to engage in cognitive activities that will make them successful in mathematics (Mochesela, 2007:111, 17).

In addition, Mochesela mentions that constructivism neither dismisses nor advocates a teaching technique; instead, it calls for the focus of the teacher to be on learners' abilities, needs, and uniqueness. She further notes that constructivism is being denoted as a proponent of "discovery and inquiry-based learning, incorporating opportunities for discussion and exchanging ideas" (Mochesela, 2007:17-18). To illustrate how constricting mathematical knowledge is more than just the learning of new concepts and entails the reconstruction of previous knowledge, she mentions the following example of the assistance provided to learners who had a misunderstanding of the utilization of an equal sign:

> To these learners an equal sign was considered as 'doing something' signal, as a result although learners understood the equation like "3x + 5 = 17" because there is a single term on the right—the result of doing something, they however; had difficulty solving an equation like '3x + 5 = x + 17". It seemed unfinished because the right side still has an addition to be performed. After a sequence of appropriate activities, learners re-organized their understanding so that an equal sign was seen as a relational symbol rather than a 'do something signal' (Mochesela, 2007:18).

Fifth, Nicola J. Mulder and her 10 coauthors talk about *Interchangeability* in terms of its connection to knowledge and collaboration in bioinformatics training initiatives in South Africa. In addition to mathematics, the training is also done in "biological science, computer science, physics, and statistics." While only a few universities offer formal degree programs in bioinformatics, many other universities either have full-scale research projects on bioinformatics aspects or are part of other disciplinary projects. Students in the programs are instructed in bioinformatics courses, subject-related activities and modules offered at their universities, and related activities offered by other bioinformatics research institutions (Mulder et al., 2016:6).

In addition to the span of bioinformatics subject matter to which mathematics and computer science students are exposed, they also receive training in basic molecular biology, and the biology students in turn receive training in computer programming. Students' knowledge is enriched by the many networking opportunities available to them and the sharing of information that takes place via the bioinformatics activities. The courses and activities are so popular that there is not enough space to accommodate many of the students who are interested in the programs. As Mulder and her colleagues underscore the problem with the limited space, "The substantial number of applications from bench scientists to attend these courses highlights the current gap in the training of bioinformatics users" (Mulder et al., 2016:6-7).

A fitting three-pronged question here has to do with (1) the reason South Africans jumped on the bioinformatics bandwagon in its early stage, (2) how they deal with its challenges and promote it, and (3) the benefits they have accrued from the initiative. The following extract from Mulder and her colleagues addresses this question:

> Bioinformatics is now a critical skill in many research and commercial environments as biological data are increasing in both size and complexity. South African researchers recognized this need in the mid-1990s and responded by working with the government as well as international bodies to develop initiatives to build bioinformatics capacity in the country. Significant injections of support from these bodies provided a springboard for the establishment of computational biology units at multiple universities throughout the country, which took on teaching, basic research and support roles. Several challenges were encountered, for example with unreliability of funding, lack of skills, and lack of infrastructure. However, the bioinformatics community worked together to overcome these, and South Africa is now arguably the leading country in bioinformatics on the African continent (Mulder et al., 2016:1).

Sixth, in the work dealing with European mathematical didactics traditions edited by Werner Blum and his four colleagues, we get a perspective of *Interchangeability* vis-à-vis mathematical didactic interactions between France and African countries. But before getting to the explication of this perspective, we are first introduced to the notion of *interchange* via the lens of validation and proof of the French tradition. For the French, *Interchange*, which facilitates the exchange of ideas within the community and others, operates hand-in-hand with *cohesion*, which indicates that members of the French community share knowledge, experience, and theoretical models which allow them to talk about a "French Didactique," and *dissemination*, which refers to ideas in mathematical education across the globe (Blum et al., 2019:31).

Specifically about *interchange*, we learn that the phenomenon was promoted between France and African countries via the *Colloque Franco-Allemand de Didactique des Mathématiques et de l'Informatiqueis*, with its book series associated with *Recherches en Didactique des Mathématiques*. A number of other on-going international exchanges are also noted to be part and parcel of the endeavor. The tenth Summer School of Didactics of Mathematics convened in Houlgate in 1999, for instance, served as a venue for researchers to study the works of their colleagues. While young researchers are invited to participate in the *interchange*, they are nonetheless not the focus of the activities. The *interchange* is consistently nurtured through the participation of mathematicians from many countries across the globe and by some of the articles being published in multiple languages, specifically English, French, and Spanish (Blum et al., 2019:32-33).

In terms of the didactic interactions between France and African countries, we are informed that during the past several decades, mathematicians across Francophone Africa have collaborated significantly with French didactics mathematicians. The French initiative started early after the Francophone African countries gained their independence from France. For example, the didactics mathematics collaboration between France and Senegal dates back to 1975. When doctoral programs in the didactics of mathematics were launched in French universities in 1975, African students were admitted into the programs and wrote their doctoral theses under the supervision of French researchers. By 1985, 11 students from four Francophone African countries had already written and successfully defended their doctoral theses; and, by the end of the 20th Century, 20 more students from nine Francophone African countries had accomplished the same (Blum et al., 2019:44).

In addition, we garner that significant progress has been made in the areas of co-tutoring and co-supervision of doctoral students in the field of Mathematics by mathematicians from France and Francophone Africa. By the year 2000, 25 doctoral theses had been successfully completed and defended due to these collaborations. Other fruitful collaborative efforts include (a) the biannual Summer Schools of Didactics Mathematics in which many African didactics mathematicians participate; (b) the tri-annual conferences of the Francophone Mathematical Space (EMF), which began in 2000 with the goal of establishing the EMF structure that would include more Francophone researchers from non-affluent countries into the international community of mathematics education; and (c) alternating EMF conferences between France and Francophone African countries (Blum et al., 2019:44).

Furthermore, we are apprised by the cases of Benin, Mali, Senegal, and Tunisia that the *interchange* between the French and Francophone African didactics mathematicians has made it possible to establish several master's and doctoral degree programs in didactics of mathematics in Francophone African countries and support the growing community of didactics mathematicians in those countries. An obvious indication of this growth is the recent establishment of the Association of African Didacticians of Mathematics (ADiMA), whose first conference was convened by the Ecole Normale Supérieure (ENS) of Yaoundé in Cameroon in August of 2016 and the second one held at the Institute of Mathematics and Physical Sciences of Dangbo in Benin in August of 2018. We also learn that Tunisia is an exemplar of the fruitful *interchange*, beginning with the first collaborations in the 1970s involving the Tunisian Association of Mathematical Sciences (ATSM) to the French Association of Mathematics Teachers (APMEP) today. The collaborations that have been birthed by the *interchange* have fostered the growth of a Tunisian community of didactics mathematicians which is today anchored in the Tunisian Association of Didactics of Mathematics (ATDM) and enriched the mathematical research perspectives in both France and Tunisia (Blum et al., 2019:44-45).

Conclusion

In the Ancient Kemetian language, the idea of *Interchangeability* is reflected in the many terms used to explicate the ability and the means employed to transmit and receive messages. First, there are two terms for "message." One term is *åab*, which means "decree." The other term is *uput*, also meaning "embassy, order, decree, errand, command, mission, duty, and commission." Second, there are six words for "messenger": (1) *åupti*, which also means "envoy"; (2) *theftenn*, which likewise denotes "envoy"; (3) *unges*, which similarly means "envoy"; (4) *pehreri*, also denoting a "runner" [as in a person who smuggles specified messages into or out of a country or area] and "envoy"; (5) *hebu*; and (6) *khunnu*. Third, the "divine messenger" or "herald of the god" is referred to as Neter uhem or Uhem neter, meaning "envoy of the gods," Fourth, "The messenger-god" is called Åp-t or Åpu-t. And fifth, the "king's messenger" is called *uputi nesu* (for these translations, see Budge, 1978).

Also, we learn from Teodor Lekov that *Interchangeability* underlies the connection between the Ancient Kemetian denotation of the heart (*ib*) with that of the stomach (*rA-ib*), which literally denoted "entrance (mouth) of the heart." The *rA-ib* is the frame of the torso—the cavity in which the heart is found. Thus, the *ib* is more than the physical heart; it is the inner core of the body and the person. He adds that another complexity for comprehending the position of the heart in this and other similar milieus is the fact that there are two different words for heart in the Ancient Kemetian language. "First, and older *ib* roots in Semitic stratum of the language and has parallels in the languages as Akkadian and Arabian. Its usage implies nuances of 'internal,' 'hidden,' 'invisible,' 'inner self,' etc. On the contrary, the second word for 'heart'—*HAtj* is descriptive and means literally something like '(that) which is in front,' in sense of 'outer,' 'visible,' etc., for example as a designation for the pulse—a 'movement of the heart-*HAtj*'" (Lekov, 2004:33).

In addition, according to Nili Shupak, in "The Tale of the Eloquent Peasant" (one of the masterpieces from the golden age of Ancient Kemetian literature during the Middle Kingdom—i.e. from 2040 to 1782 BC), the object spread by Nemty-nakht (an Ancient Kemetian official who had the position of a scribe and astronomer of Amun during the reign of Thutmose IV in the Eighteenth Dynasty—i.e. from 1549/1550 to 1292 BC) on the path is at certain times referred to as *ifd* (a trimmed piece of cloth with four corners) and other times as *hbs* (straightforwardly a garment). The *Interchangeability* of the two terms leads Shupak to suggest that "what is meant here is the usual man's garment in Egypt, a skirt consisting of a length of material wrapped around the body and tied in front. Nemty-nakht did not place merely a piece of cloth on the road, but his own garment" (Shupak, 1992:7). The essence of this act hinges on the fact that, as Shupak points out, "According to the ancients, a man's spiritual forces are hidden in his clothes. A man's garment contains his identity and symbolizes it; it has magical properties and symbolic legal significance. Damaging, grasping, or tearing a man's garment were considered acts which caused its owner shame" (Shupak, 1992:7). What all this indicates is that Nemty-nakht was not only equipped to utilize linguistic and extra-linguistic tokens for the transmission of messages in the Ancient Kemetian language system, he was also well equipped to receive messages in the same system articulated by the eloquent peasant. As Shupak puts it, "a symbolic, perhaps even magic, significance can be related to the high official's garment [i.e. Nemty-nakht's garment] (called an "apron"), which is mentioned in connection with its owner's functions of protecting the lower class: "You are the apron (*šndyt*) of the motherless," meanings that the judge defends and spreads his patronage over the orphan" (Shupak, 1992:7).

Chapter 7

Cultural Transmission

Cultural Transmission, as I state elsewhere (Bangura, 2014), culling from the writing of Anouschka Bergman, Kathleen Curie Hall and Sharon Miriam Ross (2007), "refers to the fact that there are aspects of language which people can acquire only through communicative interactions with other users of that language system. Even children, whose ability to learn languages may seem innate, must still learn all of the specific signals of those languages by interacting with other speakers. The language we acquire as children is not influenced by our genetic or hereditary background" (Bangura, 2014:23). In this chapter, I examine the subject of *Cultural Transmission* as it relates to African languages and African mathematics. Before discussing the relationships between the former and the latter two aspects, a bit more will be said about the linguistic conceptualization of *Cultural Transmission.*

To start with, John Lyons makes two important points in his conceptualization of *Cultural Transmission.* The first point is that differentiation should be made between *Cultural Transmission* and *Genetic Transmission* (the transfer of genetic information, i.e. relating to genes or heredity, from one generation to another), since the former concept deals with the notion that the capacity to speak a certain language is passed down from one generation to another not by instinct, as the latter concept denotes, but by teaching and learning. Thus, Lyons urges analysts to transcend the question about whether knowledge of the more general features of the grammatical system of languages is genetically instead of culturally transmitted to explain why a significant component of the system of certain languages is obtained through learning. Such an analysis, he believes, would allow for the systematic testing of the hypothesis that children are born with the knowledge of particular truths that dictate the system of language (Lyons, 1977:82).

The second point Lyons makes has to do with the discovery that as opposed to the once widely-held belief that the signaling-behavior of non-human species was solely instinctive, the behavior is actually obtained by an amalgamation of instinct and learning. He cites the example of the chaffinch (a Eurasian and North African finch—i.e. a seed-eating songbird that typically has a stout bill and colorful plumage—typically with a bluish top to the head and dark wings and tail). He points out that while the general pattern of the bird's song may be determined as instinctual, the song's more intricate buildout and rectification hinge upon the bird's ability to hear and learn from the singing of the more grown-up chaffinches. Hence, for him, in terms of *Cultural Transmission*, there is no definite distinction between language and non-human semiotic systems (Lyons, 1977:82-83).

Next, Richard Nordquist starts off by pointing out that linguists define *Cultural Transmission* as the process by which a language is transferred from one generation to another in a linguistic community, and that this process is also referred to as "cultural learning" and "sociocultural transmission." He then goes on to describe the following three general types of *Cultural Transmission* as delineated by Tao Gong (2010): (1) "*horizontal transmission*—communication among individuals of the same generation"; (2) "*vertical transmission*—in which a member of one generation talks to a biologically-related member of a later generation"; and (3) "*oblique transmission*—in which any member of one generation talks to any non-biologically-related member of a later generation" (Nordquist, 2019a:1).

As was done in most of the preceding chapters, the rest of this chapter is also divided into three major sections. The first section comprises a discussion of various linguistic expositions of *Cultural Transmission*. The second and third sections entail African linguistic and African mathematical evidence on the subject, respectively.

Linguistic Expositions on Cultural Transmission

A reading of a sample of works on *Cultural Transmission* as a language design feature makes it possible to delineate the following eight thematic categories in which the discussions can be subsumed: (1) evolution of human behavior, (2) cultural vocabulary, (3) cultural level of usage, (4) language transfer, (5) cultural variation, (6) cultural context, (7) cultural identity, and (8) cultural language styles. As was also done in most of the foregoing chapters, these themes are presented separately in this section for comprehensibility.

Evolution of Human Behavior

In their review of a series of essays on the nexus between *Cultural Transmission* and the evolution of human behavior, Kenny Smith, Michael L. Kalish, Thomas L. Griffiths, and Stephan Lewandowsky address a number of issues. First, they note that in many diverse areas, humans learn from one another. Thus, according to these scholars, knowledge and behavioral structures are culturally transferred among humans. They therefore suggest that in order to properly investigate the evolution and outcomes of *Cultural Transmission*, researchers should inquire about (a) how extensive *Cultural Transmission* is among animals, (b) how *Cultural Transmission* operates within various groups of human, (c) the byproducts of cultural evolution, and

(d) how our human species has been molded by the interplay of culture and biological evolution (Smith et al., 2008:3469).

Second, Smith and his colleagues recount that the use of evolutionary approaches to investigate developments and outcomes of *Cultural Transmission* has a well-established history. The investigation of undeviating equivalents between biological developments and the cultural development of words continues to attract serious researchers. Using formal frameworks of evolutionary processes, a great deal of these researchers' works have been either theoretical or empirical focusing on actual cultural occurrences. Smith and his fellow scholars contend that although these frameworks constitute vital approaches for investigating cultural developments, there are other possibilities. They suggest the experimental approach as one such possibility for investigating the processes and properties of *Cultural Transmission.* This is because the experimental approach has the advantage of allowing a researcher to establish a connection between a formal model's generality and control and the naturalism of empiricism of actual behavior in actual cultural settings. Smith and his partners state that a longstanding and effective experimental approach, albeit currently undergoing a rebirth, is to examine cultural evolution straightforwardly in laboratory populations under controlled states in order to determine what actually takes place when people learn from one another and when non-humans learn from one another (Smith et al., 2008:3469).

Third, according to Smith and his coauthors, a basic question that underlies the investigation of humans' ability for culture has to do with the identification of the evolutionary roots of culture. The question is usually phrased in terms of whether this human ability is a recent phenomenon or a very old one that has reappeared in a different manner among us humans. Smith et al. also inform us that in order to provide an answer to this question, researchers employing the *diffusion chain experimental* method (i.e. an approach "based around a pairwise learning interaction, whereby one individual produces a behaviour for observation by another individual, who attempts to learn or reproduce that behaviour…the second is then used as the model, producing behaviour for a third individual and so on—a multiculture is created in the laboratory") discern *Cultural Transmission* and social learning from other processes with the ability to yield similar group-level behaviors. They further apprise us that impressive findings have been generated by these researchers indicating that many non-human species such as birds, fishes, primates, and rodents engage in *Cultural Transmission.* In addition, they note that these researchers have demonstrated that a behavior can be planted within a group and the behavior would spread through that group in an unforced and unregulated manner (Smith et al., 2008:3470).

Fourth, Smith and his partners explore the reality that a major incentive for comprehending the ability of humans for culture is that it constitutes some of the most astonishing attainments of humanity. They include extensive religious and social ceremonies, highly advanced sciences, and revolutionary technologies as being among the inventions that have resulted from "a cumulative process of cultural evolution, whereby each generation builds on the achievements of their predecessors in a gradual, approximately monotonic ratcheting up of complexity and functionality" (Smith et al., 2008:3471). Furthermore, they mention the following foci on which researchers investigating this aspect have focused:

a) The extent to which *Cultural Transmission* "yields products that are well designed, and we can use experimental techniques to delve into the processes that produce these functional outcomes" (Smith et al., 2008:3471).

b) Cumulative cultural evolution's "taxonomic spread (its presence in non-human is contentious); the mechanisms underpinning it (it remains unclear whether sophisticated imitation is required for cumulative cultural evolution or whether more basic social learning mechanisms will suffice); and the types of outcomes it yields (e.g., whether cumulative cultural evolution can deliver behaviours that are universal cross-culturally) (Smith et al., 2008:3472).

c) The degree to which *Cultural Transmission* "improves traditions, suggesting that children are prone to over-imitation—they copy both task-relevant and task-irrelevant (and therefore non- and a-functional) behaviours" (Smith et al., 2008:3472).

d) The extent to which "adult human participants negotiate communication systems through repeatedly playing a graphical communication game similar to the parlour game Pictionary" (Smith et al., 2008:3472). *Pictionary* is a word-guessing game whereby a team moves a piece on a game board formed by a succession of squares. On each square is a letter or shape representing the type of picture to be drawn on that square.

Fifth, for Smith and his fellow scholars, biology, which has been ascribed to distinctively human social learning systems, must be the fundamental root of the extensively cultural essence of human perception. They note that biology and culture influence each other. They mention as an example of this mutual influence our discernment capabilities having been greatly shaped to work with our human cultural inheritance. Therefore, they assert that the existence of design in human behavior hinges upon biological or cultural evolution; consequently, in order to comprehend which inheritance structures transmit and configure which behaviors, as well as to comprehend how these inheritance systems interact, the roots of complex and adaptive human behaviors need to be explicated. In addition, Smith and his colleagues make references to the following aspects to which researchers interrogating the biological strand of *Cultural Transmission* have paid particular attention: (a) "human behavior involving cultural evolution and the popular Evolutionary Psychology approach—i.e. the account that sees human behaviour as governed by a set of hard-wired, task-specific ecological challenges posed by the ancestral human environment"; (b) explanations of the "variation in handedness in human populations"; and (c) "situations where culturally transmitted traits (mate situations…and niche-constructing capacity) impact on or change the course of biological evolution" (Smith et al., 2008:3473).

Cultural Vocabulary

Cultural Vocabulary can be generally denoted as a body of information shared by most of the people in a particular culture. Linguists tend to treat this phenomenon in relation to linguistic or lexical "borrowing": i.e. the process by which a word (referred to as a "borrowing," a "borrowed word," or a "loanword') from one language is adapted for use in another language. In the rest of this subsection, I present how Joseph H. Greenberg (1957) and John Lyons (1971) broach the concept of *Cultural Vocabulary.*

Beginning with Greenberg, he tells us that *Cultural Vocabulary* will absorb a very large number of word resemblances and resemblances in items with grammatical function, especially if they are repetitive via a number of languages, which is a reflection of a genetic connection. These resemblances would also congregate in particular semantic domains that mirror the cultural essence of the contact and point toward one, two, or three languages as the source or sources. The forms would be quite similar to those that exist in these specific tongues given the significant differences in other respects and the resulting historic distance of the connection, if one exists. Greenberg cites the example of how Romance borrowings in English are almost all similar to the French forms, including the fact that they barely enter the basic vocabulary of English. In essence, English would have displayed roughly equal similarities to all of the Romance languages if it were actually a Romance language (Greenberg, 1957:39).

In addition, Greenberg suggests the following basic probability considerations when determining whether the resemblances between two languages are simply serendipitous and/or symbolic:

> The likelihood of finding a resemblance in sound and meaning in three languages is the square of its probability in two languages. In general the probability for a single language must be raised to the (n – 1th) power for n languages. Thus if five languages each showed a total of 8 percent sound-meaning resemblance to one another, on a chance basis one would expect $(0.08)^4$ or 0.0004096 resemblances in all five languages. This is approximately 1/25,000. In other words, were one to compare sets of one thousand forms from all five languages, one would have to do this twenty-five times before a single instance of a resemblance in all five languages would occur. Even recurrence in three languages would be rare on a chance basis, 0.0064, that is, less than 1 percent. Hence the presence of a fair number of recurrent sound-meaning resemblances in three, four, or more languages is a certain indication of historical connection (Greenberg, 1957:39).

Also, according to Greenberg, other considerations exist in terms of the phonetic form. Adding to the merit of the comparison of these considerations is the existence of recurrent correspondences—i.e., correspondences that are not unique. This issue leads to the question about whether forms that exist in many languages imply that they are altered forms of a prevalent origin. By considering languages that are closely connected on each side, an analyst would be able to determine the history of the tentative reconstructions that have intersected. However, this procedure will not work if only two languages are to be compared (Greenberg, 1957:39).

Furthermore, Greenberg suggests the possibility that should the preceding factors lead to the determination of a historical link, then borrowing would explain the identified resemblances. Even though in certain and infrequent cases borrowing may seem like an implausible explanation, it is still possible to ascertain whether borrowing explains the multitude of resemblances between two languages. Greenberg also proposes that "A basic consideration is the a priori expectation and the historical documentation of the thesis that borrowing in culture words is far more frequent than in fundamental vocabulary and that derivational, inflectional, and pronominal morphemes and morph alternations are the least subject of all borrowing" (Greenberg, 1957:39). Concomitantly, he adds that "While it cannot be said that any single item might not on occasion be borrowed, fundamental vocabulary is proof against mass borrowing" (Greenberg, 1957:39).

Next, preferring the related term *Cultural Words*, Lyons makes four salient observations. The first observation is that compared to other words, cultural words are more susceptible to borrowing, albeit no set of words that are fundamental in a language are unsusceptible from borrowing or replacement. Nonetheless, Lyons asserts that this aspect is less vital than it might appear to be. This is because, according to him, given the languages which have been comparatively and thoroughly studied, the vague idea of resemblances has been replaced with the more exactly denotable concept of *systematic correspondence* (Lyons, 1971:26). "Systematic correspondence" generally refers to a relation or mapping between two linguistic structures, such that if the same lexical content occurs in the two structures, then there exists a constant meaning relation between them.

Lyons' second observation concerns the possibility that much of the resemblances researchers find in the vocabularies of various languages may hinge upon those languages having borrowed words either from one another or from a third language. He points to the large numbers of words from the Greek and Latin languages that exist in modern European languages as proof of this possibility (Lyons, 1971:25-26). He further notes:

> ...if we use the term 'borrowing' here to refer, not only to words which have been taken directly from the classical languages, but also to words which have been created in recent times by the deliberate combination of parts of Greek and Latin words, then we must say that most modern scientific terms, as well as the names of such modern inventions as the telephone, television, automobile, cinema, etc., have been indirectly borrowed from Greek and Latin. Borrowing from the classical languages explains many of the more obvious similarities in the words of different modern European languages (Lyons, 1971:26).

The third observation made by Lyons deals with the fact that modern European languages have borrowed from one another in various degrees and at different historical epochs, and continue to do so in contemporary times. Given what is known about these European languages and the general evolution of European culture, not much effort is needed to determine most of the borrowed words in the vocabularies of the modern versions of these languages. Nonetheless, there is a greater likelihood that one would not be able to determine the exact degree of the borrowing among these languages if s/he could not discern the borrowed words and then exclude them from his/her calculation (Lyons, 1971:26).

Lyons' fourth observation relates to the fact that while early comparative philologists had an idea that languages borrow from one another, they, however, did not have methods to help them decipher borrowed words from the rest of a vocabulary. Thus, their analyses were based on grammatical resemblances as proof of linguistic linkages and were very careful in stating resemblances between words when these were not easily identified in the "basic vocabularies" of the relevant languages; the "basic vocabularies" being the most indispensable words learned earliest and utilized always (Lyons, 1971:26).

Cultural Level of Usage

To begin with, Andrew MacLeish describes *Cultural Level of Usage* as a phenomenon that reflects the educational and socioeconomic backgrounds of the speakers of a language. He identifies two general levels of cultural usage: (1) *Nonstandard*, which is characterized by "illiterate speech, ungrammatical speech and writing, excessive and unskilled slang, careless use of vocabulary"; and (2) *Standard*, which denotes "cultivated language, clear, grammatical writing" (MacLeish, 1971:32).

In order to grasp the essence of *Cultural Level of Usage*, MacLeish recommends that it be contrasted with *Functional Variety of Usage*, which hinges upon the intended utilization of the function. Here, he also pinpoints two types of *Functional Variety of Usage*. They are (1) the *familiar-colloquial*, which is indicated by "familiar conversation, private correspondence, informal conversation, and familiar public addresses"; and (2) the *formal-public*, which is indicated by "pulpit or platform speech, public reading, worship, belles-lettres, and legal, scientific, or other expository prose writing" (MacLeish, 1971:50).

In addition, MacLeish reveals that there are "four usage-scales of native central English" that influence how the language is employed in speech and writing. These four scales are (1) Age, (2) Style, (3) Breadth, and (4) Responsibility (MacLeish, 1971:50), which he did not define. So, I denote these scales as follows: (a) *Age*—the length of time that a speaker or writer has lived or a thing spoken or written about has existed and a distinct period of history; (b) *Style*—the manner a speaker speaks or a writer writes about something; (c) *Breadth*—the range or extent of the speech or writing; and (d) *Responsibility*—the state of being held accountable or blamed for a speech or writing.

Also, MacLeish alludes to the following five styles of usage: (1) "frozen," (2) "formal," (3) "consultative," (4) "casual," and (5) "intimate." He provides detailed and quite comprehensible descriptions of these styles that are worth restating here verbatim as follows:

> The frozen style is characteristic of print and declamation. It is characterized by the absence of authoritative intonation in the text and by the fact that the reader or hearer cannot cross-question the author.

> The purpose of the formal style is to inform. It is characterized by the absence of participation on the part of the hearer and by the fact that neither the speaker nor the text is "involved"; the style is an objective one. The consultative, casual, and intimate styles are all characteristic of the colloquial mode. The consultative style is the norm for coming to terms with strangers. It is characterized by the fact that the speaker assumes no background information on the part of the listener; he supplies all this information. Further, the person who is addressed in this style is continuously participating in the discussion (MacLeish, 1971:50-51).

MacLeish continues:

> The casual style is used with friends; it is the style for "insiders." This style uses ellipses and slang. The intimate style is that used in an intimate group, usually between two people. It is characterized by the extraction of small units of wording or intonation from what in other situations would be longer lengths of discourse. These small units or wording or intonation are frequently a kind of code between two people (MacLeish, 1971:51).

Furthermore, based on MacLeish's exposition, it can be posited that the quintessence of *Cultural Level of Usage* and *Functional Variety of Usage* is that both concepts allow us to comprehend from a linguistic perspective the general concept of *Usage*, defined by him as "The customary manners of speaking or writing a language; the lexical, grammatical, and phonological choices we make" (MacLeish, 1971:130). He states that *Usage* also means a certain precedent of a usual way of speaking or writing such as employing the double negative. He argues that although describing the levels and types of usage is a quite complicated process, there are, as stated and denoted earlier, two common types of English usage: i.e. (1) the *Standard* and (2) the *Nonstandard*. MacLeish cerebrates about the fact that the distinction between the two usages is not clear-cut, as usage is often shared between them. Yet still, Nonstandard English is not often employed by Standard English speakers, as some of the former's expressions, such as the double negative, socially and economically disadvantages the user. But since the distinction between the two usages is not always clear-cut, one must be cautious about claiming that the Standard is superior to the Nonstandard. MacLeish adds that there are three general varieties of Standard English: (1) the *formal*, which is utilized in formal speeches and serious writing; (2) the *informal*, which is employed in books and magazines geared toward the general reader; and (3) the *colloquial*, which is used in discourses that desire conversational locution (MacLeish, 1971:130).

Language Transfer

Language Transfer, which is also referred to as *Transference*, can be generally defined as the effect of a language's linguistic features on another language in such environments as language learning and bilingualism. Kenji Hakuta explores many aspects of *Language Transfer* apropos *Cultural Transmission*. For starters, he describes the various approaches that have been used to study the phenomenon. The first approach to emerge was the behaviorist framework that postulated the notion of habit as the causal factor for language acquisition. Undergirding this postulate was the perception that if first-language acquisition was the result of building up of habits, then second-language acquisition must be a process of getting rid of those first language habits where differences exist between the two languages and maintaining the good habits from the first language where similarities exist between the two languages. Hakuta recalls that a dynamic propinquity existed between behaviorists and linguists who engaged in contrastive analysis, whereby the similarities and differences among grammatical systems of language usage were investigated. Based on the belief that second-language acquisition must be different from first-language acquisition, the main objective of this propinquity between behaviorists and contrastive linguists was to determine aspects that made acquiring a second language challenging by providing detailed contrastive analyses of various languages (Hakuta, 1986:116).

Hakuta also recounts that the behaviorist postulate of language acquisition did provide some useful explanations; nonetheless, as the postulate came under scrutiny, the contrastive analytical approach with its attendant idea of language transfer was also scrutinized. He believes that the contrastive analytical approach would have come under scrutiny anyway even without the presence of the habit theory due to the serious doubts that emerged over the approach's inability to account for second-language acquisition data. Its use of habit theory simply made the contrastive analysis more prone to scrutiny, especially when it failed to predict all of the errors that actually occurred and falsely predicted those errors that did not occur (Hakuta, 1986:117).

Given these problems with contrastive analysis, Hakuta suggests that a theory that would account for language transfer would be not one tied to habit theory, but one that is compatible with second-language acquisition, with the exception of the radical varieties. This is because any theory that would discount the place of prior experience in the attainment of new knowledge, for instance, would not explain language transfer. Also, a variant of radical nativism would suggest that humans naturally have all of the systems of human languages and, thus, exposure to a second language simply makes a reality of the already existing system. In essence, when confronted with a new language, a person would draw upon the language with which s/he is familiar (Hakuta, 1986:123).

Next, Hakuta articulates the nexus between individual variation and language transfer. He expresses the view that the characteristics of an individual learner might explain language transfer. He cites the example of the English language learning experience of a Japanese child called Uguisu and argues that he is not convinced that Uguisu's wrong usage of the word *mistake* and the preposition *with* in relation to reflexives would exist among all other Japanese children learning English. He believes that while the mistakes could have resulted from Uguisu's knowledge of the Japanese language, it is also likely that they were the result of Uguisu's personal "insight" of the connections between the two languages (Hakuta, 1986:124).

The foregoing idiosyncratic occurrence, says Hakuta, is analogous to Chomskyan linguists (who proffer the view that humans were born with the predisposition to learn language) sitting around and sharing with one another their hunches about the English language. While all of them would be quite knowledgeable about the language, they would, nevertheless, come up with some different arguments and analyses for even the appropriate description of a sentence. This is due to the fact that the part they select to work on will be influenced by the various features of their individual makeup, even though their descriptions would come from the same knowledge source—the English language, and may even share the same form of linguistic argumentation. Similarly, maintains Hakuta, the points at which language transfer occurs in second-language acquisition may be due to individual variation (Hakuta, 1986:124).

In addition, Hakuta considers linguistic relativism as a framework that explicates language transfer. According to him, at the level of grammatical concepts, every aspect of transfer shown in second-language learners is an aspect of confirmation for the idea that the essence of the native language has an impact of the worldview about language. As an example, he mentions the fact that to the native speakers of the Chinese, Japanese and Korean languages, the pervasive linguistic features *a*'s and *the*'s that appear before nouns in the English language may sound nonsensical; but to the native speakers of French, German and Spanish, these linguistic features are normal and well known. He therefore proposes that if the acquisition of a second language is perceived as an act of problem-solving, then it is only logical to assert that the native language would affect the manner in which that problem is solved (Hakuta, 1986:126).

Also, Hakuta discusses morpheme acquisition order vis-à-vis language transfer. He delineates three generalizations from discoveries of studies that have investigated (a) whether second-language learners resemble first-language learners and (b) whether second-language learners with different native-language backgrounds undergo the same process. The first generalization is that the order of appearance of morphemes among second-language learners is significantly different from that among native learners of English. About this finding, Hakuta is, however, quick to caution that it "does not mean that the process of first- and second-language acquisition are different...since the two types of learners are different in respects other than their level of language development. It may be that general cognitive development influences the order found in first-language children. For example, the past tense form may be acquired late because its mastery requires a developed concept of time" (Hakuta, 1986:120-121). The second generalization is that across children with different language backgrounds, the order of acquiring English morphemes appears to be quite similar. The third generalization is that for some of the morphemes by native-language transfer, the common order of English grammatical morpheme acquisition is modified (Hakuta, 1986:121).

Furthermore, Hakuta explains how the social context of second-language acquisition can influence language transfer. He tells us that a significant number of transfer errors can be accredited to the structure of the second language. Hence, he posits that the most significant discoveries about language transfer would be those that inform us about the sociolinguistic contexts in which the transfer takes place and those in which it does not take place (Hakuta, 1986:125).

Cultural Variation

The concept *Cultural Variation* can be generally defined as the rich array in social customs that different societies across the globe manifest. The discussion of the concept here as it relates to *Cultural Transmission* is in line with two of the features examined by Stephen C. Levinson: i.e. (1) "universals" and (2) "sociolinguistic variation" (Levinson, 1983). *Universals* refer to linguistic patterns that occur systematically across natural languages and perceived to be potentially true for all of them. For instance, all languages possess nouns and verbs, and all spoken languages possess consonants and vowels. *Sociolinguistic Variation* denotes the manner languages differ and evolve in communities of speakers in terms of the interaction of social factors such as age, ethnicity, gender, race, degree of integration into their community, etc. and linguistic structures such as grammatical forms, intonation features, sounds, words, etc.

Starting with universals, Levinson begins by arguing that in order to address problems concerning the ideas of context and grammaticalization upon which the definition of pragmatics hinges, we must make an important distinction between *universal pragmatics* and *language-specific pragmatics*. This is because the former is the fundamental theory of what aspects of context get encoded and how, and the latter pertains to individual languages. For example, as Levinson explains, "the pragmatics of English might have relatively little to say about social status (beyond what we need to describe the appropriate contexts for the use of *sir, your honour* and the like), while in contrast, the pragmatics of Japanese would be greatly concerned with the grammaticalization of the relative social ranks of participants and referents" (Levinson, 1983:10).

Next, Levinson contends that a functional explication of the principles of the social organization of universals, or near universals, is imperative in order to render as valid the ubiquitous assertion that almost all of the languages in the world have the three fundamental sentence-types: (1) declarative, (2) imperative, and (3) interrogative. This is because, according to him, given that these sentence-types appear to be employed paradigmatically to assert, question, and order, respectively, there is no point to look for internal linguistic inducement for them, since they occur repeatedly in the languages of the world because humans are, probably, particularly interested with these three functions of language: (1) put in order other people's actions, (2) obtain information, and (3) communicate information. Noting that this explication is obviously *post hoc*, he insists that we must demonstrate that these three activities are foremost in social life. Alternatively, adds Levinson, most languages possess some, and many languages possess elaborate, ways to encode relative social status among participants (Levinson, 1983:40).

In addition, Levinson suggests that the maxims of *conversational implicature* (discussed in detail in an earlier chapter) are only tenable if their application entails universals or, at least, what other culture-specific constraints obtain in social interactions. The reason for this, as he explains, is that the maxims produce inferences that transcend the semantic content of sentences that are uttered in all types of cooperative exchanges (Levinson, 1983:103). These inferences are, as Levinson puts it so eloquently,

> ...by definition, conversational implicatures, where the term *implicature* is intended to contrast with terms like *logical implication, entailment* and *logical consequence* which are generally used to refer to inferences that are derived solely from logical or semantic content. For implicatures are not semantic inferences, but rather inferences based on both the content of what has been said and some specific assumptions about the

> co-operative nature of ordinary verbal interaction. These inferences come about in at least two distinct ways, depending on the relation that the speaker is taken to have towards the maxims. If the speaker is *observing* the maxims in a fairly direct way, he may nevertheless rely on the addressee to amplify what he says by some straightforward inferences based on the assumption that the speaker is following the maxims...Another way in which the inferences may be generated by the maxims is where the speaker deliberately and ostentatiously breaches or flouts the maxims (Levinson, 1983:103-104).

Also, Levinson posits that an accounting for universals is necessary for a fundamental understanding of generalized conversational implicatures. This is due to the fact that, according to him, universality is anchored on the postulate that if maxims of conversational implicatures are extractable from thoughts involving rational cooperation, then they should be universal in their use, at least in interactions that are cooperative. He adds that another prediction from this postulate is empirical verifiability that could produce a test for implicature, since implicatures are said to emerge from the notion that cooperation underlies conversation (Levinson, 1983:120-121).

Thereafter, Levinson addresses the question concerning the extent to which universals reflect the features of conversational organization. For him, this question is essential for at least the following four reasons: (1) "linguists may be able to explain significant linguistic universals by pointing to universal functional pressures exerted by basic patterns of language use"; (2) "general patterns of child language acquisition may be explained by reference to a single basic learning situation, which is conversational"; (3) "pedagogical programmes in second language acquisition can take certain pragmatic parameters for granted"; and (4) "there are distinct limits to the kind of social variation in the use of language that has been explored in the ethnography of speaking" (Levinson, 1983:368-369). He also mentions that such universals would highlight a fundamental attribute of human nature: i.e. we humans are as much distinguished by conversational activity as we are by our various cultures and intricate social structures (Levinson, 1983:369).

Finally, in considering the role of universals in the application of pragmatics to difficulties in second-language learning, Levinson states that although procedures such as implicature have a probable universality, there seem to be abundant variations in both the systems of languages and how they are employed. He also says that there is a significant likelihood for cross-cultural misinterpretation even when there are underlying universals of usage, as evident in the composition of polite expressions. Citing the example of German speakers who seem to be significantly more direct, or less polite, when making complaints or requests compared to English speakers, he mentions the likelihood for misinterpretations. Thus, he suggests that this is a fertile area for linguists interested in systematic contrastive pragmatics to conduct research on aspects of misinterpretation that could emerge from a learner's belief that a linguistic construct in a language s/he is learning will entail the same conversational uses, illocutionary force, implicatures and presuppositions as those that exist in his/her native tongue (Levinson, 1983:176).

As it concerns sociolinguistic variation, Levinson broaches two important aspects. The first aspect is in regards to the variable phonological realizations related to social dialects and how pragmatics addresses them. He begins by proffering the view that the most restrictive perspective that pragmatics deals with linguistically encoded features of context would make phonological realizations related to social dialects appear to be outside the scope of pragmatics. Such a rejection, he says, would be based on the confined thinking of *encoding* that needs, among other things, that the significances being considered are intentionally transmitted and normally linked with the related linguistic forms. This is because, according to Levinson, the association of certain accents, evident by portions of phonological variables, with certain geographical or social communities is mostly not a component of an intentional communication (Levinson, 1983:29).

Furthermore, Levinson suggests that if one is to take the wider purview of pragmatics that connects language comprehension to context, then cases will exist whereby sociolinguistic variables are relevant to language comprehension. These variables would then be employed to adduce spheres of interpretation such as the marking of transitions from chat to business. Citing the example of a comedian telling a joke about an Englishman, an Irishman and a Scotsman, Levinson states that the comedian would have to mimic features of their different accents in order for an audience to follow which of the character is talking. Thus, he argues, it is quite challenging to draw a dividing line between sociolinguistics and pragmatic occurrences. This reality, he adds, is a result of the different ranges that have been declared by sociolinguistics and also because sociolinguists are interested in the interconnections between language and society however these occur in grammatical structures. As he further notes, sociolinguistics is not an element or level of grammar as are phonology, semantics, syntax and, quite credibly, pragmatics (Levinson, 1983:29).

The second aspect of linguistic variation broached by Levinson concerns what he refers to as the "variationist paradigm," which he characterizes as the notion that the systematic operationalization of sociolinguistic variables is partly conditioned by discourse. But since the fields of pragmatics and sociolinguistics share many general concerns, Levinson is convinced that there exists no actual peril for the absence of cross-fertilization, particularly among sociolinguists interested in language comprehension (Levinson, 1983:375).

Cultural Context

Cultural Context can be designated as the circumstances that form the immediate physical and social setting in which people live and how their way of life affects their behaviors and opportunities. Here seven themes pertaining to *Cultural Context* limned by Min-Sun Kim are discussed.

The first theme discussed by Kim is about communication theory and cultural context. He starts by mentioning that communication theory has a strong Western partiality as it is a Eurocentric enterprise, particularly in the United States, while Eastern notions that could add significantly to the theory have been left out of the Western theorization. The corollary to this development is that communication theory as developed in the United States is emerging as the world's communication theory (Kim, 2002:1-2).

Next, Kim informs us that the increased focus on individual-level cultural dimensions was a good step toward expanding communication theory beyond the realm of Eurocentrism, as some scholars began inquiring about the possibility of developing novel or reciprocal theories, methodologies and conceptual elements that could be generated from a collective global perspective. He makes the case that the "center of gravity" of communication theory would be balanced if non-Western communication researchers could develop scientific innovations from their own cultural legacies, instead of simply translating Western research instruments or applying Western communication theories to their research on their non-Western cultural settings. He believes that such an approach will broaden our knowledge of *Cultural Context*, as we will learn that the "self-related variables in cultures in which people usually have independent self-construals may be very different from the consequences observed in cultures in which people have more interdependent self-construals" (Kim, 2002:24-25).

In addition, Kim apprises us that the individualistic framework of the self is either directly or indirectly utilized by the majority of the human communication discipline. He points out that this framework, which is based upon Western philosophical constructs of human nature and many coatings of praxis and institutions, exhibits an aura of objective reality. He therefore argues that while the framework is adequate for explaining American and European empirical work on human communication that operates on individualistic maxims, it is inadequate for explaining the objective empirical realities of other cultures that do not operate on such individualistic maxims (Kim, 2002:26); i.e. they operate on collectivistic maxims.

The second theme examined by Kim deals with interventions and cultural context. For starters, he states that the success of an intervention and also for developing a bond between a client and a counselor, for instance, requires cultural sensitivity. He mentions the following aspects to support his assertion: (a) a client could encounter difficulty establishing confidence and affinity with a therapist when misinterpretations emerge due to cultural differences in values or communication styles; (b) an individualistic value orientation undergirds the value-laden process of teaching assertiveness or argumentative skills; since the dominant culture and that of psychology see this orientation as the most conducive, it could lead therapists to employ a "let's fix the minority" method that would prevent them from seeing other conducive alternative methods—a situation that could in turn prompt the minority client to reject the values of the dominant culture; (c) by eschewing the notion that there is a one best way, therapists would be better able to help clients to make choices in their communication patterns; (d) in order to build and increase the range for responding effectively in various contexts and to identify clues that discern different situations, therapists would be best served by developing a comprehension of the assertive or argumentative behaviors in different contexts; (e) by therapists giving clients the power to choose, the clients would obtain a range of culturally suitable skills; and (f) the most basic attributes of cultural sensitivity for clients include therapists' openness to the clients' experiences and a sincere respect the choices they make (Kim, 2002:53).

Also, Kim tells us that more than any other time in its history, the United States today has a more composite blend of students from many different cultures. The increase in the population of minority and international students is well reflected in national and regional demographic alterations and distributional movements. As a result, educational institutions and teachers have been left scrambling to accommodate the different cultural values, communication styles, customs, languages, and religious orientations these students bring to the classroom. Culture underlies the problems that many students and teachers encounter in the culturally diverse classrooms. And, as Kim argues, "in fact, interactions between teachers and pupils from different cultures are fundamentally problematic, and cross-cultural misunderstandings often occur because classroom interaction is an archetypal human phenomenon that is deeply rooted in the culture of a society" (Kim, 2002:53-54). He therefore calls upon teachers to draw from pertinent intercultural communication resources to aid the learning process (Kim, 2002:54).

Furthermore, Kim enlightens us that a key component of the Western communication field for over two millennia has been teaching students to argue constructively and effectively. Helping to maintain this tradition is the solidity of the evidence supporting the constructive nature of argumentativeness. Thus, verbal communication is regarded as vital for attaining educational objectives. Associated with this belief is the notion that the school, operating in the individualistic culture, is an oral domain in which speech is the main vehicle for passing on knowledge. As Kim succinctly puts it, "The more a person talks, unless the talk is overly negative or morally reprehensible, the more positively the person is perceived. In U.S. classrooms, the individual is expected to be an active participant in class discussions, and speech is essential...Instructional approaches offer short-term relief from what are considered the debilitating effects of communication avoidance, and therapeutic approaches offer long-term so-called cures for the problem" (Kim, 2002:54).

The third theme Kim explores pertains to meaning systems and cultural context. About this aspect, he makes a couple of observations. One observation is that cultural meaning can developmentally impact human psychological structures to such an extent that the latter can become an intrinsic component of the former. As an example, he notes that people's ascendant selfways (i.e. "communities' ideas about being a person and the social practices, situations, and institutions of everyday life that represent and foster these ideas") have a greater likelihood of being reiterated, maintained, and subsequently regularized to become a somewhat independent or interdependent psychological system when people's behaviors, feelings, and thoughts become consistent with those ascendant selfways. Seriatim, the psychological structure that is birthed would have the potential to lead to answers that bode well and, therefore, build up features of the cultural structure itself. In this manner, everyone's psychological procedures and systems would be slowly included into the greater cultural context. Contrastingly, behaviors that are deemed unsuitable with the selfway would not be perceptively developed (Kim, 2002:9).

The other aspect about meaning systems and cultural context observed by Kim relates to the fact that *self-concepts* (i.e. selfhood, selfways) serve as the cerebral renderings of the idiosyncratic qualities employed by people to characterize and modulate their own behaviors. He adds that the concept *self-system* is also utilized to denote the entire series of concepts employed to define the self. He also says that self-concepts are perceived to entail information relating to the characteristic attributes of the self in particular moods, situations, temporal contexts, in addition to the links among these aspects. He further mentions that perspectives about the accession and system of self-concepts have been derived from two schools of thought with competing explanations of the nexus between the notions of self and social relationships. While one school postulates that identity is built from social relationships and that these relationships with others make up identity, the other school proffers the view that identity is built as a person disconnects from initial relationships and that attributes and encounters unique to that person comprise identity (Kim, 2002:9-10).

The fourth theme treated by Kim deals with multicultural perspectives of cultural context. He starts off by informing us that some scholars have been questioning the generalizability of Western models of communication theories to non-Western societies and minorities in the United States. These queries have led to calls for the generation and inclusion of non-Western theories in American communication studies and to broaden diversity in the field across ethnic, gender and racial demarcations. Kim argues that the discipline of communication, a la the North American tradition, needs serious restructuring. This is necessary because, as he eloquently points out, "The field of communication which has been built on top of the individualistic model of self-identity, precludes an adequate appreciation of the social basis of communication—an ironic state of affairs that should cause alarm to those in the field" (Kim, 2002:3).

Next, as stated by Kim, a majority of what we know about the character and impacts of self-related variables in Western societies is based on the unuttered belief that people have to a certain extent individualistic self-construals. Thus, he posits that the ostensible laws and rules of human communication may be irrelevant or appear differently in societies where people perceive themselves as being interdependent. He adds that this is a very important matter even in the United States whose combined racial and ethnic minorities are projected to make up the majority population in the year 2020 (Kim 2002:3).

Additionally, Kim insists that it is imperative for the field of communication to move away from an Anglo-centered to a multipolar discipline. He is hopeful that the multiculturalism movement serving as the "fourth force" in the United States will encourage the development of more alternative conceptual systems. Concomitantly, he believes that communication theories must be released from the shackles of the Euro-American perspective of the autonomous individual. Kim also thinks that the utilization of alternative viewpoints to investigate human communication will show how cultural models greatly shape both conventional and scientific knowledge (Kim, 2002:3-4).

The fifth theme Kim examines is in regards to research trends and cultural context. He begins by suggesting that in order to get a sense of the cultural underpinning of communication research it is helpful to look at the trends in related fields that reveal some advancement in this regard. For example, he indicates, psychologists, being more self-conscious of their field, have been propelled attain a wider outlook of the "discipline qua discipline" (Kim, 2002:4).

Thereafter, Kim indicates that despite the fact that various American, Asian and European theorists have pointed out that the individualistic framework of self is inadequate for explaining social behavior, the observation has been mostly ignored in empirical and theoretical investigations of communication behavior. He therefore enjoins human communication scholars to heed the following three recommendations that have been made in the social psychology discipline: (1) "cultural decentering away from Euro-American theory," (2) "recentering the discipline within the cultures of interest," and (3) "integrating the different cultural perspectives to move toward a truly universal theory of human communication" (Kim, 2002:4).

Subsequently, Kim notes that there is a growing sense of what may be referred to as the cultural foundations of communication theories. He is convinced that in order to harness this awareness, a systematic theoretical framework must be established. While he is not certain about how this endeavor will come to fruition, he is nevertheless certain that communication scholars taking on the task will gain significantly if they are to employ a cultural perspective to peruse existing theories (Kim, 2002:4).

The sixth theme broached by Kim is about self-construal and cultural context. In his lengthy discussion of this aspect, he starts with the observation that individualism for the West and collectivism for the East are the factors that influence these cultures' communication styles. These styles, he indicates, have been criticized because they miss the many cultural and social explanatory variables that underlie the explication and determination of behavior. He notes that the comparative analytical approach employed to analyze different cultures is also commonly utilized to analyze the effect of culture on communication behaviors. He argues that while this approach has some utility for determining the existence of cross-cultural differences, it nonetheless has little usefulness for determining why culture has an impact on those differences. Thus, he recommends that the arbitrating character of self-construals be employed to gauge the impact of culture on communication behavior (Kim, 2002:22).

Next, Kim apprises us about the few noteworthy tries that have been made to operationalize self-construals at the individual level and to connect the discrete variations in self-construal with variations in communication behaviors. The following are findings from recent research on the topic mentioned by him that he believes are germane for appreciating the essence of self-construal vis-à-vis cultural context: (a) self-concept is a vital arbiter of cultural behavioral systems, (b) the concept of self has been connected to a number of communication styles that were once perceived to be linked to cultural facets like collectivism and individualism, and (c) the idea of self-concept as an intervening variable has been shown to help us better identify the exact position of the self in controlling choices for conversational approaches. Based on these discoveries,

Kim believes that identifying variables that are impacted by culture and that in turn impact communication styles will increase our knowledge about the intricacies of the impact of culture on communication (Kim, 2002:22-23).

In addition, Kim edifies us that individual-level procedures have at various times been adduced by many intercultural communication theoretical frameworks used to examine cultural contrasts. He notices, however, that empirical evidence about the specifics of the self-systems of individuals from different cultural settings is missing in these investigations. He discerns that a key factor that effectively influences the choice of a conversational technique appears to be the type of self that plies in a certain culture. He points out that culture has been determined to be a concept that is unwieldy to perceive and define, and defies measurement. He adds that a number of major cultural dimensions, such as collectivism versus individualism, come from various scholarly works as high-level psychological ideas that can assist in explicating cross-cultural variations in behavior across a wide spectrum of circumstances. In a number of cases, a broader idea is employed to characterize significant features of the behavior of a whole cultural community, albeit high-level concepts such as collectivism and individualism are usually not precisely denoted. Thus, the soundness of these concepts cannot be ascertained since their limits remain opaque (Kim, 2002:23).

As also alluded to by Kim, self-construal seems to be a significant intervening variable that can aid in the explication of how culture impacts communication behavior. This is due to the fact that aspects of the self-system such as "independent and interdependent selves" can assist us in comprehending the exact procedures through which culture affects communication behavior. Since self-concept supplies a connection between a culture's norms and its values and people's everyday behavior, it has the potential to provide explanations for intercultural conversational approaches. These explanations would in turn help to ease the difficulty of classifying people as either individualistic or collectivistic at the individual level (Kim, 2002:23).

Thereafter, Kim states that operationalizing individualism and collectivism on the individual level has advantages, as it would make it possible for us to describe the individualism-collectivism dispositions of various groups and to investigate the respective significance of individualism and collectivism within each group. He says that by employing such an operationalization, a researcher would not have to make a premature assumption that the groups being studied are individualistic or collectivistic; instead, s/he would simply provide an empirical explanation of the characteristics of the groups. He adds that in order to adequately demonstrate how psychological variations among individuals from different cultural settings influence their perceptions and behaviors, it is imperative for researchers to get a better grasp of the effect of culture (Kim, 2002:23-24).

Furthermore, Kim mentions that contemporary between-culture studies that incorporate self-construals have tendered proofs of the essential part self-construals perform in communication deportment. He proposes that self-construal give us a simple or economical explication for variations among cultures in terms of self-connected variables and communication conduct. He also notes that the level of the "independent and interdependent construal of self" methodically impacts how people think about the essence of "conversational constraints" within a cultural entity and across cultural entities. He adds that culture affects intervening procedures (self-construals) that impact the stimuli for oral communication. Or, as he puts it alternatively, "cultural variability does not have a direct impact on verbal communication motivations per se but, rather, its influences are indirect through other processes" (Kim, 2002:24).

Moreover, as illustrated by Kim, the idea of self-construals as "individual-level correlates of individualism and collectivism" has the possibility of illuminating and extending the muddled notion of culture by yielding more succinct, logical, desegregated, and empirically verifiable features of cultural differences and also connecting psychological elements to a cultural feature. He believes that it is only via a reasoned nexus between culture-level analysis and individual-level analysis that we can develop a cogent theory to explain the connection between self-construal and cultural context (Kim, 2002:24).

The seventh and final theme Kim deliberates over is sociocultural context. He commences with the assertion that the deficit perspective is ethnocentric in that it characterizes the self as a vital force that is self-reliant, delimited, and individualistic. He proposes that others' assessments comprise another way a person's social function can be menaced. In such a situation, instead of emerging from self-doubt about the romanticized role-identity as an individualistic self that a person carries to an interaction, an increased sense of assessments can emerge from a broader reactivity to social context. Ceteris paribus, individuals who are intimately connected with the social selfhood that encircles them might be more apprehensive due to their belief that what occurs in the social setting is increased. One of the core features of the collectivistic self is a broader reactivity to others' assessments and trying to fit into a group (Kim, 2002:39).

Next, according to Kim, by avoiding oral expression, a person could restrain his/her feelings and may occasionally make it possible for him/her to express passive sentiments, thereby eliminating the hazard of a direct dare (Kim, 2002:39). This is because, as Kim states,

> Just as messages of silence (extreme forms of indirectness) might be evaluated differently within different relationship contexts, avoidant styles (or withdrawal) can be seen as positive or negative by members of different cultural orientations (as they are measured against what is expected in that context). In individualistic contexts, there often appears to be a demand to interact that characterizes much of dyadic communication—a built-in assumption that when people are engaged in focused conversation, it is their responsibility to keep verbal communication active. Silence or avoidant communication styles might, at times, represent a threat to this responsibility... Low motivation for verbal output among independents may be seen as the failure of positive politeness—the need to be involved with others. This failure can occur in any culture but seems to be especially common in cultures that may be characterized as relatively individualistic (Kim, 2002:39-40).

Also, Kim informs us that in an investigation involving subjects from Hawaii, mainland United States, and Korea, it was discovered that there is a negative correlation between a person's independent cultural orientations and his/her susceptibility to being uncommunicative in "both first- and second-attempt" entreaties. In this case, there is no need for a person to state his/her meaning in words in order to be understood as there already exists a greater understanding between communicators who share similar views, suppositions, and closeness (Kim, 2002:40). Kim goes on to enlighten us about other investigations that have found correlations between (a) interdependence and "embarrassability," (b) high communication avoidance and pathology, and (c) compulsive communication and pathology (Kim, 2002:40-41).

In addition, Kim directs our attention to the verity that while there is abundant knowledge about the correspondences and outcomes of communication avoidance in the United States, there is little knowledge about the outcomes of atypically "high levels of communication approach." He therefore posits that just as a "particular degree of communication approach tendencies" is imperative for adequate interpersonal communication in certain cases, so can a problem emerge when the communication styles are excessive or hostile (Kim, 2002:41).

Finally, as Kim notifies us, various degrees of communication approach and circumvention may be engendered by diverging ideas of self, which are partially the outgrowths of the various cultural surroundings in which people live. Thus, he suggests that in order to keep away from misconstruing the communication behaviors of people from other societies, it would be beneficial to learn about their preferred levels of oral communication. He adds that a profitable approach for explicating intercultural misconception is the notion that verbal communication motivation may operate variably across societies. Consequently, according to him, it is the existing culture at a particular time that shapes "what level of verbal communication will be perceived as normal; who will be required to adapt to whose level of verbal communicativeness; and whose verbal style will be seen as deviant, irrational, or inferior" (Kim, 2002:41).

Cultural Identity

A general definition of *Cultural Identity* is a person's feeling of belonging to a group. For the discussion of this topic, I cull from the work of William B. Gudykunst (2004). It behooves me, however, to mention here that Kim (2002) has provided a detailed analysis of the topic covering such aspects as the uni- and bi-dimensional models, the ego identity model, various paradigms, committed ethnic identity, complementarity, dichotomous categorization, "dynamic betweenness," individualistic vs. collectivistic orientations, mechanical systems, and yang communication behaviors.

Gudykunst discusses three attributes of *Cultural Identity*. First, he talks about how people's cultural identities represent their social identities which rest on their memberships in their cultures. These cultural identities, he proffers, have a great effect on people's communication in quotidian life, although they mostly are not significantly aware of the effect of these identities. He also says that people get to know the effect of their cultural identities on their communication styles when they find themselves in other cultures or when circumstances in their own cultures compel them to interact with people from other cultures (Gudykunst, 2004:66, 68).

Second, Gudykunst talks about the strength of cultural identity, which he says involves the degree to which people perceive their culture as vital in the manner they denote themselves. For an example, he cites the cultural identities of United States Americans as being the degree to which they identify themselves with being "Americans." He states that people frequently perceive their cultures as being more vital when they are in other cultures compared to when they are in their own cultures. He adds that unlike people who have not traveled beyond the borders of their countries, those who have traveled to and/or live in other cultures tend to view themselves as more similar to those who share their cultures and rate their cultural identities more positively. He also mentions that African Americans identify less with United States culture compared to Asian, European and Latino/Latina Americans (Gudykunst, 2004:68).

Third, Gudykunst delineates the content of cultural identities. On this aspect, he makes several reflections. One reflection is that individualistic and collectivistic propensities tend to affect our communication with strangers the most. These tendencies, as he explains, can be found in every culture. Given that everyone possesses both individualistic and collectivist thoughts, he therefore suggests that there is the likelihood for an individualistic culture to have collectivists and a collectivistic culture to have individualists. He identifies the following three factors as undergirding people's collectivistic inclinations: (1) individual goals are ranked lower than group goals, (2) the in-group is perceived as an appendage of the self, and (3) the in-group discernment is strong. He also contends that since individualistic and collectivistic inclinations do not necessarily collide, people can thus possess both of these inclinations (Gudykunst, 2004:70). Based on the views of Kim (2002) presented in the preceding section, Gudykunst and Kim provide opposing postulates on individualism and collectivism. While Gudykunst sees hardly any conflict between the two tendencies, Kim sees many.

Another reflection of Gudykunst pertains to how the strength with which people identify with their cultures impacts the thrusts of their cultural discernments. For Americans in the United States who identify strongly with their culture, he says that they esteem freedom, independence, and social recognition more so than those Americans who weakly identify with the culture and even more so than Japanese who identify with their own culture strongly. The Japanese who value their culture strongly, he notes, hold in higher esteem the principles of accepting traditions, harmony, and self-sacrifice compared to Japanese who identify weakly with their culture and Americans who strongly identify with their own culture. He also mentions that Asian Americans who identify strongly with the United States culture accentuate independent self-construals and harbor individualistic principles more so compared to those Asian Americans who identify weakly with the culture of the United States (Gudykunst, 2004:70).

The other reflection by Gudykunst relates to why when people from a particular culture communicate with people from other cultures, they begin from the premise that they share a similitude with people from the other cultures. For instance, he mentions that people from individualistic societies tend to believe that people in collectivistic societies do things collectively, thereby using that belief as the basis to communicate with people from the collectivist societies. For exemplification, he says that the belief makes it possible for people to envision people in collectivistic societies place greater significance on their social identities compared to their personal identities when they communicate. He cautions, however, that some people in collectivistic societies can be quite individualistic and, thus, it is important to determine whether an individual identifies strongly with his/her dominant cultural tendencies (Gudykunst, 2004:70).

Cultural Language Styles

Cultural Language Styles can be generally described as the choices of words, symbols and gestures used by specific groups of people when they speak, write and use sign language in their particular societies. Ronald B. Adler and Neil Towne provide succinct descriptions of six *Cultural Language Styles.* Before doing this, however, they begin by indicating that the utilization of language is more than merely selecting a specific category of terms to transmit a notion. Every language possesses its own style that differentiates it from other languages. The key elements for communicating competently include issues such as brevity or detail, formality or informality, and precision or vagueness. Problems tend to occur when a person uses a communication style from his/her culture in a different culture (Adler and Towne, 1993:184).

The first *Cultural Language Style* described by Adler and Towne is the *low-context* variety. According to these authors, this style is employed to plainly and rationally convey feelings, ideas, and thoughts. They mention North Americans being an example of users of this style in that citizens of Canada and the United States prefer straight talk and become impatient when someone is "beating around the bush" (Adler and Towne, 1993:184-185).

The second *Cultural Language Style* described by Adler and Towne is the *high-context* type. This style, as they point out, is utilized to preserve social harmony. Instead of annoying others by communicating plainly, users of this style are capable of finding meaning from the setting in which a message is transmitted. The contextual variables at play in this style include (a) the general social rules that underscore interaction among people, (b) the history of the connection, and (c) the non-oral behaviors of the talker. Adler and Towne cite most Asians and Middle Easterners as employers of this style as they value the maintenance of harmony more than clarity directness (Adler and Towne, 1993:184-185).

Before discussing the rest of the styles, I must mention here that Adler and Towne also allude to the fact that even within a society, various subcultures can have contrasting ideas about the essence of direct speech. They invoke the example of Puerto Ricans in the United States whose language style is much similar to the Korean or Japanese high-context style than it is to the low-context style of the European Americans. Puerto Ricans, in general, cherish social harmony and, thus, shun conflict, which makes them to methodically speak in an indirect manner so as not to offend other participants in a conversation (Adler and Towne, 1993:185-186).

The third *Cultural Language Style* described by Adler and Towne is the *elaborate* kind (i.e. involving many carefully arranged parts or details; detailed and complicated in design and planning). They reference Arabic speakers as an example of people who employ this style, as these speakers use language that is richer and more eloquent compared to what most speakers of the English language use. They point out that a familiar attribute of Arabic includes strong declarations and overstatements that would seem amusing in English. When used with people in a different culture, this style can lead to misinterpretations by conversational participants (Adler and Towne, 1993:186).

The fourth *Cultural Language Style* described by Adler and Towne is the *succinct* variety (i.e. something written or spoken briefly and clearly expressed). These scholars mention that this style is maximal in cultures that rate silence highly. Citing the example of Native Americans as users of this style, Adler and Towne note that these speakers privilege silence over talking in social situations they perceive to be equivocal. As these coauthors put it so interestingly, "When you contrast this silent style to the talkativeness that is common in mainstream American cultures when people first meet, it's easy to imagine how the first encounter between an Apache or Navajo and an Anglo might feel uncomfortable to both people" (Adler and Towne, 1993:187).

The fifth *Cultural Language Style* described by Adler and Towne is the *formal* type (i.e. speaking or writing using a conventionally recognized form, structure, or set of rules). As these scholars mention, this style, which emphasizes proper speech, is predominant in African and Asian societies. The style, they say, is more about denoting social position than it is about employing correct grammar. They cite the case of Korea where the language conforms to the Confucian structure of relational pecking orders and, thus, the language has specific terms to indicate different sexes, levels of social status, degrees of intimacy, types of social occasions, etc. "For example," they add, "there are different degrees of formality for speaking with old friends, non-acquaintances whose background one knows and complete strangers. One sign of being a learned person in Korea is the ability to use language that recognizes these relational distinctions" (Adler and Towne, 1993:187).

The sixth and final *Cultural Language Style* described by Adler and Towne is the *informal* kind (i.e. utilizing a style of writing or conversational speech characterized by simple grammatical structures, familiar vocabulary, and use of idioms). According to these coauthors, this style is predominant in countries such as Australia, Canada, the United States, and also Scandinavian countries. They further state that adherents of this style use casual friendliness even when they talk to strangers, as they are uncomfortable with what they perceive as a stiff and unfriendly communication style (Adler and Towne, 1993:187).

African Linguistic Evidence

To begin with, Richard Hayward (2000) proffers the view that when it comes to linguistic *Cultural Transmission* in the African context, linguists ought to be particularly cognizant of the intellectual indebtedness of the Westerners to speakers of Afroasiatic languages because of the significant revolution they attained in the representation of language in written form and made possible the writing of alphabets. This revolution, he asserts, has made it possible to transmit cultural beliefs and practices across all human undertakings.

Next, Christopher Ehret reminds us that through most of human history, as long as there was a society to which a language belonged, that language would be used as a medium of social and cultural communication. He also notes the verity that the language can die as people lose the sense that they belong to a community that is different from those of other communities and refuse to pass the language down to the generations that come after them. He mentions the example of the Dahalo, a language of the Southern Cushitic subgroup of the Afro-Asian language family spoken in Kenya. He recounts that over 2,000 years ago, the Dahalo who were gatherer-hunters spoke their own language; but after periods of close relations with their neighboring, dominant Southern Cushitic-speaking family society, the Dahalo gave up using their Khoisan language. He also informs us that the Dahalo have done their best to maintain many of the words from their old language and continue their ancient ways of collecting food (Ehret, 2000:275-276).

Also, Øyvind Dahl presents six respective linguistic *Cultural Transmission* "frames of reference" in the Malagasy language. Frame of reference number one has to do with truth being expressed in social relations, not in the accuracy of propositions. Accordingly, the Malagasy word for truth is *marina*, whose root word is *arina*, whose literal meaning in turn is to state something plain, without ups and downs. Frame of reference number two is to eschew open confrontation. Frame of reference number three is for an individual to strive to adapt to circumstances. Frame of reference number four is for a person to endeavor to achieve harmony (*mampitovy tantana*), balance (*mandanjalanja*), and the golden middle (i.e. circumventing confrontation). Frame of reference number five is that one must not express his/her feelings openly. And, frame of reference number six is that there is nothing good about self-praise (Dahl, 2000:126-127).

In addition, Joseph Holloway and Winifred Vass demonstrate that American English is pregnant with African linguistic *Cultural Transmission*. For example, they identify the following eight African culture clusters in South Carolina alone: (1) Akan, (2) Mande, (3) Mano River, (4) Niger Delta, (5) Bakongo, (6) Ovimbundu, (7) Yao, and (8) the Unknown (Holloway and Vass, 1993:xxvii). This denouement is due 'largely' to the Trans-Atlantic Slave Trade which took place from the 16th to the 19th Century and saw more than 12 million enslaved Africans brought to the Americas and the Caribbean. I use the term 'largely' because there is an abundant amount of evidence which shows that Africans were already in the New World almost 200 years before Christopher Columbus' voyage to the area, as King Abu Bakri/Bakar of the Malian Empire had spearheaded a series of voyages to the area in 1310 (for a listing of some of the sources containing this evidence, see DawaNet, 2013).

Furthermore, in *Culpability of the Trans-Atlantic Slave Trade* (2017), I discuss many aspects relating to African linguistic *Cultural Transmission*. I will only list them here for the sake of brevity; the reader interested in the details can consult the book. These aspects include the following: (a) one out of every five Africans lives outside the continent of Africa and contributes in some way to the languages and other linguistic systems of his/her host society; (b) the Trans-Atlantic Slave Trade was the most significant catalyst for the dispersal of Africans and their linguistic culture to other parts of the world; (c) even though African languages were brutally oppressed during slavery, Africans still managed to preserve the phonological, syntactic and grammatical core of their languages and use them to contribute to the languages of the Western oppressors; (d) many enslaved Africans wrote about their experiences which became sources of knowledge for non-Africans and inspiration for the revolutionary Africans that followed; (e) as Black nationalism intensified, more attention was given to African linguistic origins; (f) linguistic evidence of how African languages influenced major world languages and being the roots of Semitic languages is continuously being unearthed; (g) how Ebonics, or African American English, has served and continues to serve not only as a means of communication, but also as transmitter of African culture; and (h) how the pidginization of African languages and the evolution of African American linguistic features have and continue to benefit European languages and other linguistic systems (Bangura, 2017:115-152).

Moreover, Rujuta Patil states that the seven principles of the Kwanzaa festival are an example of African *Cultural Transmission*. She adds that the festival is being utilized by African Americans to honor their African traditions and establish a connection with the African cultural heritage that hinges upon seven principles which include creativity, faith, purpose, self-determination, and unity (Patil, 2018: 1 & 4). Besides the fact that Patil only lists five of the principles and, thus, a complete listing is warranted, a bit more also needs to be said to give the reader who may not be familiar with the Kwanzaa principles a better idea of why the festival is a paragon of African *Cultural Transmission*.

What then is Kwanzaa? In my *Falolaism: The Epistemologies and Methodologies of Africana Knowledge* (2019), I begin by indicating that Kwanzaa was delineated by Mwalimu Maulana Karenga as part and parcel of what he calls *The African Code*, a Pan-African concept that advocates unity via diversity based on the following seven principles: "(1) *Maat*—truth, justice and equality; (2) *Kujichagalia*—self-determination; (3) *Umoja*—unity; (4)

Mila—culture, (5) *Kulea*—education; (6) *Jamii*—family; and (7) *Ujamaa*—economics" (Bangura, 2019:13). I add that *The African Code* has been translated into more than 30 languages and employed as a "non-political, non-religious cultural commonality for African people seeking self-determination everywhere...employs the Ge'ez alphabet and treats Kiswahili as the official Pan-African language, and subsequently Ge'ez as an African script to replace all forms of Latin to write all African languages" (Bangura, 2019:13).

Next, I point out that Mwalimu Toyin Falola asserts that African Americans treat the Kwanzaa holiday as a practical Pan-African project utilized to establish an African consciousness and connection with their Motherland. I then quote him to support his proposition as follows:

> ...whether political, economic, or cultural, [the Kwanzaa project has] been undergirded by the intellectual ideas around a set of interrelated issues: the generation of knowledge (utilitarian knowledge to be sure), the building of institutions to spread knowledge, the ideology to organize the facts and opinions, and the structures to use the knowledge. Both in Africa and the United States, the intellectualization of the Black experience is one of the most successful academic achievements of the twentieth century. Indeed [Kwanzaa is among the practical projects that] have also led to vigorous debates on a variety of opinions to emancipate Black people (Falola, 2013a:74; quoted in Bangura, 2019:192).

As I also mention, Ronald W. Walters tells us that Kwanzaa is a celebration by African Americans to demonstrate the common cultural features that bond people of African descent everywhere in the globe. I restate his explication of the raison d'être for Kwanzaa as follows:

> Africans in the Diaspora are not part of the daily particularities of African history and so cannot possess the identity which flows from these experiences. However, they do share certain aspects of this history, and the basis upon which they do so is (1) their affirmation of an African heritage; (2) their participation in the Diasporic aspects of Pan-African political struggles; (3) their continuing concern with the status of Africa and their efforts to improve it; and (4) their relationship to other hyphenated Africans in the Diaspora (Walters, 1993:385; also quoted by Falola, 2013b:74 and Bangura, 2019:410).

Indeed, as I further posit, it was only fitting that Kwanzaa led to the emergence of *Kawaida*: a concept that was "conceived and crafted in the midst of the liberation struggle of the 1960s as an emancipatory philosophy dedicated to Cultural Revolution, radical and social change, and bringing good in the world" (Bangura, 2019:410-411).

Before proceeding to the next section, it behooves me to mention here that *Cultural Transmission* has been employed by some analysts to discuss other linguistic and extralinguistic features pertaining to Africa. For example, some of the authors who contributed to a World Bank-sponsored work titled *Culture and Development in Africa* (1992) and edited by Ismail Serageldin and June Taboroff discuss *Cultural Transmission* in terms of cultural theory and development practice; culture and civic society; cultural institutions; conservation of the built historic environment; development, archaeology, and the environment; culture and cause or consequence of economic development; and future directions for culture and development in Africa. Also, Larry M. Hyman (2011) looks at *Cultural Transmission* in his areal linguistic reconstruction of Macro-Sudan Belt and Niger-Congo languages focusing on verb morphology and word order. In addition, Claertje Frieke-Kappers (2007) talks about the phenomenon in her study of "the creative use of genre features continuity in patterns of language use in Budu, a Bantu language of Congo (Kinshasha)."

African Mathematical Evidence

To begin with, Paulus Gerdes teaches us that while the *sona* sand tradition was developed mainly by the Chokwe, kindred groups such as the Luchazi and Ngangela did so as well. These groups occupy eastern Angola, neighboring zones of the Congo, and northwest Zambia. Gerdes gives us many examples of *Cultural Transmission* through the *sona* sand tradition that is pregnant with mathematical operations. A good example is the monolinear "chased chicken" design of dimensions 5 X 6, 9 X 10, and 3 X 8, respectively. When the same algorithm is applied to a grid of dimension 10 X 5, the three "chased chicken" lines are needed to embrace all points on the grid (Gerdes, 1999:182). Gerdes also asserts that one can then ask about how the number of "chased chicken" lines are related to the dimensions of a rectangular grid: [$g(r, c) = ?$]. Given this, one can then extrapolate the following mathematical equation, according to Gerdes (1999:184):

$$G(r, c) \text{ is a common divisor of } \frac{r+1}{2} \text{ and } \frac{c+2}{2}$$

or further still

$$f(2m, 2n + 1) \text{ is the greatest common divisor}$$

$$\text{(gcd) of } \frac{r+1}{2} \text{ and } \frac{c+2}{2}$$

Next, Gerdes informs us about the *Cultural Transmission* that took place during the early centuries between North Africa and Africa South of the Sahara through the exchange of mathematical ideas. *ᶜIlm al-Hisab* (mathematics) of pre-colonial Central Sudan was part of the Islamic sciences introduced in Kanem-Borno, Nigeria sometime after the 11th Century and later in Hausaland, Nigeria in the 15th Century. This Arithmetic was taught in both secular and Muslim (*Islamiyya*) schools to be used in the courts to calculate inheritance, in the distribution of almsgiving (*zakat*), and for business transactions and surveying land. In both Kanem-Borno and Hausaland, scholars used Coptic solar calendars from North Africa to determine economic activities, particularly in the area of agriculture. The transmission also yielded a great deal of literature in the fields of Mathematics, Medicine, Astrology, and other affiliated sciences written in Arabic, Fulfulde, Hausa and other languages, all of which provide proof that these African scholars had researched the mathematical, agricultural, astronomical and medicinal sciences way before colonialism was foisted upon Africa by imperialist Europe (Gerdes, 1994:360).

Gerdes also provides numerous examples of the earlier mathematical and other scientific advancements in Africa. The following is a small sample of them:

> (a) Muhammed ibn Muhammed al-Katsinawi from Katsina (now northern Nigeria) on chronograms and magic squares; (b) the study of formal logic in Timbuktu and Lapousterle (Bamako, Mali) and three mathematical manuscripts, written in Arabic, that belong to the Ahmad Baba Library in Timbuktu; (c) one of the manuscripts, whose calligraphy is typical of Africa South of the Sahara, was written by a mathematician from Mali named al-Arwani; (d) the other two manuscripts contain references to medieval mathematicians from the Maghreb; (e) an example from the eastern coast of Africa is provided by the comments of the late 15th-Century traveler, Ahmad Ibn-Madjid, who reported on the counting of the Wac-Wac, probably not a Bantu but a Khoi-San people, living at that time in southern and central Mozambique (Gerdes, 1994:360-361).

Gerdes adds that other possible reciprocal links between the development of mathematics in Africa South of the Sahara and across the Indian Ocean, the possible influence of mathematical ideas of enslaved Africans from the African continent, and from India and Indonesia, on the development of mathematics on Madagascar, Mauritius, and other islands, need to be investigated (Gerdes, 1994:361).

Also, as C. R. Guglielmino, C. Viganotti, B. Hewlett, and L. L. Cavalli-Sforza point out, mathematical treatment demonstrates that *Cultural Transmission* is significantly successful when African parents transmit culture to their children, which is their cultural inheritance. They characterize cultural inheritance as a process of modification that is facilitated by communication, which has achieved its highest growth in humans and can facilitate a long period of maintenance of quick metamorphosis of culturally transmittable characteristics based on contexts and necessities. They postulate that methods of sociocultural transmission mostly regulate "conservativeness" or "flexibility" (Guglielmino et al., 1995:7585).

In addition, Edward B. Eastwood and Benjamin W. Smith allude to the notion of *Cultural Transmission* in their examination of geometric and hand-printed rock art in the Central Limpopo Basin located in Southern Africa. Their numerical calculations reveal that the specific collection of forms selected for utilization in the Geometric Tradition art of the Basin was not generated there; instead, it had its cultural origin in the "geometric hunter-gatherer rock art" of Central Africa. Thus, they agree with the earlier suggestion by other researchers that the art is most probably the legacy of an invasive cognitive system instead of an indigenous knowledge system developed in the Basin. They also mention that this conclusion is buttressed by the results they obtained from their scrutiny of the pigments employed, the application techniques utilized, and the places and painting surfaces selected. Furthermore, their results also indicate how the unique characteristic of Geometric Tradition art and its represented subject indicate a cognitive incompatibility between its painters and the painters of the traditions of the San and Northern Sotho. Moreover, they state that also evident is that the "Geometric Tradition has a significant period of overlap with San rock art...it shows a local spatial preference for ecological zones with good pasture, and is part of a distinctive tradition that extends across a similar area of southern Africa to that postulated for herder migrations [which leads one to] conclude in favour of a migratory Khoekhoe authorship for Geometric Tradition art" (Eastwood and Smith, 2005:73).

Thereafter, we glean from Daniel Thanyani Rambane and Mashudu C. Mashige an idea of *Cultural Transmission* in terms of the part played by mathematics in African scientific thinking as it relates to human, intellectual and sociocultural advancement in a select group of countries in the continent they investigated. They provide evidence to debunk the myth that Africans lacked an "intelligible sense of numeracy" before the intrusion of Westerners in the continent and show that the myth was part and parcel of the colonialists' agenda to marginalize Africans and "other-ise" African knowledge systems. They call for the discovery and advancement of the early contributions to mathematics and scientific thinking by Africans as a way to move away from the conventional Western-based method of teaching the subject to embracing an African cultural method that is more accessible to African learners. They believe that such a shift will help to foreground African epistemologies in producing knowledge, spurring human and social advancement, and leading to the realization of the African Renaissance (Rambane and Mashige, 2007:183).

Rambane and Mashige apprise us about a number of ways this African mathematical and scientific, cultural heritage was transmitted. The first was the emphasis placed on the use of the knowledge gained to uphold cultural values that would ensure a harmonious future. The second was to get learners to develop an intellectual affinity for the history of a cultural-educational connection with mathematics—for example, the African roots of Pythagoras theorems in Ancient Kemet/Egypt. The third was African mathematicians establishing a contextual cultural reality and anchoring knowledge within a cultural praxis that facilitates historical continuity. The fourth was devising mathematical ideas (such as in riddles and puzzles) according to a particular cultural, ethnic and social milieu. The fifth was connecting mathematical ideas to those of other disciplinary areas such as archaeology, history, psychoeducation, etc. (Rambane and Mashige, 2007:185-197).

Furthermore, Alexander O. E. Animalu and Jonathan Okeke Chimakonam delineate a *Cultural Transmission* feature of a perennial African traditional mathematical motif. They summarize the procedure they employed and the phenomenon so elegantly that I must allow their words to speak for themselves as follows:

> ...we rely...on the production of "light caustics" by quasi-crystals which is a recurrent motif of traditional African (Igbo) cultural symbol and the Arewa sign as well as the most recent *ex tempore* observation of 4x4 matrix diffraction pattern of white light by a glass (~conical pyramid) of water (~quasi-crystal) to construct a 4x4 magic square representation of the complementary duality of African three-valued thought/logic. In technological and geometric terms, the three-level logic provided the paradigm behind the technological feat of topological transformation of a sphere mould of alloy into a bronze torus recorded by Africans at Igbo-Ukwu (c.10,000-5,000BC). [There exists an] implication of the 4x4 Magic square representation for the complementary duality of the torus and cube-hexagon genospace and the consequential elaboration of three-level logic as a paradigm for lattice-gauge geno-Einstein's general relativity theory...(Animalu and Chimakonam, 2012:141).

Finally, as Kori Lloyd Hugh Maxwell and Iman C. Chahine explicate the oral *Cultural Transmission* in mathematics teacher education in Morocco and South Africa, the techniques utilized enable learning and the discovery and identification of how the local craftspeople think. The thinking of these people successively indicates how they comprehend mathematics and use it to make a living. Maxwell and Chahine also point out that story-telling and apprenticeship are the modes via which knowledge—cultural, mathematical, and otherwise—is generally transmitted. The people operate orally and learn mostly through "apprenticeship, discipleship, and systematic strategies to support the memory." Thus, according to the coauthors, orality establishes an "ontology based on interrelating forces rather than on tangible objects" (Maxwell and Chahine, 2013:65-66).

Conclusion

First, the concept of *Cultural Transmission* is well represented by the Ancient Kemetian terms *hab* and *ānsh*, which mean "to transmit" and "to send a message." Second, the concept manifests itself in the terms *āankh*, denoting away of "life" or "to live" by Kenetians; and *ānkh*, meaning a way of "life of prosperity, all stability, all health and joy of life, a formula of good wishes which follows each mention of the king's name in official documents" (for these translations, consult Budge 1978).

As Fayza M. H. Haikal elucidates, Ancient Kemetian cultural continuity is evident in every aspect of modern Egyptian culture, despite the changes that have taken place in the country's circumstances, language, and religion throughout its very long history. Still present in today's Egyptian society are Ancient Kemetian modes of expression, funerary religion, the effect of the Nile and its regenerative forces, conception of the universe on the origin of certain beliefs, and folklore. To buttress his assertion, Haikal provides a plethora of evidence which includes the Ancient Kemetian beliefs in (a) the larger universe or the realm of the Divine, (b) monotheism, (c) the invisible/intangible world that is the origin for most of the superstitions or taboos transmitted over time, (d) the Creator and all the invisible forces he created that surround us all the time, (e) the complexity of the mortal being with his/her multiplicity of facets, (f) the soul as the sacred life force, and (g) rituals to avert evil forces (Haikal, 2011).

In addition, Joshua J. Mark mentions how Kemetian *Cultural Transmission* manifested itself through trade, especially via the Silk Road (see the preceding chapter for details on this), and impacted other civilizations and faiths. He cites the example of Kemetian burial rites, which date back to 4000 BC, that spread the belief of eternal life and the inevitability of a personal state of living beyond death. The perception is even said to have inspired the Christian belief of eternal life (Mark, 2013:2).

Chapter 8

Arbitrariness

In "Domesticating Mathematics in the African Mother Tongue," I state the following based on the observations of Anouschka Bergman, Kathleen Curie Hall and Sharon Miriam Ross (2007):

> This feature of human language entails five aspects. First, the words of a language are generally recognized to represent a nexus between a group of sounds or signs that give the words their forms and the meanings, which the sounds are said to represent. Second, evidence of *arbitrariness* can be observed in the fact that the inner core of an object can be given a name. This evidence of arbitrariness can also be observed in cross-linguistic comparisons when words with the same meanings usually have different forms in different languages, and similar forms usually express different meanings. Third, *onomatopoeia* (i.e. words that imitate natural sounds or possess meanings that are associated with such sounds of nature) provides evidence of arbitrariness being the norm in languages, at least in terms of the connection between the forms of words and their meanings. Fourth, a counterexample of arbitrariness is *sound symbolism*: i.e. certain sounds occur in words not because they directly imitate some sounds but because they are evocative of particular meanings. Fifth, *non-arbitrariness* or *iconicity* plays a minimal role in language (Bangura, 2014:25).

The immediate question here, then, is this: How has the idea of *Arbitrariness* been conceptualized by linguists? Beginning with David Crystal, he characterizes the notion as the nonexistence of any tangible connection between gestures, actions, or sounds that are used to convey information and the distinct and independent things in the universe to which they allude. He cites as an example the word *table*, which he says resembles nothing tangible in the universe as the thing itself in terms of the way it is pronounced or written. He adds that evidence from onomatopoeic and other symbolic utilizations of sound is often invoked to denote the opposite of *Arbitrariness* (Crystal, 1992:26).

Next, William O'Grady, Michael Dobrovolsky and Mark Aronoff provide a simpler definition of *Arbitrariness* as the lack of an innate nexus between a linguistic symbol and the entity to which it refers. They also note that with the exception of a small number of onomatopoeic words, arbitrariness is a more predominant feature in human languages. Among other species, as they point out, (a) bee dancing exhibits a degree of *Arbitrariness*, since there is no known relationship between the type of dance and the amount of space from the hive; (b) noises made by bees concerning the quality of food source and the path to the food are not arbitrary; (c) calls by some birds are very appropriate for purposes such as warning about dangers, which cannot be easily pinpointed and are, therefore, arbitrary; (d) the vocalization of a majority of the nonhuman primates is to a certain extent adjustable; and (3) the alarm calls by vervets (a common African guenon with greenish-brown upper parts and a black face) have been found to be arbitrary (O'Grady et al., 1989:407).

Given the preceding backdrop, the rest of this chapter, whose aim is to show the connection between *Arbitrariness* and African linguistic and mathematical ideas, respectively, is divided into three major sections and a conclusion. The first section broaches linguistic expositions on *Arbitrariness*. The second and third sections entail discussions of African linguistic and mathematical evidence on the subject, respectively.

Linguistic Expositions on Arbitrariness

Many different features of *Arbitrariness* have been investigated by linguists. I have placed these features into the following seven categories for the sake of cohesiveness: (1) nominalism; (2) sound symbolism or onomatopoeia and phonesthetics; (3) sound-meaning relations; (4) characteristic of language; (5) grammatical considerations; (6) linguistic theoretical considerations; and (7) other linguistic considerations—creativity, duality, and iconicity.

Nominalism

Analyses of *Nominalism* have focused on three aspects: (1) providing a general definition of the concept, (2) the debate between nominalists on the one hand and Platonists and realists on the other, and (3) issues pertaining to the subject in contemporary ontological discussion. The rest of this section discusses these aspects sequentially.

For starters, Crystal, for example, defines *Nominalism* as the linguistic, philosophical perspective which argues that no intrinsic connection exists between word forms and their referents. Instead, the relationships between word forms and their referents are the outcome of how a linguistic community uses them, and are therefore arbitrary (Crystal, 1992:268). Michael, J. Loux denotes *Nominalism* as a reductionist framework of investigating issues concerning the actuality and essence of metaphysical things with distinct and independent existence (Loux, 1998:1).

Next, apropos the debate between nominalists on the one hand and Platonists and realists on the other, a number of observations have been made. First, *Nominalism* was one of the two contending postulates that engaged early Greek thinkers, with the other being *naturalism*: the postulate that there is an inherent relationship between words and the things to which they refer (Crystal, 1992:268). While Platonists employed the ontological approach to argue that attributes such as kinds, relations, properties, propositions, sets and states of affairs are primitives (i.e. words, bases, or roots from which others are historically derived) that cannot be reduced or simplified, the nominalists counter that there are no "abstract entities" and that so-called "abstract entities" are actually analyzable based on concrete things (Loux, 1998:1).

Second, during various epochs, the debate between nominalists and Platonists concentrated on different aspects. During the Middle Ages, the focal point was on universals. While Platonic realists insisted that universals are composed of "real beings" (*Encyclopædia Britanica*, 2018:1), the nominalists countered that all that exists are particulars and that universals are linguistic expressions with general application. To support their proposition, nominalists drew upon the semantics of general terms with which universals could be identified. Classical empiricists concurred with medieval nominalists and the former sought to advance particularism by making connections between mental representations and general terms (Loux, 1998:1). Also, during this period, Platonic realism (and also Aristotelian realism) was linked to orthodox religious conviction while *Nominalism* was condemned as heretic (*Encyclopædia Britanica*, 2018:1).

In the 17th Century, the moderate nominalists that emerged agreed that there is some similarity between particulars and the general terms used to refer to them. If this were not the case, they reasoned, speech would be out of the question. Simultaneously, "by explaining thought and speech through the use of symbols, such as mental images or linguistic terms, *Nominalism* seems to imply some form of conceptualization that involves more than the mere correct use of symbols and thus is not clearly distinguishable from conceptualism" (*Encyclopædia Britanica*, 2018:1).

From the 20th Century to the present, some linguists began to make a distinction between the doctrinal attribute and the praxis of *Nominalism*. The importance of this development is very captured by Michael Shapiro when he states the following:

> The use of terms and concepts from the history of philosophy to make headway in linguistic theorizing may be interesting but also possibly confusing, the latter for the following reason. The linguistic

> phenomena classified might include linguistic universals (the Peircean 'types') as well as linguistic events (the Peircean 'tokens'). And one who is familiar with the nominalist/realist distinction as a matter of doctrine only might naturally suppose that by 'nominalist linguist' is meant one who denies the reality of linguistic universals. That, of course, would be an application of the nominalist doctrine to linguistic phenomena; but that, one can see now, is distinct from nominalist linguistics as a practice or method. Nominalism as a practice would not necessarily deny that universals are real; rather, it consists in deciding their classification arbitrarily—both their classification into subtypes, if they are segregated from individuals, and whether to so segregate them. Even their classification as real or unreal would be quite arbitrary (Shapiro, 2013:1-2).

The aforementioned "Peircean" terms refer to linguistic concepts developed by Charles Sanders Peirce (1839–1914), who is considered to be the founder of American pragmatism.

In addition, regarding issues pertaining to the subject of *Nominalism* in contemporary ontological discussion, many of them have been noted. To begin with, since the work of Ferdinand de Saussure on semiotics/semiology in the 20th Century, there has been a considerable examination of the arbitrary connections between words and things from a naturalistic perspective—for example, the study of onomatopoeia (Crystal, 1992:268). Other issues that have attracted the attention of scholars dealing with the subject include the existence and status of events, propositions, sets, and state of affairs; and set-theoretical systems which are said to be comprised of concrete particulars (Loux, 1998:1). Also, "in modern logic a nominalistic concern is reflected in the form that is given to the universal quantifier. Instead of saying 'man is mortal,' or even 'all men are mortal,' the modern logician circumvents the universal by saying 'for any x, if x is a man, it is mortal.' Neopositivism, in reproducing metaphysics, has often been explicitly nominalistic, insisting that there exists only 'the facts' of observation and experiment" (*Encyclopædia Britanica*, 2018:1).

Sound Symbolism or Onomatopoeia and Phonesthetics

Linguistic discussions on *Sound Symbolism* or *Onomatopoeia* and *Phonesthetics* (also spelled *Phonaesthetics*) have focused on the following four themes: (1) definition of the concept, (2) genesis of human speech, (3) the idea of symbolism, and (4) discontinuous transmissions. These themes are explored in the rest of this section in the order they are stated.

Commencing with the definition of the concept, Crystal sates that *Sound Symbolism* refers to the undeviating relationship between the form of language and its denotation. This occurs when phonetic sounds mirror those in the external universe (i.e. *Onomatopoeia*) as in the English words *splash* and *cuckoo*. Also, aesthetic attributes such as light and size (i.e. *Phonesthetics*) have been noted in such English words as *glitter* and *slimy* (Crystal, 1992:360).

Also, focusing only on *Onomatopoeia*, O'Grady and his coauthors define the concept as indicating terms that have been developed to sound like their referents. English words like *buzz* and *sizzle* are among the cited examples. Onomatopoeic terms with the same things to which they refer can vary among languages because terms are not precise phonetic duplicates of their referents (O'Grady et al., 1989:107).

Regarding the genesis of human speech vis-à-vis *Sound Symbolism*, Paul A. Gaeng begins by asseverating that since the only authoritative sources in concerns of historical inquiry which are written records are nonexistent and, thus, there is no scientific basis to recompose the remnants of original language, we will not transcend the domain of mere conjecture on the origins of language. He points out that all of the theories that have been proffered to explain the origins of language (discussed earlier) have been challenged; in fact, only small portions of their explications have been accepted. For illustration, he mentions the Bow-wow Theory, which was among the first attempts to develop theoretical explanations for the origins of language, advancing the view that speech developed gradually from sounds imitating those in nature. Most of the evidence on *Onomatopoeia*, however, shows that a relatively little amount of the lexicon of any language represents such sounds, and some Native American languages have none. Gaeng recalls that the Linguistic Society of Paris became so uninterested in theories on the origins of human speech that in 1866 stated in its bylaws that it would not accept any papers on the topic (Gaeng, 1971:7).

Next, Gaeng mentions how it has been suggested numerously that we could imaginably acquire some knowledge about the initial endeavors of humans to speak by investigating how children learn their native languages. A number of experiments have been done throughout history with newborn babies to determine whether they would develop a language of their own if they are isolated from the sounds of a language. An example of this is the experiment by the Kemetian King Psammatichos secluding two infants in a mountain discussed earlier (Gaeng, 1971:7).

In reference to the idea of symbolism, we glean from Gaeng's exposition on the notion that the core of language is symbolism; and, since symbols are reflections of conventions and are essentially arbitrary, particular sounds and series of sounds entail particular denotations only by concurrence in a specific speech community and are meaningless outside of it. He states that while words are symbolic representations of notions and objects, there is, however, no correlation between a symbol and the transmitted message. Adducing to the English word *horse* which has almost the same meaning in French (*cheval*), German (*Pferd*)), Hungarian (*ló*) and Spanish (*cabballo*), albeit these languages have different words for the animal, he says that the correspondence between the English word and its referent is only possible because of the agreement among members of the English speech community. Thus, according to him, there exists no elemental need for a word to denote what it does, since nothing exists in the discernable material nature of the symbol that could be thought of as relating to any attribute of the symbolized thing or notion. Nonetheless, he argues, the concept *arbitrary* does not suggest that the selection of a world symbol is at the whim of a speaker or that s/he can alter it willy-nilly. Instead, language, he asserts, is a "socially contract" through which the sounds of speech and their relationships to the messages they convey are shared among members of a speech community (Gaeng, 1971:15), otherwise coherent communication would be impossible.

In addition, Gaeng considers onomatopoetic words (i.e. "sound-imitative or echoic words") as constituting a special category of words that appear to mimic the sounds they depict. Examples of these words he cites include *bang, boom, whisper,* whose sounds are directly connected to their meanings. He also notes that even in cases where the same natural sounds appear, they tend to sound differently to different people: for example, the rooster sounds like *cock-a-doodle-do* to an English language speaker's ears and *cocorico* to those of a French-language speaker. Obviously then, according to him, even echoic words are somewhat arbitrarily selected and are relatively conventionalized, their direct phonetic mimicking notwithstanding. Consequently, he contends, it is imperative for a speaker to learn the exact mimicking employed in his/her language (Gaeng, 1971:15-16).

Furthermore, Gaeng indicates that akin to *Onomatopoeia* are *Interjections*—i.e. abrupt expressions made as asides or interruptions, especially as parts of speech (e.g. *ah!* and *dear me!*). He demonstrates that even with *Interjections,* when two or more languages are contrasted, they differ across the languages. For instances, he says that while on the one hand when a speaker of the English language feels mild pain, s/he exclaims *ouch*; the speaker of the French language, on the other hand, exclaims *aïe*. In short, he proposes, unlike, for example, groans, *Interjections* are not reflexively or naturally expressed; instead, they must be learned like all other conventional sound sequences of a language. He adds that *Onomatopoeic* words and *Interjections,* even combined, while they are an important component of any language, still constitute only a small portion of a language. Also, he asserts, both linguistic features are not exempt from the arbitrary relationship that manifests between symbol and meaning (Gaeng, 1971:16).

As it pertains to *Discontinuous Transmissions*, Anthony Arlotto broaches several aspects about how this linguistic feature is important for understanding the history of a language. One aspect he talks about is the reality that *Onomatopoeic* words and *Interjections* will always be obvious no matter how much we construct and reconstruct sound laws or other linguistic processes. He adds that it has long been determined that these words are not amenable to the functioning of sound laws (Arlotto, 1972:196).

Another aspect deals with how *Discontinuous Transmissions* of language can serve as a solution to the preceding problem by investigating how languages are conveyed from one generation to another. When children learn their languages, they are not given grammar books from which they commit to memory clearly-stated forms and rules of grammar; instead, they hear people around them speaking and by some internal procedure referred to as a *language acquisition device* they extract the rules of the grammar of their native tongues from what they hear. As a result, most of the words of a language are conveyed across generations, with each generation probably appending alterations which may significantly or insignificantly change how the words are pronounced. The subsequent generation then starts with the alterations and may also expand them (Arlotto, 1972:197).

The other aspect concerns the point at which *Onomatopoeic* words and *Interjections* begin to differ from the rest of the vocabulary. These linguistic elements are not as entirely arbitrary compared to the majority of the other words in a language. They are also delimited by what is called *appropriateness* and by the customary system of mimicking or phonetic symbolism of that language. As such, the worth of these words transcends their being transferred from generation to generation, and they are also not solely the results of history; rather, they are birthed by each generation of speakers as they learn the language (Arlotto, 1972:197).

Sound-Meaning Relations

To start with, Joseph H. Greenberg inculcates us that *Sound-Meaning Relations* are arbitrary since any meaning can be depicted by any mixture of sounds. He cites the example of the English word *dog* which is represented by the combination of the sounds *Hund* in German, *cane* in Italian, *sabaka* in Russian, and *kalb* in Arabic. He also points out that the thousands of meaningful forms of any language are generally independent. He further proffers that with the exception of a small number of occasional homonyms, there is a common principle that just because we can call a *dog* a "dog," we are not compelled to call a *cat* a "cat." Nonetheless, he notes, it is improbable that we will call the *cat* a *dog*. Therefore, he suggests, similitudes that transcend chance in form and meaning call for a historical elucidation, either via appropriation or via common roots (Greenberg, 1957:35-36).

Next, as suggested by Greenberg, when engaging in such a historical analysis, we must make a distinction between "lexical resemblance" and "grammatical resemblance." While the former refers to similar sounds and meanings based on root morphemes (e.g., the English *'hand* and the German *'hant*, both meaning *hand*), the latter refers to similar sounds and meanings that are based on non-root morphemes (e.g., the English *-ər* and the German -*ər*, both designating the comparative of adjectives). Both types of similarities, he indicates, are therefore imperative for providing historical explanations of *Sound-Meaning Relations*. Nonetheless, he submits, similarity in denotation not followed by similarity in sound or vice versa may be treated as having insignificant merit. Therefore, for him, to be left out of a line of reasoning for historical relationship is the existence of sex gender exhibited by morphemes without phonetic similarities (Greenberg, 1957:36).

In addition, Greenberg suggests that to be possibly deemed as a formal characteristic like sound is the sequence of meaningful linguistic features. He identifies only two possibilities of occurrence in the arrangement of forms vis-à-vis syntactic considerations: either (1) *A* precedes *B* or (2) *A* follows *B*. Consequently, he asserts, lines of reasoning based on word sequence are of little value, especially when the types of constructional denotation—such as genitive or actor-action—which are potentially significant are essentially small. Thus, the "historically unconnected occurrences" of such similarities are abundantly probable and, accordingly, over-chronicled (Greenberg, 1957:36).

Also, in his early work on Theoretical Linguistics, John Lyons illuminates the arbitrariness of the substantial realization of substance and form in terms of *Sound-Meaning Relations*. He begins with the analogy that just as the shape of chess-pieces is immaterial to the working of the game of chess, so too is the specific differentiations of shape or sound whereby the expression-elements of language are discerned immaterial. He then goes on to posit that the link between a certain sound or letter and expression-element is a case of arbitrary practice (Lyons, 1971:63).

In a later work dealing with Semantics, Lyons posits that numerous symptoms (i.e. changing states of speakers—emotional, health, intoxication, etc.) and numerous individualistic indicators are either inherently or traditionally motivated, while others are arbitrary. He therefore stresses the point that language-signals are very intricate in that while some parts of an utterance-signal seem to be indexical, others are not. Thus, according to him, the concept "indexical" includes numerous attributes of social meaning, and that it is only by the efficacy of the interpersonal connections that are developed within social collectivities that we can assert our own individual identifiable personalities (Lyons, 1977:109).

Characteristic of Language

On the topic of *Arbitrariness* as a *Characteristic of Language*, we glean from Francis P. Dinneen's work several strands. One strand deals with the *arbitrary* and *conventional* nature of language being among the several characteristics linguistic researchers have discovered to be the basis for the precise illustration of language. This characteristic, he says, is about the meaningfulness of language, as the sounds uttered by speakers are related to factors besides language itself. Another characteristic is that because languages are *sound*, they can therefore be treated as *linear*. The other characteristic concerns the fact that all languages are *systematic* in terms of the acceptable blending of the sounds of a language and in the blending of the meaningful attributes of the language (Dinneen, 1967:6).

Another strand has to do with the verity that even when the message of any two languages is indistinguishable, exchanging information through speech alone between two speakers of different languages cannot occur since there is no inevitable link between the sounds employed in each language and the message that is conveyed. Thus, according to Dinneen, the assertion that language is *arbitrary* is simply the reality that in order for more than one language to exist there must be no unswerving, compulsory link between the characteristic of the things or notions language contends with and the linguistic elements or amalgamations by which these elements or notions are conveyed. He adds that *Onomatopoeia*, which has the potential to invalidate this proposition, is inhibited by the fact that such words are relatively few in any language and the exactness of the mimicking hinges upon the sounds that exist in a language (Dinneen, 1967:8-9).

The other strand is in regards to minimal pairs or the parallel distribution of sounds in relationship to *Arbitrariness* as a *Characteristic of Language*. Dinneen makes two points about this issue. The first point is that an arbitrary characteristic of a sound system is that not all of the discernible phonemes of a language can appear in all positions. This characteristic allows us to identify the systematic and regular features of both the allophonic variations and functional differences that exist in any language. The second point is that while it is difficult to provide complete evidence for the arbitrary, systematic composition of sounds in a language, we can, nevertheless, show mostly how some of the sounds in the language exist in parallel and, thus, contrastive, distribution in a small number of contexts (Dinneen, 1967:36).

Grammatical Considerations

Here, three grammatical considerations regarding *Arbitrariness* are presented. The first is Lyon's consideration of grammatical structure. He apprises us that by glossing linguistic productivity exclusively with regard to a language's grammatical structure, this type of productivity can, to a certain degree, be characterized by the attribute of *Arbitrariness.* To give an example, he talks about how the simple attributive adjective mostly comes before the noun it modifies in the English and German languages, but it commonly comes after the noun in the French Language. Another example he provides is how the verb generally comes at the beginning of a sentence in the Irish language, but it comes at the end in the Turkish language. He therefore argues that word order is to this degree arbitrary, just as much as any other aspect in the grammatical structure of languages (Lyons, 1977:77).

Having made the preceding observation, Lyons states the axiom that not everything in grammar is arbitrary. For instance, the fact that in the sentence *Thomas Sankara entered the National Assembly of Burkina Faso and he spoke,* but not *He entered the National Assembly of Burkina Faso and Thomas Sankara spoke,* Lyons would say that 'he" can refer to Thomas Sankara is not simply an arbitrary and unexplainable fact of English. It can be explained in accordance with the proposition that except they appear in subordinate clauses, pronouns apply either to elements that are present in the setting or to previously mentioned elements, and this successively hinges partly upon the fact that "language-utterances, in their natural medium and transmitted in the normal vocal-auditory channel, are produced and processed in real time and partly upon the fact that what is said earlier serves to amplify or modify the context for what is said later" (Lyons, 1977:77). He adds that many other cases of non-arbitrariness in grammatical structure exist and enjoins us to always bear this reality in mind when we analyze the characteristics of a language (Lyons, 1977:77).

In addition, Lyons informs us that there exist certain utterances whose originality does not exist only because they have never appeared in the previous encounter of a speaker or writer, but in their recognized inventiveness of style. It is for this type of originality that the concept "creativity" is most suitable, albeit linguists do not uncommonly employ it to represent productivity. It is indeed arguable whether creativity is a characteristic of language or a characteristic attribute of the utility of languages by certain speakers and writers in certain instances. No matter one's position on the issue, creativity as it occurs in the metaphorical utility of lexemes or their normal, or characteristically constructive, amalgamation, evidently hinges upon the semantic construction of the language-system, and it has attracted the attention of semanticists for a long time. As such, the question about whether creativity, like productivity, is arbitrary or rule-governed and to what extent remains unanswered (Lyons, 1977:77-78).

The second is Dinneen's consideration of the grammatical method of Dionysius Thrax (born in Kemet/Egypt in 170 BC and died in 90 BC). To begin with, according to Dinneen, during Thrax's time, this Greek thinker's technique was the most didactic to the best of his capability, albeit in modern times it was considered to be quite restricted. Thrax attempted to instruct us about what to inspect and hearken to so that we can determine what the structurally significant "parts" of a language are. For him, the "parts" are significant because of their formal differences; a correlation exists between the differences of the meaning variances in his language (Greek); and there exist different characteristics of form and content which occur repeatedly and vary either in terms of their morphological makeup, syntactic functioning, or both. As a result of their variations, the "parts" can be employed in the Greek language to signal either variegated denotations or variations in denotation (Dinneen, 1967:106).

Next, as Dinneen explains, for native speakers of the Greek language, it would probably suffice to simply identify the "parts" of a sentence on the basis of semantics. This type of technique suggests that an analyst comprehends the sentences to be examined methodically. Consequently, the analyst must differentiate two quite different types of criteria employed in language investigation and assess their functionality individually (Dinneen, 1967:106).

One of these criteria comprises grammars which can allow an analyst to either speak about or speak the language. A purely formal grammar allows the analyst only to speak precisely with a particular level of enculturation when people can utter the noises that are varacious in each recognizable social context. By possessing the knowledge of a culture's semantic system, the analyst can function as a constituent of that culture. Thus, purely formal grammars so denoted constitute a squandering of time for a person who is simply interested in employing the language as it is used by its native speaker (Dinneen, 1967:106).

The other criterion is that by considering the four types of knowledge the Greeks identified—i.e. (1) "practical mastery without reflection," (2) "mastery with the ability to compare techniques," (3) "intuitive mastery of art," and (4) "demonstrative knowledge of our subject," an analyst can discover that a "grammar" of denotation will be futile for him/her go past the initial type of knowledge. Also, inquiries about language that are formulated according to "meaning" will yield no beneficial answer besides the non-contemplating type, unless the analyst has the ability to ask questions about language, since the connection of form and content is arbitrary in any language. Dinneen, therefore, suggests that linguists need a method that will allow for the analysis of the inner structure of a language, since any language will differ from other languages because it is an arbitrary system and the basic approach for contrasting linguistic meanings differ in application for all languages, whether morphologically, lexically, or syntactically (Dinneen, 1967:106-107).

Also, as Dineen notes, Thrax's grammar was principally about the morphological model of the Greek language, with the major limitation of the scholarship being the absence of information about syntactic structures. It was not until 100 AD that Appollonius Dyscolus (also born in Egypt, flourished about 138 AD, and died in 140 AD) partly filled the hole in Thrax's method by cataloging some attributes of grammatical connections among form classes and also among the procedures for "the case government of nouns in construction with prepositions" (Dinneen, 1967:107).

In addition, Dinneen mentions that one would have expected other analysts that succeeded Thrax to have utilized his method since he was the first to recognize linguistic elements in terms of their formal differences and then provide a semantic relationship for each of them. Interestingly, as Dinneen recounts, while some of the successors followed Thrax's footsteps, others did not, mainly due to the "anomaly-analogy" disagreement that polarized the thinkers of his time. Dinneen also recalls that as the dispute over the formal connection between form and meaning in language subsided grammarians shifted their greater deliberation to the semantic classes and paid less attention to formal distinctions (Dinneen, 1967:107).

The third is Dinneen's consideration of the grammar of Priscianus Caesariensis, commonly known as Priscian (born in Mauretania, now Cherchell, Algeria and flourished about 500 AD), via criticism of its phonological component. Before discussing this aspect, it makes sense to first provide a brief description of the grammar itself, according to Dinneen, for those readers who may not be familiar with it.

For starters, Dinneen educates us that Priscian, a Latin, while teaching Greeks in Constantinople in the 6th Century AD, wrote the most comprehensive and definitive representation of the Latin language during the Middle Ages. According to Priscian, his grammar for Latin was based on that of Dyscolus for the Greek language. Mimicking Dyscolus' model, Priscian also divided his work into 18 books of asymmetrical lengths. The first 16 books, referred to by the medieval scholars as *Priscianus Major*, covered Latin morphology; and the last two books, referred to as *Priscianus Minor* by the same scholars, dealt with syntax. Priscian's grammar and the small instructive work of Aelius Donatus (a Roman grammarian who was born in 320 AD and died in 380 AD) emerged as the authoritative sources for the Latin language and for dialogues about language within scholarly circles during the Middle Ages (Dinneen, 1967:114).

Next, Dinnen teaches us that the work of Priscian is significant for two reasons. The first reason is that it is the most comprehensive and precise representation of Latin by a native speaker of the language. The second reason has to do with the fact that Priscian's grammatical theory is one of the bases of the conventional approaches for analyzing language. Even though Priscian averred that semantic guidelines were the fundamental standards for analyzing language, he also routinely employed formal guidelines in his analysis. The opaqueness of the semantic guidelines and the transparency of the formal guidelines were justification for utilizing a different and sequential application of formal and semantic guidelines for the analysis of any language. The possibility for doing this is absent from Priscian's grammar, since he insisted that semantic guidelines are the core for distinguishing the "parts of speech." Following the Aristotelian approach, Priscian privileged the noun over the other parts and the nominative form in nominal inflection. But even more, he was not consistent when applying both of his own basic principles (Dinneen, 1967:114).

Also, Dinneen instructs us that in conceptualizing elements of language, Priscian suggested the following four types of sounds accommodated in a language, although, according to Dinneen, not all of them are worthwhile for providing an account of language: (1) "*vox articulata*—an articulated vocal sound—is a sound that is limited or associated with a meaning by a speaker"; (2) "*vox inarticulata* is a sound not uttered in order to manifest a meanings"; (3) "*vox literata* is a sound that can be written, whether *articulata* or *inarticulata*"; and (4) "*vox iliterata* is a sound that cannot be written" (Dinneen, 1967:114). Dinneen adds that Priscian defined a *letter* as the smallest part of a sound that we can write; a *syllable* as a sound that can be uttered in one breath with an accent and also be written—it has between one and six letters; a *word* (*dicto*) as the smallest part of a composite utterance and a part of its entire denotation; and an *oratio* as a stipulated ordering of words that designate an entire thought (Dinneen, 1967:114-115).

Furthermore, Dinneen edifies us that Priscian defined eight "parts of speech." They are as follows: (1) *noun* gives a common or proper quality to each of its subjects, bodies, or things; (2) *verb* comprises tenses and moods, albeit lacking case, indicating acting or upon which an action is taken; (3) *participles* which are not clearly denoted, but are said to come in third place and justifiably so because they have in common with nouns case as well as voice and tense with verbs; (4) *pronoun* can take the place of the real name of any person and designates a specific person; (5) *preposition* is indeclinable and is placed before other words, either next to them or becoming an amalgam of them; (6) *adverb* is indeclinable and its denotation is added to verbs; (7) *interjection* which is not directly denoted, but is said to be different from adverbs with which it is identified due to its syntactic independence and emotive denotation; and (8) *conjunction* is indeclinable and connects other parts of speech, and its importance hinges upon its capacity to clarify their denotations or connections (Dinneen, 1967:115).

With the preceding background, the question here then is the following: What does Dinneen say about the phonological aspect of Priscian's grammar as it pertains to *Arbitrariness*? According to Dinneen, while the words utilized by Priscian indicate explicit contrasts, a careful examination reveals that he was chaotic. Priscian's initial binary designation was between sounds that are articulated and those that are not, suggesting dissimilarity between sounds such as groans and sighs and language sounds; however, his definition of sounds of groans and sighs are akin to Aristotle's denotation of *vox*: i.e., a sound made by an animal connected to a particular image. In essence, this definition is a reflection of Aristotle's binary designation of meaningful and meaningless language sounds. This cannot be clearly discerned from Priscian's definition, as "meaning" in this case rests on the intensions of the speaker. The next group of Priscian's definitions also indicates that a sound that can be written (*vox literata*) can either be meaningful or meaningless based on, again, the intensions of the speaker. In addition, he also did not enumerate why certain sounds cannot be written; in fact, it is a contradiction to the principle of articulatory phonetics which states that any sound can be written as long as symbols can be developed to represent the movements needed to make the sound. In addition, his denotation of the syllable, while it is not invariably appropriate, is non-semantic because it is *arbitrary* and is also productive for illustrating Latin because it is propitious (Dinneen, 1967:118).

Linguistic Theoretical Considerations

About linguistic theoretical considerations as they relate to *Arbitrariness*, Dinneen's work on general linguistics is quite profitable. He broaches the following four themes: (1) linguistic sign, (2) linguistic value, (3) linguistic theory, and (4) Edward Sapir's language. His analyses of these themes are presented in the following paragraphs in sequential order for clarity. I must further mention here that Lyon's view on the first theme is also included.

First, as it pertains to *linguistic sign*, Dinneen starts off by defining the concept according to Saussure's use of it as a broad statement alluding to what other linguists discern as morphemes, words, phrases, clauses, or sentences. Nonetheless, Saussure's utilization of the concept "morpheme" is limited to inflectional and derivational affixes instead of roots and stems. Dinneen, however, says that there are two basic kinds of signs: (1) a *simple sign* which cannot be analyzed into constituent signs and (2) a *syntagme* which is comprised of two or more meaningful constituents. These two kinds of a sign, he adds, account for a significant variation in the analysis of the elements of the sign (Dinneen, 1967:203).

Next, Dinneen points out that any sign has two core attributes: (1) *arbitrary* and (2) *linear*. He notes that Saussure's idea of the *Arbitrariness* of signs was the conventional variety that no "motivation" for the phonetic composition of a sign is to be discovered solely in simple signs. For syntagms such as complex or compound words, syntactic composition, etc., there exists "relative motivation": e.g., inflected forms are correspondingly composed to signal the same denotational connections; syntactic compositions employed in similar contexts are correspondingly composed. Thus, syntagmes are inspired reciprocal to one another (Dinneen 1967:203).

Also, as Dinneen indicates, since the concept "symbol" insinuates a gauge of motivation to Saussure, he favored the concept "sign" to explain statements in language. He notes that Saussure was not convinced that the number of onomatopoeic signs was large enough or significant in language to invalidate the fundamental proposition that signs are arbitrary (Dinneen, 1967:203).

In addition, according to Dinneen, the linearity of the linguistic sign is evident mostly in the *signifiant* (i.e. the acoustic image of what is meant) which can be divided into components following one another in temporal sequence. The sequence constitutes what Saussure referred to as the *chain of speech*. Dinneen suggests that Saussure envisioned the outcomes of this expected reality to be "incalculable" in linguistics, as linearity is the total foundation of linguistic systems and the guideline by which language can be differentiated from other systems of sign (Dinneen, 1967:203-204).

Thereafter, Dinneen informs us that in addition to the arbitrary and linear characteristics of language, Saussure also assigned two obviously antithetical characteristics to sign—(1) *immutable* and (2) *mutable*, although from different perspectives. The fundamental rationale for immutability of language and the signs that comprise it is the result of each generation inheriting its language and the signs that go with it, and both the linguistic community and the individual are acquiescent in receiving them. This reality indicates that the "conventional" essence of language is an anomalous kind and renders obsolete the idea that there exists an unambiguous "contract" about meanings among speakers. Hence, according to Dinneen, Saussure is quite correct in asserting that "a law accepted by the community is a thing that is tolerated, and not a rule to which all give their free assent" (quoted by Dinneen, 1967:204).

Furthermore, Dinneen tells us that Saussure tenders four rationales for why linguistic signs are immutable and, consequently, of *la langue* (i.e. language), in contrast to *la parole* (i.e. speaking) and *le langage* (i.e., the ability to express a thought and communicate with signs). The first rationale is that since any sign is just as good as another, signs are essentially arbitrary. Thus, there is no need for speakers of a language to discuss or have a preference for signs. The second rationale concerns the fact that even if people want to change the arbitrary writing system of a language, because the number of elements is small and are open to criticism, language signs are boundless, and this reality discourages linguistic change. The third rationale concerns the verity that language is a quite complicated system about which only a small number of experts are well informed. Even these experts have not been able to change languages notably. The fourth rationale is in regards to language being the only social system that everyone uses, which explains why speakers are generally prudent about changing their linguistic customs (Dinneen, 1967:204).

Moreover, Dinneen posits that the mutability of linguistic signs and, consequently, of language, is an inevitable historical reality. He argues, however, that this mutability is from a different perspective: i.e. language is mutable historically due to shifts from *signifiant* (the acoustic image or the signifier) to *signifié* (what a "thing meant" or is signified) as a result of the evolution of sound and analytical shifts. From the perspective of the present-day speaker, he suggests, all the reasons for the reluctance of language to change are evident. Therefore, in order to advance scientific analysis, he calls for the defense of the "fiction" and necessity of a stable language. He nevertheless insists that a more objective approach that allows an analyst to investigate the cumulative impacts on the language system by its speakers over time be utilized (Dinneen, 1967:204).

For Lyons, based on the conventional elucidation of the principle of *Arbitrariness* by linguists, in Saussure's analysis of the linguistic sign, the *signifié* is identified with the concept instead of signification. Lyons therefore suggests that it could be argued that the principle of *Arbitrariness* should be utilized in conjunction with Saussure's model if an analyst is to be able to combine the sign and the concept. This issue, he points out, has received a great deal of coverage in scholarly journals in the field of Linguistics. What is evident from these investigations, he says, is that the idea of *Arbitrariness* is more complex than it appears at first glance, but what is abundantly evident is that "arbitrary" and "conventional" are not coequal (Lyons, 1977:101).

Second, in regards to *linguistic value*, Dinneen begins with the observation that there exists no externally foisted structure that regulates the amalgamations of forms and meanings. He also observes that linguistic values remain completely relative, which explains why social fact by itself can establish a linguistic structure (Dinneen, 1967:209).

Next, Dinneen mentions that Saussure employed the following three terms to conceptualize value based on observations he had made concerning meaning: (1) *signification*—the relationship between a particular sound and a particular concept; (2) *value*—an opposable term to other terms, both paradigmatically and syntagmatically; and (3) *content*—the sum total of signification and value. Dinneen adds the following technical descriptions of the terms, although he makes it very clear that Saussure himself did not use them in these ways but that they are consistent with his thinking: (a) *signification* basically suggests the conventional idea of referential meaning; (b) *value* is in accordance with Saussure's notion of *la langue* as a system of interrelated terms in which the value of one term is contingent on the values of the other terms; and (c) *content* is neutral in terms of the other two terms—i.e. signification and value. Dinneen also states that while Saussure wrote about signification in a positive manner, he wrote about value and content in a negative way in order to distinguish the former term from the latter two (Dinneen, 1967:209).

In addition, Dinneen talks briefly about how Saussure utilized the idea of value to explore both the conceptual and the material (or phonetic) attributes of language. The lesson from the investigation is that all that is needed of the elements of language is that they be differentiated from the other elements of the same kind. In terms of phonetics, it matters not what kinds of sounds are employed in a language; what is important is that the sounds are different from one another. This is also the case for writing, as letters can be formed differently based on the positions they occupy in words (Dinneen, 1967:210).

Third, as it concerns *linguistic theory*, Dinneen commences with the position of *Arbitrariness* in "linguistic theory and reality" tendered by Louis Hjelmslev in his Glossematics. Dinneen tells us that *Arbitrariness* is one of the two factors (the other factor being *appropriateness*) Hjelmslev identified that allows linguistic theoreticians to introduce particular premises from previous experience to fulfill the requirements for analyzing particular empirical data. These premises, Dinneen says, entail the greatest possible generality and, thus, fulfill the requirements for analyzing a great amount of empirical data (Dinneen, 1967:331).

Also, Dinneen talks about Hjelmslev asking about whether the object dictates and impacts the theory or vice versa. He states that Hjelmslev proffered the view that if a theory is to be defined as "a set of hypotheses," then a theory can be said to influence an object; if the definition is not acceptable, then it can be said that an object influences a theory. Thus, according to Dinneen, from Hjelmslev's perspective, a theory is independent in itself or an experience, since in itself it indicates nothing about its utilization and connection to empirical data. He adds that a theory does not include an existing postulate; instead, it is comprised of pure deduction in that it could be utilized solely to calculate the likelihoods that ensue from its premises (Dinneen, 1967:331).

Next, Dinneen considers the standing of *Arbitrariness* in Hjelmslev's "entities of the analysis" in his Glossematics. On this aspect, Dinneen explains that names are selected in order to make it possible to arrange the information pertaining to their manifestation in the simplest possible manner because their *Arbitrariness* depends on the reality that they do not at all involve the manifestation. He also notes that Hjelmslev analyzed elements that are different from those analyzed by other linguists or by conventional grammar. For Hjelmslev, according to Dinneen, linguistic theory is to be used to analyze a *text*—an approach that will allow a linguist to identify (a) the form supporting the "substance," (b) the comprehensible empiricism supporting the text, and (c) a system of *categories* that allows an analyst to extrapolate the possible elements in a language by using the definitions of those categories (Dinneen, 1967:348).

In addition, Dinneen maintains that the preceding approach is purely formal and done independently of the relevant linguistic or nonlinguistic "substances." In essence, the approach does not evolve into either a phonetics or semantics; instead, it develops into a *linguistic algebra* that provides the formal justification for making inferences. Since such "algebraic" elements lack innate classification, Dinneen recommends that a linguist must arbitrarily and appropriately ascribe names to them (Dinneen, 1967:348).

Fourth, apropos *Edward Sapir's language*, Dinneen initiates his exposition with the assertion of Edward Sapir (a United States anthropologist and linguist born in 1884 and died in 1939) that a distinctive and psychologically arbitrary and symbolic relationship exists among all conceivable elements of awareness on the one hand and some elements in the "auditory, motor, and other cerebral and nervous" areas on the other. Dinneen then reveals that by using this assertion as support for his definition of language, Sapir stated the basic conventional denotation of language as "a purely human and non-instinctive method of communicating ideas, emotions and desires by means of a system of voluntarily produced symbols" (quoted by Dinneen, 1967:221).

But, as Dinneen also points out, Sapir elucidated the preceding conventional definition of language by adding two amendments. The first amendment is that language consists of a visual image, a category of images, or a perception of connection. Still, insists Dinneen, this amendment does not imply that meaning can be so straightforwardly talked about since, like language, it cannot be isolated in any specific section of the brain (Dinneen, 1967:221).

The second amendment formulated by Sapir is that the interconnection between speech and meaning is one that could but need not exist. This is because while the movement of language is continuous, its internal meaning, its cognitive worth or potency differs without interference vis-à-vis consciousness. As such, Dinneen argues that while thought and speech are strongly interconnected, they are not to be deemed as having the same boundaries or extent in space, time, or meaning. He adds that given that there exist numerous delineations or proxies for language as generally envisaged (i.e. "primarily an auditory system of symbols"), one can talk about "thought without language" in terms of a procedure in which there is a lack of recognition of auditory symbolism. In such pursuits as thinking without being cognizant of any auditory imagery, the motor and other systems that are diametrically connected to speech production may be utilized (Dinneen, 1967:221-222).

Together, the two amendments, as Dinneen asserts, make it possible to postulate that thought is indispensable and that the external form of language alone stays consistent. These truisms, he argues, led Sapir to conclude that (a) language form deserves to be investigated for its own purpose; and (b) at each stage of the formal analysis, meaning must be addressed at least as the highest quiescent possibility, in light of the fact that the entire impetus of language is the communication of meanings. Thus, according to Dinneen, the formal analysis necessitates identification of the attributes that are the conduits of communication and the types of meanings to be conveyed by language (Dinneen, 1967:222).

Other Linguistic Considerations: Creativity, Duality, and Iconicity

The discussion here on *Creativity, Duality,* and *Iconicity* as they relate to *Arbitrariness* is based on the analyses of Lyons who has, to the best of my knowledge, done the most definitive work on the topic. First, on the matter of *Creativity* as it relates to *Arbitrariness,* Lyons tackles the question about whether *Creativity* is rule-governed and the extent to which it is arbitrary, as opposed to being inspired. Conceding that he has no solid answer to the question, he, nonetheless, says that there exist utterances whose newness does not depend only on the fact that they have never transpired in an antecedent encounter of a speaker or writer, but in their accepted inventiveness of style; and it is for this type of newness or inventiveness that the concept *Creativity* is most fitting. He also, however, acknowledges that the issue of whether *Creativity* is an attribute of language or a typical aspect of the utility made of languages by speakers and writers is arguable. He adds that no matter which position one takes on the issue, the way *Creativity* reveals itself in the metaphorical employment of lexemes or their untypical, but stylistically effective, combination obviously rests of the semantic organization of the system of a language, and it has conventionally been of interest to semanticists (Lyons, 1977:75-76).

Second, about *Duality,* Lyons begins by stating that from a semantic perspective, *Duality* has the benefit of allowing an analyst to differentiate an immense number of forms, and incidentally an enormous number of lexemes and statements, by integrating a rather small number of lower-level features in a number of different ways. He adds that combined with the attributes of grammatical productivity, *Duality* represents the verity that sine die many formally different utterances can be composed in any natural language using a somewhat small set of phonological features (Lyons, 1977:74).

Next, Lyons tells us that as it works in language, *Duality* is tied to *Arbitrariness* in that should each phonological feature in a particular form has to have a noticeable iconic relationship, whether conventional or innate, to certain features of its denotation, then it is apparent that there will exist strong limitations on the potential to amalgamate phonological features with one another. This will be the case when there is a perpetual, but non-iconic, connection between a particular phonological feature and the denotation of the forms in which it manifests itself. The supposed dominance, vis-à-vis stability and rationality, of a language in which all forms connected to position in space started with the same phonemes, all forms having to do with union ended with the same phoneme, etc., will be more than valuable by its lack of pliability and the restricted quantity of word-forms of a particular length that would be constituted from a small collection of attributes (Lyons, 1977:74-75).

In addition, according to Lyons, only to a limited extent in natural languages in cases dealing with sound-symbolism and onomatopoeia can one find a disconnection between *Duality* and *Arbitrariness.* He also notes that it is not unusual, nonetheless, to find phonological features of particular kinds or phonological systems to have a particular grammatical function, which limits the *Arbitrariness* of the language-system, although it does not expand its *Iconicity* (i.e. relating to or of the nature of a sign whose form directly reflects the thing it signifies; for example, the word *snarl* pronounced in a snarling way). For illustration, he mentions languages in which (a) there is a variation between the phonological structure of the forms of nouns and that of verbs; (b) there is a variation between the phonological structure of the forms of pronouns, prepositions or conjunctions and that of nouns, verbs or adjectives, etc.; (c) there is also the context of boundary-signals or *Grenzsignale*—i.e. phonemes or prosodic attributes that mark the boundaries of forms in order to make them more distinguished in the effusion of speech: examples include how stress functions in the Czech and Turkish languages and how vowel-harmony works in the Hungarian and Turkish languages (Lyons, 1977:75).

Furthermore, Lyons apprises us that the disconnection between *Duality* and *Arbitrariness* therefore places particular restraints on the merger of phonological features in certain languages. In addition to these restraints, he also points to systematic and random ones that are free of grammatical function and denotation. He cites as an example the nonexistent in the English language, with the exception of the forms for certain proper names, word-forms beginning with *tv-* or *sr-* as in the Russian language. These types of blending of consonants, he says, are not essentially arduous to enunciate; the fact that they do not appear in the English language is evidence that all languages have these types of restrictions or hindrances that are said to possess a level of phonological structure instead of a list or collection of phonological features. Therefore, he argues, the more restrictions there are, whether they are systematic or random, on the fusing of phonological features, the more we are certain about the "redundancy" in the signals, and "redundancy" is imperative for thwarting channel-noise (Lyons, 1977:75-76).

Third, concerning *Iconicity*, Lyons alludes to many aspects. To start with, he conceptualizes *Iconicity* by contrasting it with *Arbitrariness*. Correspondingly, he states that an iconic signal, or part of a signal, is one which in somehow "geometrically similar" to what it denotes or represents; if it is not, then it is arbitrary (Lyons, 1977:70).

Next, Lyons considers *symptomatic* signals (i.e. those in the mind of a speaker) to be appealing when viewed from the perspective of *Iconicity*. He posits that the connection of intensifying loudness of voice and rising pitch with intensifying anger or excitement, for instance, is iconic. Even more, the connection is innate instead of customary—i.e., it is governed biologically instead of culturally. Thus, according to Lyons, what may be referred to as innate *Iconicity* is definitely an indispensable and chief attribute of language-behavior, albeit it may be portrayed as paralinguistic as opposed to linguistic, and is quite typical of other types of signaling behavior employed by humans and other animals (Lyons, 1977:71).

Also, on the report of Lyons, *icon* (i.e. a sign which would have the feature that makes it notable, even though its object lacks actuality) is the concept used for non-arbitrary signs by Charles Sanders Peirce (United States logician, mathematician, philosopher, and scientist born in 1839 and died in 1914). Lyons mentions that Peirce employed the term "interpretant" in his theory on semiosis or pragmatic semiotics to denote the mental influence engendered by a sign. Thus, for Lyons, the normality, or *Arbitrariness*, of symbols, in contrast with what he refers to as "the naturalism of iconic signs," hinges upon the utilizer's comprehension or cognizance of the norm (Lyons, 1977:102).

In addition, as Lyons points out, it is unclear how icons and symbols differ based on Peirce's account. This is because for Peirce, *Iconicity* hinges upon some innate geometrical or functional resemblance between a sign and the object it represents. Lyons doubts the entire idea of resemblance being unconnected to our identification of the pertinent attributes by strength of two entities being similar. Add to this doubt the fact that our identification of a resemblance between a sign and what it represents hinges upon our grasp of particular cultural habits of interpretations. Correspondingly, "iconic" cannot be regarded as the same as "natural." While a difference can be discerned between the natural and the cultural on the one hand and between the arbitrary and non-arbitrary on the other, it is the case that icons are a sub-category of non-arbitrary signs in which the resemblance may be either cultural or natural. Furthermore, *Iconicity* is, for the most part, a complicated and diverse characteristic in that there are numerous and various types of icons in languages as well as in other semiotic systems. Moreover, only marginally iconic are many of the onomatopoeic forms of spoken language and of the characters and hieroglyphs of the ideographic writing systems in that if we can know their meaning, we can also notice the resemblances of their form and meaning, although we cannot infer what their meaning is based only on their form (Lyons, 1977:102-103).

Thereafter, Lyons explains that there are two types of *Iconicity*. The first type he calls *Primary Iconicity*, which he says is "medium-dependent." He mentions the English word-form *cuckoo* as an example given that it is iconic in the phonic medium, but not in the graphic one. From an analytical point of view of the lexicon of natural languages, the idea of resemblance, or relatedness, between different meanings connected with some or a group of forms is the most important. A more complicated kind of *Iconicity* may exist between form and meaning, moderated by what, from a historical perspective, is an expansion of meaning from a fundamental to a conveyed, or metaphorical, perception. The second type Lyons calls *Secondary Iconicity*, which hinges upon the primary iconic connection with the characteristic of the referent, as far as the forms of the word is concerned, and the commonly agreed-upon connection between the word and what it represents, as far as the denotation of the words is concerned (Lyons, 1977:103-104).

Subsequently, Lyons apprises us that what is referred to as "secondary onomatopoeia" is actually a particular aspect of *Secondary Iconicity*. Secondary onomatopoeia, he says, refers to when the sound of the spoken word-forms is perceived to be suitable to the denotations of the lexemes of which they are forms, although the words do not really represent sounds or their sources—a phenomenon that is referred to as "sound symbolism," also called "phonaesthetics." He maintains that while it is true that in certain languages particular sounds or blending of sounds are tied to features of meaning (e.g., in poetry), what remains unclear is the level at which the postulates of sound-symbolisms are used jointly by separate languages. He adds that sound-symbolism, as it obtains, seems to be limited to a somewhat small portion of the lexicons of languages (Lyons, 1977:104).

Thereupon, Lyons tells us that a major feature of the connection between the conventional and the iconic is that they are not opposed to each other. He cites the example of the perceived connection between owls and wisdom, which may or may not hinge on anything in nature. But, in a culture where such a perception exists, the relationship is obviously not arbitrary. In essence, the perception, like many others like it, does not yield easily to a demarcation between what is biologically (nature) or culturally (nurture) ascertained (Lyons, 1977:104-105).

Furthermore, as a cautionary note, Lyons proposes that while the concept "iconic" has earned its place in linguistic literature, although it is just a designation for a somewhat heterogeneous collection of non-arbitrary relationships of forms and meaning, and it is a functional technical concept, it should not be regarded as being the same as "natural." He also suggests that unlike the concept "symbolic," the concept "iconic" is expedient because it does not have a different non-technical meaning outside semiotics (Lyons, 1977:105).

African Linguistic Evidence

This section is composed of summaries of findings derived by some scholars who have explored the notion of *Arbitrariness* as it pertains to languages in Africa. First, in her investigation of the categorizations of languages within the Niger-Congo phylum, Cynthia Hallen (1998-1999) culls from the work of Gwenyth Lafleus to indicate that when Malcolm Guthrie (1948) would find it challenging to utilize the techniques of dialect geographers in order to categorize the Bantu languages, he would tweak the technique by taking into consideration that the *Arbitrariness* attribute is critical to the technique. As Hallen further points out, Lafleur notes that Guthrie would start with one language and move outwards, grouping with the language all the neighboring languages that possess similar characteristics. When Guthrie arrives to a point where he has moved into another group, he would start the process over from the center.

Second, according to Ellen Contini-Morava (1983), even though rules exist in Kinyarwanda (a Bantu language of Central Africa) for pronoun usage inspired by *Arbitrariness*, there are other rules shaped by extra-linguistic factors. Accordingly, she also insists that *Arbitrariness* in language must therefore be perceived as a scale, given her findings on the order and interpretation of pronominal object prefixes in Kinyarwanda. She additionally finds that some pronoun sequences in the language are acceptable while others are not. Among the pronoun sequences that are acceptable, some provide information about the respective participants' roles of the pronouns' referents while others are ambiguous as to the roles. Hence, she suggests the following:

> ...both the restrictions of ordering possibilities and the interpretations of acceptable sequences can be explained in terms of two alternate strategies for the ranking of participants in a reported event: ranking according to degree of emphatic distance from ego (egocentricity) and

> ranking according to degree of potency in bringing about the event (contribution). Ambiguous pronoun sequences are cases where there is a potential conflict between these ranking strategies; unacceptable sequences violate the ranking according to egocentricity (Contini-Morava, 1983:425).

Third, in their examination of *Arbitrariness* vis-à-vis the theoretical cognitive explications of the isiZulu noun class system, Britta Zawada and Mtholeni Ngcobo (2008) assess the oft-cited axiom that the order in which the noun class system of the Bantu languages of Southern Africa has developed suggests a morphologically arbitrary system, without any conceptual or semantic underpinning or purpose. Their finding, however, indicates that the isiZulu noun class system is grounded on conceptual notions such as case of acquisition, frequency, and prototypicality.

Fourth, while cogitating about iconicity in spoken languages, Pamela Perniss, Robin L. Thompson, and Gariella Vigliocco discovered that African languages entail depictive forms that utilize iconic mappings which had been identified since the early part of the 17th Century and had been observed more studiously in grammars since the mid-19th Century. Nonetheless, these scholars note that despite their apparently distinctive configurations and tendencies, in addition to their depictive and expressive features, these iconic mappings eluded traditional linguistic analysis. Due to the fact that iconicity can be identified in only a small quantity of words in spoken language, the subject is not receiving adequate attention from contemporary linguists. Perniss and her colleagues agree that "direct imitative form-meanings mappings in spoken language are only possible for acoustic properties and events, which are arguably far less frequent than visual properties and events in our experience"; therefore, they argue, it is not accidental that "onomatopoeic words make up only a small portion of spoken languages' total lexica, although they appear consistently across all spoken languages" (Perniss et al., 2010:2).

Also, Perniss and her fellow scholars note that onomatopoeic imitations in spoken languages are generally derived from the sounds of other animals and those produced by objects in motion or upon impact on other objects. They say that many of these words, especially the latter variety, also display another kind of iconic mapping referred to as "phonesthesia." For the attendant "phonesthemes" of this variety of iconic mapping, they propose that there is a correlation between a similarity of configuration, usually in "word-initial or word-final consonant clusters," and a similarity of denotation. Perniss and her partners further observe that as the convergence between certain phonesthemes and onomatopoeic configurations reveal, the existence of phonesthesia in an African language, as in other languages, may be partly due to generalization from the denoted mappings of acoustically iconic configurations (Perniss, 2010:2-3).

In addition, as Perniss and her colleagues point out, almost all African languages in Africa South of the Sahara have been reported to have lexica with large representations of acoustic or sound symbols. This finding seems to impugn the Eurocentric linguistic claim that the Indo-European language phylum enjoys predominance on iconic mappings being employed to communicate all types of sensory occurrences (Perniss et al., 2010:3).

Fifth, Harald Hammarström in his interrogation of "linguistic diversity and language evolution" discusses a notion of *Arbitrariness* in the Gula Sara, a Central Sudanese language spoken in Chad. He begins by speaking about certain languages separating all referents into classes/genders demonstrating contrasting agreement configurations. He notes that while certain languages, on the one hand, compel a speaker or a writer to mark the gender openly everywhere applicable in a sentence, other languages, on the other hand, require the speaker or the writer to make only a few repetitive markings in the sentence. He adds that a systematic distinction is made for natural gender (masculine/feminine) in many class/gender linguistic systems, although they differ in their approaches of doing it. At certain times, he also indicates, referents lacking gender possess their own class; these elements are designated as either all masculine, all feminine, or arbitrarily categorized (Hammarström, 2016:24).

Hammarström then goes on to say that probably, the most discernible distinction has to do with "speech-act participants": i.e. "you versus me," etc. Generally, as he restates a well-established fact, grammars possess simplex and mandatory contrasts for first, second, and third person, and some mandatory contrasts for the number of participants as well. He mentions that while it is tempting for an analyst to suggest that this is the only type of "sensible system of oppositions," he states that many languages would render the assumption untenable. As proof, he cites the case of Gula Sara which "has a special simplex form for 1 + 2 person singular (you and me) as well as an inclusive/exclusive distinction...It also seems that, historically, a 'plural' ending containing a *g* might have been added to simplex forms of the categories person 1, 1 + 2, 2 and 3 ('minimal') to form the 'plurals' 1 exclusive, 1 inclusive, 2 and 3 ('augmented')" (Hammarström, 2016:24).

Sixth, from David Robinson's work, we are edified that ideophones (i.e. words that bring to mind ideas in sounds) can be found across Africa. The earliest formal investigations on ideophones were by anthropologists who conducted research across the continent in the late 19th and early 20th Centuries. For example, a study of Bobangi, a language spoken in the Congo, led to the discovery that "colouring words...were among the most graphic in the language (and) often...have such force that sentence after sentence can be constructed by means of them, without the use of a single verb" (Robinson, 2019:5-6).

In addition, Robinson mentions that the vigor of ideophones hinges upon the way the words themselves are composed, which includes being uttered in a dramatic tone and visual gestures to assist the hearer to envision the phenomenon being mentioned. To illustrate this aspect, he states that the "Ewe word *dabodabo*, for instance, translates as 'duck'; in addition, the repetition of its basic syllables evokes an uneven walk, and the sounds are accompanied by an exaggerated waddling motion. 'It's a really good example of an ideophone as a word that imitates not just sound but a whole sensory scene'" (Robinson, 2019:7). Other interesting findings on ideophones discovered in West Africa are the following:

> Comparing the ideophones of half a dozen West African languages, [it is found] that certain 'front' or 'closed' vowels—such as the [i] sound in the English word 'cheese'—tended to be used to represent concepts that were light, fine or bright, for instance; while 'back' or 'open' vowels—such as the [ɔ] sound in 'talk' or the [ɑ] in past—tended to be associated with a sense of slowness, heaviness and darkness. Meanwhile, voiced consonants such as 'b' or 'g'—so-called because they require the vocal cords to resonate—were associated with greater weight and softness, while voiceless consonants, such as 'p' or 'k', tended to be used to represent lighter weights and harsher surfaces. So, for example, in Ewe, someone *kputukpluu* is thinner than someone who is *gbudugbluu* (Robinson, 2019:7-8).

Also, similar connections were discovered among linguistic tones as follows:

> In many languages, the pitch at which a syllable is spoken can change a word's meaning [and] words representing slowness, darkness and heaviness tended to have lower tones than those depicting speed, agility and brightness, which were formed of higher tones. Many also include the repetition of syllables, which can be used to signify number and a continuous action or state. So, for instance, in Ewe, *kpata* is used to denote one drop falling, but *kpata kpata* depicts many drops falling (Robinson, 2019:8).

Nonetheless, as Robinson further points out, there is the peril that these evolutionary perspectives ensconce the notion of present-day ideophones being antediluvian. He argues that this prejudice, which has existed since the first detailed studies of ideophones in Africa were conducted, is misinformed.

This is because ideophones provide us with an opportunity to comprehend how people can use sounds to exchange sensual experiences temporally and spatially. They evoke the reality that language is strongly embedded in the "body" and that each word is like a "performance-piece" that utilizes many of our feelings. And, like poetry, ideophones in many of the African languages, as in many of the other languages across the globe, assist a person to envision a novel illumination, assist him/her to relish words, and they are reminiscent of our sensory vistas (Robinson, 2019:19).

Seventh, we also learn from Kimi Akita and Mark Dingemanse that ideophones (also referred to as "mimetics" and "expressives") are prevalent across Africa, just as they are across Asia and the Americas. This type of linguistic feature, they say, is comprised of marked words that represent sensory resemblances. They add that ideophones manifest themselves in onomatopoeia and many sensory areas like manner of motion, texture, and mental states. They also state that "Across languages, ideophones stand out as marked words due to special phonotactics, expressive morphology including certain types of reduplication, and relative syntactic independence, in addition to production features like prosodic foregrounding and common co-occurrence with iconic gestures" (Akita and Dingemanse, 2019:1). In addition, they provide many examples from the following African languages to illustrate their observations: Ngbaka, a Ubangian language; Gbaya, a Niger-Congo language; Somali, an East Cushite Afro-Asiatic language; and Siwu, a Kwa Niger-Congo language (Akita and Dingemanse, 2019:2).

Furthermore, Akita and Dingemanse evince the possibility of a boundary case in Luhya languages such as those of the Bantu and Niger-Congo phyla in which ideophones are always low in terms of expressiveness but high in terms of morphosyntactic integration. They also indicate that additional evidence for this phenomenon can be found in gestures, which tend to "co-occur with ideophones when they are at their most depictive, as marked by high levels of expressiveness and low levels of grammatical integration" (Akita and Dingemanse, 2019:7-8). Based on this evidence, the two scholars suggest that researchers studying ideophones in Africa should consider both linguistic and extralinguistic attributes in their investigations (Akita and Dingemanse, 2019:8).

African Mathematical Evidence

For starters, as Ron Eglash inculcates us, the notion of a spectrum progressing from more orderly to less orderly is quite evident in certain African material designs, which tend to reveal periodicity and aperiodicity—often moving from order to disorder—along the other. He adds that similar geometric visualizations of this type of spectrum are utilized by scholars in the field of Computer Science (Eglash, 1999:172-173). He then describes the *Arbitrariness* of the random music of the Birom, a form of musical instrument indigenous to Nigeria that has something of a white noise distribution of sounds, this way:

> ...a flute ensemble designed to allow each musician to express individual feelings through whatever idiosyncratic noise (even silence) he or she chooses, resulting in an indeterminate process in which the sounds produced by the players are not obstructed by a conscious attempt to organize the rhythms and harmony (Eglash, 1999:174; partly quoting Akpabot, 1975:46).

The Birom musical instrument's formation of sounds to represent the words that are uttered, imitating the Rössler attractor utilized by Eglas as feedback (Englash, 1999:169), can be depicted in this fractal mathematical expression:

Musical expression = Z

Total system: $x' = -(y + z)$

$y' = x + 0.15y$

$z' = 0.2 + z(x - 10)$

Next, Dirk Huylebrouck describes the *Arbitrariness* feature of one of the African mancala-type games called *ayo*: the full name, *Ayo Olopon,* is a traditional game played by the Yoruba people of Nigeria. The player must possess a lot of cognitive strength and mathematical skill. It is played in a carved wooden box, containing 12 holes (six holes on each side), and 48 *Ayo* seeds (four seeds in each hole). Huylebrouch begins by explaining the procedure of the game as follows:

> At first, the pawns are sowed in an arbitrary hole. Next, they are put one by one in the successive holes in the direction of the cala. If there are too many, the player continues from the hole opposite to the cala. Again, if the last pawn falls in the cala, the player gets an additional turn, but if it falls in an empty hole, the game stops. If the last pawn falls in a hole with pawns in it, the player sows these pawns further one, all together, making a "bridge" (Huylebrouck, 2006:155).

Thereafter, Huylebrouch describes the mathematical feature of the procedure as follows:

> ...if s(n) is the smallest number of pawns necessary for the nth hole to lead to a game the player can win (if he plays cleverly), the number of pawns leading to a winning position approaches n^2/π, for large values of n (asymptotically, mathematicians would say). Their computations

> involve high-level analysis, including the so-called gamma function Γ and the hypergeometric function F. The result is impressive, because the mathematical constant π = 3.14159265...pops up in the context of an African game (Huylebrouck, 2006:155).

In addition, we learn from Paulus Gerdes and Ahmed Djebbar about an aspect of *Arbitrariness* in *Talkhîs* (the famous work on the "Science of Arithmetic") penned by the Moroccan scientist Ibn al-Bannâ (1256—1321 AD). In addition to demonstrating the meticulousness of the cubic root and square of arbitrary whole numbers, the work also entails "the demonstration of the famous rule of signs…the demonstration of the existence of solutions of quadratic equations by a procedure that… permits it to express the number of combinations of *n* objects taken *p* at a time, with the help of an arithmetic formula" (Gerdes and Djebbar, 2007:16-17).

Also, Judah Paul Makonye demonstrates that *Arbitrariness* in learners' obsession with the mathematical concepts and procedures they learn early explains their *arbitrary errors* in introductory calculus work in South Africa's senior school certificate examinations. He describes an *arbitrary error* as one that "occurs when a learner changes a question to a form acceptable to him or her. It could be that the question seems not to make sense to the learner (and) thus wants to make it in a form familiar to him/her. The change of a question is actually declining the mathematical demand of the task and is often never condoned in mathematics" (Makonye, 2011:222). The *arbitrary errors* he discovered were in relation of equation balancing, legally invalid inference, executive differentiation, dropped exponents, and meta-cognition. The errors were caused primarily by the lack of competence in language to decode certain aspects in the questions (Makonye, 2011:202-222). Makonye goes on to provide a plethora of evidence culled from the learners' examination answers to demonstrate the *Arbitrariness* that underlie their mathematical errors.

Conclusion

In the Ancient Kemetian/Egyptian language, the notion of *Arbitrariness* can be identified in the terms for sounds that give words their forms and meanings. The following are eight examples: (1) *kheru em pe-t*—the sound from heaven; (2) *kheru remm*—the sound of weeping; (3) *kheru qa*—singing voices (4) *kheru qerå*—the roar of thunder; (5) *kheru ta*—the roar of earth; (6) *kheru tau*—the whistling of the wind; (7) *kher*—to thunder; and (8) *Kheru-qerå*—the voice of Quera [i.e. the thunder of tempest or flood] (for these translations, consult Budge, 1978).

We also derive from Derek Sloan how the Ancient Kemetian pictograms or ideograms can represent "unpicturable" concepts and/or entire terms; thus, the nexus between a pictogram or ideograms and the notion exists at the *Arbitrariness* level. He cites the example of the Ancient Kemetian picture of an eye which could refer to (a) the eye itself—*ir*, (b) a part of the eye such as the pupil—*an*, or (c) or something related to the eye such as sight. He adds that Ancient Kemetians would different "ancillary marks of pictures (semagrams) to cue the appropriate meanings. Nonetheless, getting from a picture of an eye to the concept of an eye full (i.e. beauty) requires a metaphorical leap" (Sloan, 2005:3).

Furthermore, as Bjørn Jespersen and Chris Reintges posit, the Ancient Kemetian hieroglyph of the vulture is not a portrait of a numerically particular vulture. Instead, they assert, it is a portrait of a specific person, as no hieroglyph is a portrait of a certain deity or pharaoh. Therefore, for these scholars, to assert that the hieroglyph of the vulture is a portrait calls for a philosophical determination, since the vulture hieroglyph is not self-referential; instead, it is generally "representational," which means that it points to an entity that transcends itself. As they further put it, should the hieroglyphic sign be interpreted ideographically, it could be said to represent a span of possibilities: i.e. "arbitrary vulture," "each and every individual vulture," "the set of vultures," "the property of being a vulture," etc. The choice an analyst makes to characterize the sign would hinge upon the theoretical lens s/he adopts to explicate the type(s) of object(s) to which words refer (Jespersen and Reintges, 2008:2).

Chapter 9

Discreteness

Gleaning from the study of Anouschka Bergman, Kathleen Curie Hall and Sharon Miriam Ross (2007), I state the following elsewhere: "*Discreteness* refers to that feature of language that makes it possible for humans to put together discrete units to make larger communicative units. This is imperative because every language has a limited number of sounds, between ten and 100. The sounds themselves for the most part lack meanings until they are combined together. The fact that humans can develop a large number of meaningful elements (words) from a few meaningless units (sounds) is known as *duality of patterning*" (Bangura, 2014:26). How, then, have linguists explained *Discreteness*? The rest of this section attempts to provide an answer to this question.

Various linguists have offered similar explanations for *Discreteness*, albeit placing emphases on different features. Beginning with David Crystal, we learn that *Discreteness* involves linguistic features that possess distinctly determinable demarcations with no progression between them. An example of *Discreteness* he cites is a language's consonant system [p] versus [k]. He contrasts these features with those which lack distinct demarcations and refers to the latter as *nondiscrete*, and mentions as examples variations in pitch height which he says constitutes a language's intonation system (Crystal, 1992:106-107).

Next, William O'Grady, Michael Dobrovolsky and Mark Aronoff explain *Discreteness* as a communication structure that has discernible elements which can be restated. Among the examples they mention for this structure are (a) phonemes, syllables, morphemes, words, and larger combinations in human languages; (b) two or three discrete types of bee dancing, although the dances cannot be recombined; (c) distinct components in birdsong; and (d) primate call systems (O'Grady et al., 1989:407).

Also, John Lyons broaches a number of aspects in his elucidation of *Discreteness*. First, he characterizes the notion as felicitous to the signal-elements of a semiotic structure. He adds that a structure is *discrete* when the difference between its elements is fixed and does not permit ranges in terms of plus or minus, and a structure is *continuous* when the elements display the opposite characteristics (Lyons, 1977:78).

Second, according to Lyons, the verbal constituent of language is discrete when two word-forms, carefully thought about only from the perspective of their form, are completely the same (as tokens of the same kind) or completely different. A person could therefore be ambivalent about which of the two dissimilar word-forms has been produced in speech or writing, and even the signal itself could be wholly undefined in terms of the physical characteristics which would normally differentiate the two forms. Nonetheless, it must have been produced and must be explicated as one of the two forms, since it cannot be treated as a third form halfway between the two and amalgamating the meanings of both of them (Lyons, 1977:78).

Third, as Lyons argues, the rationality of *Discreteness* does not rely on *Arbitrariness*, although the two ideas interrelate with each other to expand the semiotic adaptability of a language system. Two word-forms could be different minimally (i.e. by a single discrete feature) and could be forms of lexemes that are dissimilar in denotation—e.g., *bear* and *pear*. The possibility exists, however, that minimally-distinct forms could be forms of lexemes which are significantly different denotatively and belong to grammatical categories that are different, which increases and maintains their *Discreteness* when channel-noise could have annihilated the physical variations in the signal which would normally keep them separate. In most cases, the presence of one form would mean the absence of the other form, from the receiver's perspective, that would not lead to any misinterpretation (Lyons, 1977:78).

Fourth, Lyons refers to the fact that incongruous to the verbal part of language which is discrete is the continuous semiotic system of bee-dancing. The body movements of a bee differ in terms of magnitude, both undeviatingly and unceasingly, with an analogously incessant difference in the distance to the place of nectar. Bee-dancing is also non-arbitrary given that the distance to the place of nectar is depicted by the magnitude of the motion of the bee's abdomen and the position of its body parallel to the sun (Lyons, 1977:78-79).

Fifth and finally, Lyons mentions that *Discreteness* is interconnected with *arbitrariness, duality,* and *productivity* in various ways, it exists in every language, and it is not dependent on either channel or medium. What remains uncertain is whether *Discreteness* exists in any other semiotic system besides language. Also, while it is obvious that *Discreteness* is a feature of the verbal constituent of language, it is not so obvious of the non-verbal constituent (Lyons, 1977:79).

In addition, David Minger describes the general notion of *Discreteness* as a linguistic rendering which can be divided into small, distinct elements that can then be reconnected with other small, distinct elements to form novel linguistic renderings (Minger, 2018:1). He then goes on to provide the following examples:

a) "Add a plural morpheme to something like *dog* to get *dogs* [dagz]" (Minger, 2018:1).

b) "The idea of discreteness is not limited to the morphemic level—it carries into other levels. That's why you can show that the American English phoneme /l/ is a discrete unit that contrasts with other discrete units (like /b k i n/ etc.), even though within its own phonemic range it has at least two realizations: the bright and dark versions of ell, which are again an example of complementary distribution determined by English's phonology" (Minger, 2018:1).

Furthermore, Christina Biava explicates that in linguistics, the notion of *Discreteness* primarily denotes that which language is manifested as a "stream" (such as in a "stream of speech"), it can be divided into meaningful elements (such as distinct sounds), and these elements can be put together in different ways to form bigger elements: e.g., distinct sounds like [p], [I] and [t] put together to generate the words "pit" or "tip" (Biava, 2018:1). Biava proceeds by making the following observations that are worth restating here:

> You can best understand the "stream of speech" when you listen to someone speaking a language that you don't know. If you think about it, this is the natural situation for a baby when he/she hears language being spoken by his/her mother, father, siblings, etc. in the weeks and months after birth. However, from this "stream of speech" the infant is able to gradually abstract the sounds in that language and from there combine those sounds to form words…[The] fact that a language has a small set of discrete sounds (usually between 20 and 50 or so) that can be recombined to produce thousands upon thousands of words shows how discreteness is one of the most basic characteristics of human languages, without which human language would not be possible (Biava, 2018:1).

Since the objective of this chapter is to demonstrate the existence of *Discreteness* in African linguistic and mathematical phenomena, the rest of this chapter is divided into three major parts and a conclusion. The first part entails a discussion of linguistic expositions on the subject of *Discreteness* and the second and third parts encompass presentations of African linguistic and mathematical evidence on the subject, respectively.

Linguistic Expositions on Discreteness

Discreteness is another topic in the field of Linguistics that has not enjoyed extensive coverage. Nonetheless, the coverage has involved many variegated aspects of the idea. For the sake of lucidity, I have placed these aspects into the following 10 categories and will be discussed sequentially in the rest of this section: (1) token structure, (2) contrast with semiotic systems, (3) prosodic features, (4) sociolinguistic considerations, (5) ideational function, (6) expression elements, (7) structural dialectology, (8) phonetic elements and formal linguistics, (9) Terminology Theory, and (10) Continuous Mathematics.

Token Structure

As stated by O'Grady and his colleagues, *Discrete Tokens* are differentiated from one another by identifying their "stepwise" differences. As such, no systematic metamorphosis takes place from one token to the next. As a good example of this type of tokens, these scholars cite human language. For instance, they note that the strikingly different stop and fricative phonemes in the English language are thought of as either voiced or voiceless, with no category in between. Also thought of as discrete are digital displays, as they move from one minute (or even one second) to the next with no series of successive changes, stages, or degrees. Another example is a traffic light, since it lacks an unhurried shifting from green to yellow to red (O'Grady et al., 1989:388, 453).

In addition, O'Grady and his fellow scholars mention that *iconic, symbolic,* and *symptomatic* tokens can either be graded or discrete. They provide the following distinctions with attendant examples: (a) *iconic and graded*—a threatening canine's gradual baring of its fangs; (b) *iconic and discrete*—a photograph; (c) *symbolic and discrete*—Morse code; (d) *symbolic and graded*—a dimming light indicating the start of a theatrical performance; (e) *symptomatic and discrete*—a traffic light; and (f) *symptomatic and graded*—the crying of a child or the act of blushing (O'Grady et al., 1989:388).

Contrast with Semiotic Systems

Lyons suggests that the idea that *Discreteness* (and also *Arbitrariness, Duality,* and *Productivity*) is characteristic and distinctive of language and strikingly different from other semiotic systems rests on the fact that languages, or at least their verbal constituent, can be regarded as independent of the medium in which they are essentially and normally palpable. Written languages, as one of humans' major vehicles of communication, already exhibit some level of independence. So also have the diverse gestural sign-languages utilized by people who are deaf and hard of hearing in various cultures, although, like written languages, they were first generated from spoken languages. Therefore, when comparing human signaling and other animals' signaling, or oral and non-oral signaling among humans, serious consideration should be given to the fact that humans can learn, somewhat easily and mostly very well, to transfer much of the verbal constituent of language from one channel to another and also from one medium to another (Lyons, 1977:87).

To buttress the preceding point, Lyons raises the questions about (a) whether it is possible to conceive of a group of people separating itself from the main body of a society and renouncing any use of the spoken language; (b) whether the group's written or gestural language would still be dependent upon the spoken language that elsewhere in the society would still be connected to it; (c) whether the group's written language would develop independently, similar to how other written languages have developed in partial independence of their spoken languages; and (d) whether children could be taught the new language without first acquiring a spoken language in partial correspondence to the new one. He states that while no definite answers are available for these question, there is, nonetheless, some evidence pertaining to them (Lyons, 1977:87), the exploration of which is beyond the scope of this section.

Prosodic Features

On the topic of *Prosodic Features,* relatively little and almost the same things are usually said about it. Let us look at what a handful of linguists have to say about them, beginning with Crystal. For him, *Prosodic Features* are also referred to as simply *Prosody,* which includes loudness, pitch, rhythm and tempo that occur in spoken language. He adds that the concept includes the conventional sense of the term that denotes the "metrical features of versification" (Crystal, 1992:318).

Next, Lyons talks briefly about the disagreement among linguists regarding whether *Prosodic Features* like intonation and stress entail the characteristic of *Duality* and the extent to which they are arbitrary and discrete. His position on the matter is that given their typical function, prosodic and paralinguistic attributes of language-behavior are more akin to other types of non-oral signaling than the oral constituent does (Lyons, 1977:79).

In addition, O'Grady and his fellow scholars point out that no matter where or how they are articulated, all phones possess some immanent *suprasegmental* or *prosodic* properties that constitute part of their configuration. They name and define three of these properties as follows: (1) *pitch*—the vocal attribute of a sound that allows us to position it on a scale that spans from low to high; (2) *length*—the articulation of some consonants and vowels being held longer compared to other consonants and vowels; and (3) *stress*—the combined effect of pitch, loudness and length to indicate that some vowels are more prominent than others. They also mention the concept of *Prosodic Universals*, which they say refers to how languages can be further categorized according to their prosodic or suprasegmental type—e.g., tone languages are those with phonemic pitch contrasts (O'Grady et al., 1989:35-40, 228).

Sociolinguistic Considerations

For this discussion on *Sociolinguistic Considerations* of *Discreteness*, I draw upon the insights of Peter Trudgill. His first insight pertains to the fact that *Discreteness*, as its counterpart, *continuity*, presents a serious challenge to the use of linguistic benchmarks for categorizing different languages into clear-cut languages or dialects. More specifically, he raises the question about whether the placement of linguistic and social phenomena into different categories is based on reality or is simply a "conventional fiction." He points out that while terms such as "cockney," "Brooklynese," "Yorkshire accent," and "Black dialect" are commonly employed as if they were definite and independent discrete language varieties with apparently well-defined features, it is true, however, that the phenomenon is much more complex than it is presented. He mentions the example of "American English" and "Canadian English" being talked about as if they are clearly different language systems, but that it is not easy to find one linguistic feature that is common to all varieties of "Canadian English" but does not exist in any of the varieties of "American English" (Trudgill, 1987:16-17).

Trudgill's second insight relates to the phenomenon that in almost every linguistic group, variations in social context (in its wider sense to encompass the receiver, the topic, the medium and the set of circumstances of a message) engenders the employment of various styles that are fairly discrete. He edifies us that this is the case in Javanese, unlike in English, whereby phonological styles are obviously not distinct and comprising simply moderate percentages. He states that these styles can be described via their variations in grammar, lexicon (such as address-forms and pronouns), and orthoepy. He proposes that these styles can be considered as varieties within dialects, as they occur within a person's speech due to the attributes of the social milieu and still reveal properties of the speaker's regional and social backdrop (Trudgill, 1987:111-112).

The third insight offered by Trudgill concerns the role of *Discreteness*, with its opposite, *Continuity*, vis-à-vis cases in which speakers either move along a scale of formality of style based on a particular circumstance or switch from a particular style of a dialect to another. He says that the locational varieties or styles are evidently sub-varieties of a single geographical and/or social dialect.

In other cases, he adds, locational switching must occur *between* various dialects. He also mentions that these are cases for which we must take into serious consideration the notions of *Discreteness* and *Continuity* in that the difference between the two forms of switching is more one of degree instead of kind. In these cases, he further points out, one dialect will be used in formal circumstances and the other in informal ones (Trudgill, 1987:112). He provides the following example to support his observations:

> Native speakers of Lowland Scots dialects may switch, in relatively formal situations, to Standard English (spoken with Scots accent, of course). It is legitimate to regard this situation as rather different from that of an English speaker from England who simply switches styles. In the first place, the difference between the linguistic varieties involved in the switching is much greater. Secondly, as in the case of levels in Javanese, co-occurrence restrictions are involved: it is not usual to use Lowland Scots forms when speaking Standard English, or vice versa. And thirdly, whereas other English speakers switch from one variety of their vernacular to another, Scots dialect speakers switch from their own vernacular to that of others—a linguistic variety that they normally learn only at school (Trudgill, 1987:112).

Trudgill also sees a similarity between Lowland Scots children and lower-class African American children in the United States in terms of the educational problems both groups encounter in their respective societies. In order to function in the school environment, the two groups must engage in dialect-switching instead of style-switching (Trudgill, 1987:112).

Trudgill's fourth insight has to do with how two linguistic varieties in a *diglossic* situation (i.e. one in which switching is done on a much bigger and more institutionalized scale) can be considered by speakers to be discrete, albeit it is normally not completely the case in practice and encompasses a "standardized *high* variety and a *low* variety which is also standardized but may be subject to geographical differentiation" (Trudgill, 1987:114). He further asserts that the two varieties, whose specialization function he considers to be the most significant in the diglossic set of circumstances, are openly acknowledged in a community and the labels employed to characterize them are generally known and employed (Trudgill, 1987:114).

The fifth and final insight provided by Trudgill pertains to the disagreement among linguists about the terminology used to designate what a *language* (a system of communication used by a particular country or community) is and what a *dialect* (a particular form of a language which is peculiar to a specific region or social group) is. This disagreement, he says, is the outcome of the

Discreteness and *Continuity* problem. He cites the case of Jamaica whose language is referred to by some linguists as *Jamaican English* while others call it *Jamaican Creole.* As he describes the situation in Jamaica, Standard English, as the official language, is used there by people at the top of the social scale, some educated Jamaicans, and the British. The English-based Creole is used by people at the other end of the social scale, especially the peasants in remote rural areas. The two varieties are not mutually intelligible. The linguistic differences between the two varieties are so significant that it is not farfetched to assert that while Jamaican Creole is related to English, it is also quite distinct from it (Trudgill, 1987:185).

Ideational Function

In his deliberation on *Ideational Function* (i.e. language serving as a process of communicating and explaining occurrences of the world) Geoffrey N. Leech starts off by tendering the following postulate: "In general, grammar is describable in terms of discrete and determinate categories; pragmatics is describable in terms of continuous and indeterminate values" (Leech, 1983:70). He then says that while the postulate needs to be qualified in light of the many studies that have challenged the assumptions of *Discreteness* and *Determinacy* (especially those of transformational grammar), he, nonetheless, believes that it is not farfetched to assert that grammar is a much more orderly phenomenon compared to pragmatics. Among the challenges are those studies focusing on (a) scalar phenomena in grammar, (b) gradience with an emphasis on fuzziness and grammatical categories, and (c) sociolinguistic variable rules and implicative scales employed to explain sociolectal variation in quantitative and scalar designates (Leech, 1983:70-71).

Next, Leech argues that there is no need to quickly reject the aforementioned postulate based solely on the evidence provided by studies that have challenged it because he does not believe that (a) the evidence actually invalidates the *categorical* perspective of grammar; and (b) even the evidence on the fuzziness of grammatical categories does not completely render obsolete the fact that in grammar the primary procedures are discrete, or categorical, and the secondary procedures are continuous, or non-categorical (Leech, 1983:71).

Also, according to Leech, scalar values have the potential of intruding on grammar via three modes. The first mode involves instances of syntagmatic continuity between two parts such as the movements between consonants and vowels in continuous speech. The second mode has to do with instances of paradigmatic fuzziness between two categories such as in segmental phonology whereby the articulatory difference between [p] and [b] is gradual instead of absolute. The third mode concerns the verity that a rule which refers to one or more linguistic classes may work indeterminately to yield sentences which are grammatical only to a certain extent. Should all the three continuousness appear at the same time, Leech reasons, there would be serious opaqueness about the operation of grammar. He states, however, that the grammatical system can withstand a significant amount of fuzziness and not defaulting in its work as a primarily discrete system. He supports his proposition with an analogy of the digital computer in which some tolerances are built into it to accommodate oscillations of voltage (Leech, 1983:71).

In addition, Leech tells us that the mental idea of a category is imperative for understanding the categorical essence of grammar. He also says that category can be considered as a fuzzy set of descriptive attributes, such as a vehicle being described in terms of "mobility, having wheels, locomotion, movement along the ground, carrying passengers, propelled by an engine," and so on. Nonetheless, he observes, disagreements could emerge about whether a sample of something that is distant from the original can be placed in a category or not, such as whether a child's scooter or a helicopter belongs in the category of vehicles (Leech, 1983:71-72).

Thereafter, Leech points out that in general, the notion of a prototypic category is obviously germane to cognitive and perceptual systems, in addition to linguistic and logical ideas such as ability, agency, and causation. He adds that syntactic and phonological categories can also entertain the notion of prototypic category; or, as he colorfully puts it, "some verbs are more 'verby' than others, some consonants are 'consonantal' than others, etc." (Leech, 1983:72). What is important to pay attention to, he suggests, is that categories call for two stages of experiential processing: (1) identifying the individual attributes which places an element into a category and (2) identifying the category as an entirety. For the first stage, there would be a significant amount of indeterminacy, since it is not always clear that a particular sample belongs to the original. For the second stage, it is abundantly clear that two categories would be different, since, for instance, there exists a difference between consonants and vowels, even though the real demarcations between them may not be obvious in some cases (Leech, 1983:72).

Furthermore, as Leech proffers, a notion that has been assumed on the phonic level is that the continuousness and overlap of phonetic boundaries have not stopped phonologists from developing postulates and working with discrete parts and incongruities. This foundation, he says, has undergirded the emic-etic contrast or the form-substance contrast. He notes that while there has been little agreement at the semantic level, the proposition that language involves experience in relation to categories, and that the real benchmarks for determining belonging in these categories is the domain of the psychological theory of sense is accepted. He mentions that at the syntactic level, categories are also involved, as can be seen in word classes like adjective and verb and clause features like subject and object. He admits that there is less categorical indeterminacy in syntax compared to the "outer" levels of coding, which are phonology and semantics. For this phenomenon, he adds, a logical theory of the ideational function of language would be a good fit, if it were treated as a categorical system that moderates between two spheres of mostly non-discrete facts—i.e. (1) the referential facts from models dealing with experienced reality and (2) the phonetic facts of speech (Leech, 1983:72).

Finally, Leech ends his deliberation by rephrasing his postulate that "In general, grammar is describable in terms of discrete and determinate categories; pragmatics is describable in terms of continuous and indeterminate values" (Leech, 1983:70) in the following simpler way: "Grammar is essentially categorical; pragmatics is essentially non-categorical" (Leech, 1983:73). He is, nonetheless, quick to make it quite clear that the preceding postulate does not mean that he IS oblivious to some indeterminacy in the function of grammatical rules. This is because, according to him, while we can continue to uphold the notion that grammatical rules work in an all-or-nothing fashion, we can still say that the categories which define situations for the rules are fuzzy to some extent. He adds that this in-between perspective is in line with two considerations. One consideration is that gradients do occur among grammatical categories. And the other consideration is that linguists have in the past and will in the future attempt to make a good guess about the nature of language while keeping the syntactic and pragmatic propositions of *Discreteness* (Leech, 1983:72-73).

Expression Elements

Expression Elements, according to Lyons, constitute important principles based on the difference between form and substance. To illustrate this point, he uses the difference between /*b* [b], *d* [d], *d*/ and /*p* [p], *t* [t], *i*/, which is kept going in the spoken English language by the contrast between [b] and [p]. The fact that it is this specific contrast in sound, instead of some other, is immaterial to the structure of the English language. In addition, the contrast between [b] and [p] is not absolute, but relative. In other words, what is referred to as "the sound [b]" or "the sound [p]" constitutes ranges of sound; and, in reality, there is no point at which the "[b] range" commences and the "[p] range" terminates, or the other way around. Based on the phonetic perspective, the contrast between [b] and [p] is a result of continuous difference (Lyons, 1971:67-68).

Nonetheless, argues Lyons, the contrast between the expression elements /*b* [b], *d* [d], *d*/ and /*p* [p], *t* [t], *i*/ is absolute in that the words *bet* and *pet* and all other words separated by the presence of either /*b* [b], *d* [d], *d*/ or /*p* [p], *t* [t], *i*/ do not gradually change into one another in the spoken English language as [b] is gradually incorporating a number of different elements towards [p]. While there could come a point at which it is impossible to tell whether *bet* or *pet* was intended, no word exists in the English language that is essentially midway between the *bet* and *pet* vis-à-vis its grammatical function or denotation and is recognized by a sound midway between [b] and [p]. Based on these facts, Lyons posits that the expression-plane of language is comprised of discrete elements, which are recognized in physical substance by ranges of sound within which exists the possibility of significant contrast (Lyons, 1971:68).

Therefore, according to Lyons, since the expression-units should not be conflated with one another in their meaningful actualization as sounds, a "safety-margin" must be established between the range of sounds that consummate one expression-unit and the range of sounds that consummate another. He then alludes to the fact that some differences may not disappear over time or may not be kept on all words by every speaker of a language. This reality, he says, could be explained by taking into consideration the fact that the relevant differences are at the lower end of the "threshold" of significance in the quantity of utterances of the language the speakers preserve as vital. This reality, Lyons also insists, is not a claim that the contrast between some pairs of expression-units is relative, instead of absolute (Lyons, 1971:68).

Structural Dialectology

In his essay dealing with *Discreteness* and *Continuity* in structural dialectology (i.e. the study of how a system by which a dialect retains or attains synchronic recognition evolves due to the inception of novel linguistic features), Edward Stankiewicz explores the careful choosing of phonemic yardsticks as the premise for a structural illustration of dialects. The data he used for the analysis were gleaned from scholarly works on Slavic dialectology. Given the state of dialectology at the time he did the study (the late 1950s), as a structural linguist, he had to rely on historical or phonetic descriptions which he had to elucidate phonemically. As such, the results generated from his analysis were tentative and tendered basically as an illustration of the common problems and of a methodology that could be utilized in other areas of linguistic inquiry (Stankiewicz, 1957:44-45).

The results engendered by Stankiewicz's investigation led him to make eight observations. First, he observes that the categorization of dialects in terms of their phonemic listings should be regarded as the initial procedure for a comprehensive account of linguistic domains. In order to augment the specifications for categorization, it is imperative to cull from the examination of phonemic distribution and other levels of analysis. Conventional dialectology, with its limitations, had been almost exclusively developed by utilizing phonetic principles, implying that dialects are most evidently distinguished phonemically. Whereas few data on grammar and syntax have been collected, the large amount of phonetic data that has been collected by linguists over many generations is effortlessly accessible and remains undisturbed by the all-pervading fervor for historical constructions. The available data show some consistency over wide areas. Since lexical variations which are the most obvious to the non-linguist mirror ethnic, geographic and social situations of various domains, they cannot be systematically utilized in a structural model. Thus, since they are not systematic and are essentially cultural, lexical indicators are only employed to denote social dialects such as argot, professional language, and slang. By virtue of being the most advanced branch of contemporary linguistics, phonology appears to be the most logical sub-discipline for determining the benchmarks for classifying dialects (Stankiewicz, 1957:45-46).

Second, Stankiewicz observes that some linguists insist that without the utilization of extra-linguistic principles, dialect investigations are deficient and even unfeasible. Although it is indisputable that one can use the results derived by using other approaches and from other disciplines to correlate them with those from linguistic dialectology, the former are not of primary significance to the linguist. In fact, new variables are likely to be introduced in the dialects inquiry that could obscure and be in conflict with results generated by utilizing linguistic approaches when extra-linguistic benchmarks, which are not at all better defined than the linguistic benchmarks, are employed (Stankiewicz, 1957:46).

Also, Stankiewicz informs us that in order to address or transcend the shortcomings that emerged in conventional dialectology, some German historical dialectologists developed the concept *Sprachlandschaft.* The idea was geared toward the replacement of the concept of discrete dialects with that of "dynamic cultural zones" (*geistiges Kraftfeld*) and characterizing the entire problem of dialect thresholds as a "static, rigid concept." By doing so, these German historical dialectologists moved the focus to the inquiry of geographical domains as webs of cultural, economic, linguistic and political transformation. Remaining at the core of this approach, however, is the philosophical issue of "causes" (Stankiewicz, 1957:46).

In addition, as Stankiewicz indicates, intelligibility tests also have numerous limitations for determining dialect differentiation and unity. To comprehend speakers of different speech communities, a linguist must therefore be talented and also trained in code-switching, which is much more closely connected to lexical similarity compared to structural connections among linguistic systems. Depending solely on the views of native speakers for gauging similarities and dissimilarities among dialects is also specious in that cultural and social points of view do distort a speaker's assessment of his/her own speech tendencies. This phenomenon usually explains linguistic conventionalism or the willing embrace of other linguistic structures. Stankiewicz contends, however, that by depending mostly on internal, linguistic standards, structural dialectology can supplant conventional dialect investigations (Stankiewicz, 1957:47).

Third, Stankiewicz observes that the major challenge confronting dialectology is that of categorizing different systems, spread in geographical closeness, within a superstructure of resemblances or of intersection. Consequently, dialects are perceived as varieties of what is referred to in conventional phraseology as a "language," an "overall pattern," or a "diasystem." It is therefore quite clear that this superstructure, which entails attributes common to every dialect, varies from a standard language (or languages), which is occasionally established on one prestige dialect, or which is a culturally established agreed-upon procedure of different dialects, nor can it be described according to extra-linguistic divisions such as the administrative, the cultural, or the political (Stankiewicz, 1957:47-48).

Furthermore, according to Stankiewicz, dialectology faces a serious challenge of denoting the connection between *Discreteness* and *Continuity* when differentiating dialects which are almost identical to reciprocal lucidity among speakers of adjacent regions, but vary enough to be so noticed by most speakers. He therefore calls for dialectologists to assert the soundness of the dividing lines which they set up and to adjust the degree of identicalness among the various regional varieties, if they are not to wither their analyses into a classificatory framework which contrasts systems or subsystems of most distant or significantly individuated languages simply in terms of dissimilarity or peculiarity (Stankiewicz, 1957:48).

Fourth, Stankiewicz observes that should it employ different precepts of what comprises similarity and dissimilarity structural dialectology would encounter different challenges and generate results that are in variance with those of traditional dialectology. This is because traditional dialectology had mostly utilized "atomistic" and historical approaches which involved choosing particular attributes of an antiquated historical phase and delineating the area in which the original words had been dispersed. Accordingly, since it was obvious in the traditional dialectological method which concurred with the reorganized historical unity, the stated unity of different dialects was never challenged. Dialects were perceived as purely subdivisions of an originally single protolanguage within the neo-grammarian idea of a sequence of historical "splits" (Stankiewicz, 1957:48).

In addition, Stankiewicz notes that in designating the speech of some regions to one or another "language," the availability of a standard language, political borders, and other external barometers has played a significant and pivotal function. There is a lack of consistency, however, in the choice of their categorization yardsticks, which differ from one language to another and from one analyst to another, even though dialectologists have mostly been inclined to regard older attributes as more significant compared to the contemporary ones. Indeed, historical dialectologists were quite familiar with the arbitrary tendency that hampered their method, a recognition that led to the well-known debate on whether dialect borders actually exist (Stankiewicz, 1957:48-50).

Fifth, Stankiewicz observes that structural dialectologists are well equipped to provide answers to the many challenges with which traditional dialectologists had to contend. This is due to the fact that the stress in the study of dialects moved to the synchronic stratum; and, by choosing their benchmarks in relation to appropriateness and using a systematic framework, structural dialectologists acquired impartial tools for discovering the variations among linguistic systems. On the phoneme level, if their phonemic listings vary from one another, dialects are regarded as being different. Consequently, the phonemic maps sketched by structural dialectologists differ from those sketched by traditional dialectologists: the latter featuring "primary" and "secondary" (or phonetic and distributional) attributes, and the former featuring a similitude of attributes and distinctly designated regions (Stankiewicz, 1957:51-52).

Sixth, Stankiewicz observes that while the major challenge for historical dialectologists is how to construct *Discreteness*, the major challenge for structural dialectologists is how to describe the continuity and similarity among discrete local systems and subsuming them into higher kinds. In their attempts to discover "overall patterns," some analysts have utilized a graphic device in which "cover symbols" represent different phonemes or series of phonemes. The result is that imprecisely implemented transcriptional symbols, which either over-differentiate or under-differentiate phonemic differences, can be extended not only to account for the dialects of a particular language, but a number of highly individuated languages as well. Hence, the delineated model is not a concept of attributes that are consistent in a number of interconnected dialects; it is a false idea (Stankiewicz, 1957:52-53).

Moreover, according to Stankiewicz, the extreme and purely formal structural linguists may challenge the contrasting of segments of systems on the basis that models are marked and distinctive designs. Nonetheless, this radical perspective is in practice supported by only a few linguists. Also quite limited is the practice of contrasting systems simply on the basis of their listings (Stankeiwicz, 1957:53).

Seventh, Stankiewicz observes that a gauge of the resemblance among local systems with slightly different phonetic listings is based on their use of the identical terminal phonetic parts. What's more, when contrasting the general nucleus of phonemes which exist in the dialects of larger linguistic regions, one may find that they have some phonemes that do not exist in other linguistic regions. Thus, it is not farfetched to suggest that in regions where the same distinctive attributes exist, the amalgamating likelihoods may be significantly disparate (Stankiewicz, 1957:54-55).

In addition, as Stankiewicz proposes, an analyst could gain another measure of dialectical similitude if s/he were to carefully search for the coexistence of various styles within phonemic structures. He provides a salient reason for this proposition that is worth restating here verbatim as follows:

> As structural transformations in time are attenuated by the existence of certain phonemes which are unstable in the pattern, variation in space is graduated by zones of transition in which some phonemic distinctions are not obligatory but optional, and as such endowed with social connotations. In interdialectal communication, speakers may resort to the phonemes which are "potential" in their system, but obligatory in the system of their interlocutors. As it is, these alternants are often known to them from the speech of older members of their own speech community. Thus they find projected in space the facts which pertain to the dimension of time (Stankiewicz, 1957:55-56).

Also, Stankiewicz apprises us that should analysts utilize a wider model of phonetic relevance, they would be more likely to discover *Continuity*; whereas if they employ the strict phonetic model, they would be more likely to discover *Discreteness*. Thus, some linguists have dismissed the phonetics-phonemics dualism because the attributes that are equipped with non-distinctive operations are also a component of the linguistic code and are therefore relevant. And, since the typical attributes which are imperative for the working of a sound system are well known by the speakers of a language, it would be illusory to believe that they are taught to ignore all of them in speech-sounds (Stankiewicz, 1957:56). I must add here that the *linguistic code* refers to the organization of a narrative of reality which privileges cognitive processes like creativity, imagination and symbolization that inform both literary aptitude and linguistic comprehension.

Furthermore, according to Stankiewicz, the examination of the non-distinctive characteristics, particularly with regard to the distinctive ones, shows many loci of structural intersection that were set up during a typical historical evolution and contact among dialects. Therefore, rather than masking the uniformity among phonetic contents, structural dialectologists can accentuate the various operational uses across different dialect regions. They can also use maps that show discrete phonemic systems and plot over them the phonically uniform regions (Stankiewicz, 1957:56-57).

Eighth, Stankiewicz observes that when analysts move from the synchronic to the diachronic strand, they discover that the inquiry into dialects from the perspective of *Discreteness* and *Continuity* entails additional reverberations. Being one of the major sources of information on historical processes, dialectology shows strong relations between diachronic and synchronic processes. The various uses of dialectal attributes across linguistic regions mirror the phenomena referred to in diachronic linguistics as "dephonemicization," "phonemenization," and "rephonemicization," similar to the instance of "stylistic alternants" that are related to both the "vertical" time and the "horizontal" space axes (Stankiewicz, 1957:57-58). Brief definitions of the key concepts here are in order. Beginning with the root-word, *phonemenization*, it refers to the grouping of phonemes, or explaining or transcribing sounds with reference to phonemes, or the progression of an allophone to the level of a phoneme. Thus, dephonemicization refers to the ungrouping of phonemes, etc.; and rephonemicization is the regrouping of phonemes, etc.

In addition, as Stankiewicz enlightens us, in the instance of dialectal contact whereby speakers are recognized for assuming control of their neighbors' distinctive and non-distinctive features, there is a weak correlation between zonal diffusion and temporal progression. In light of the work done by diachronic linguists concerning the effect of external factors on the history of language by stressing the import of inherent alteration, dialectologists can take into account issues pertaining to dialect interference. Such an approach that compels dialectologists to take note of the issues they share in common with diachronic linguists is more profitable than "seeking in space for a mirror-image of time." Since the configuration of culture languages or other instruments of intercommunication is usually attained through the mixing of different systems, the "mixed" dialects may extend partly over the "core" dialects at certain stages or on some stratum of their phonetic structures, albeit they do not have the general network of attributes that would allow us to observe the "core" dialects as different phonemic genera (Stankiewicz, 1957:58).

Moreover, Stankiewicz alludes to the verity that a further examination of the parallelism between dialectology and diachronic linguistics reveals that the issue of placing dialects into a general system is connected to the historical linguists' pursuit of periodization: i.e. to institute a watershed (or watersheds) in the history of a language when the systemic change ("Umbruch") of a system was far more keen and important than transformations in its listing which left the phonemic relations unimpaired. It is at these points of historical evolution that some dialects gain or lose their unique attributes or amalgamations of attributes which separate them from other associated dialects. Consequently, as Stankiewicz optimistically concludes, "As diachronic linguistics was able to avoid the dead-end of early structuralism by abandoning its concept of isolated systems juxtaposed statically in time, dialectology can resist the danger of atomization by studying dialectal varieties in their two-fold aspect of discreteness and continuity" (Stankiewicz, 1957:59).

Phonetic Elements and Formal Linguistics

The discussion in this section is based on Robert F. Port's article on the *Discreteness* of phonetic elements and formal linguistics which appears in the *Journal of Phonetics*. The essay is a response to a letter written by A. Manaster Ramer to the editor of the same periodical and published in a 1996 edition. Port divides his exploration into eight sections. Presented here are synopses of the main points in those sections in sequential order for the sake of clarity.

The first section is the introduction. It addresses the two questions raised by Ramer in his letter about (1) whether phonetic elements are discretely different from one another and (2) whether attentive hearer can effortlessly recognize speech sounds. In addressing these questions, Port begins by informing us that Ramer's concern was about how results generated on the *Discreteness* of phonetic elements in the 1980s and early 1990s may erode the methodological and theoretical foundation of phonology. Port agrees with Ramer that "incomplete neutralization" is simply a modest alternative of many other well-known occurrences from experimental phonetics and phonology, and this occurrence is quite problematic. A significant number of widely attested occurrences provide compelling proof that the assortment of sounds of human speech (or the general listing of phonetic classes) cannot be conjectured to belong to discretely separate kinds that can be accurately distinguished by a first language learner and by an edified linguist or phonetician. Indeed, speech sounds are often different, and are invariably believed to be discretely different from one another. While traditional linguistic phoneticians believe this to be the case, Port insists that it is not based on the existing evidence derived from short sound bites—for example, "Are these two instances of the same sound or different ones?"—for which one cannot presume that there will inevitably be right or unchanging responses. If this were the case, he argues, then an analyst cannot employ phonetics to explicate *Discreteness*; as such, another method is needed to explain the existence of *Discreteness* (Port, 1996:491-492).

Next, Port tells us that a large number of phonologists choose to behave as if an analyst could self-assuredly presume that the procedure of alert hearing augmented by phonetic representation done by a trained expert would yield a depiction of "all possible linguistically relevant aspects of a sample of speech." Based on the discrete symbolic account, the phonologist would then assemble different data systems "from phonological components to syllabic trees to metrical grids." Should the *Discreteness* and the expected affirmative perceptibility of these elements not exist, the phonologist and the child language-learner would seem to have no starting point for their tasks. This situation, suggests Port, raises the questions about (a) "how phonology could account for symbolic structures like language-specific consonant and vowel types, allophones, stress markings, constraint hierarchies, etc."; and (b) "from where these complex cognitive objects could possibly come" (Port, 1996:4912).

Also, as Port apprises us, there exists a group of features representing the largest collection of linguistically germane sound strains in languages across the globe. These interpretable entities can be consistently and easily discerned by every member of the human family. These fixed entities serve as elements for children's language-learning and limit the viable speech sounds of language so much so that they inherit a type of "two-dimensional matrix of feature values vs. discrete time." Consequently, to Port's astonishment, the data systems of phonological illustrations "inherit their discreteness from the phonetic atoms of which they are constructed" (Port, 1996:492).

In addition, Port asserts that the shortcoming of phonetic *Discreteness* is somewhat well-known to phoneticians. These linguists, he says, have long noticed that phonetic objects vary discretely only in archetypal situations (Port, 1996:493).

The second section addresses the following question: "What is 'incomplete neutralization'?" For starters, Port demonstrates that Ramer's concern can be expounded with the German stems *Bund* ("brotherhood") and *bunt* ("colorful") as follows:

> If the stems have a suffix, as in *Bunde* and *bunte*, then they are pronounced [*bundə*] and [*buntə*]. The orthographic difference between the D and T is clearly appropriate, and illustrates one of the most striking and profound properties of human language: the use of discretely distinct sound types for `spelling' lexical items. These phonological types, the D and T, recur in many different vocabulary items and often may appear in a range of syllable positions. However, when the /d/ or /t/ occurs as the last segment in a syllable (that is, for

these stems when there is no suffix), the contrast is often said to be 'neutralized' since both words are pronounced as approximately [bunt] with a final [t] (Port, 1996:493).

Port continues:

> The D/T distinction appears to be lost since both pronunciations (for these words and for many other pairs of voiced and voiceless obstruents in Standard German) merge into a phonetic [t] (or the cognate voiceless stop, fricative or affricate). The traditional linguistic description of this phenomenon is to say that a 'neutralization process' has applied that turned the [+voice] feature of the D into the [-voice] feature of the T. In the more recent optimality theory formulation of such data, German would be said to rank certain simultaneous constraints differently than, say, English. But the simultaneous constraints are still stated in terms of standard phonetic features (Port, 1996:493).

The flabbergasting aspect of the preceding exploratory outcome, according to Port, concerns the fact that whether an analyst engages in a production exploration or a perception exploration utilizing a good number of tokens of a minimal pair, s/he will discover that the two group of [t]-like sounds vary relatively in every case. Obviously, contrary results are invariably unchallenging to get, since not every speaker will show a difference in the frameworks selected for the exploration. Also, the analyst must have many tokens to analyze, as the impact will be on "noisy variables" such as consonant and vowel duration. While production explorations have found many statistically significant differences, not much has been found by perception explorations (Port, 1996:493).

The third section deals with linguistic phonetics as the standard theory. On this aspect, Port attempts to encapsulate the various perspectives on the theory as proffered by many linguists via four themes. Theme one concerns "competence vs. performance." Port recounts that the issue arose from Noam Chomsky's contradistinction between the two concepts: i.e. *Competence* and *Performance.* Apropos systems, the distinction is usually explained to distinguish between *Competence* as the area of *discrete variables*—i.e. symbols—and symbol structure as they are arranged differently and organized in *discrete time.* The time for processing requires "discrete jumps" among system states when a procedure is applied. Antithetically, *Performance* involves the area of *continuous variables* changing in *real/continuous time.* Included among these continuous variables are procedures connected to audition, motor control, and speech perception. In essence, whereas *Performance* concerns the physical aspect, *Competence* pertains to the cognitive or mental aspects that function as postulates that are identical to those of digital computation, logic, and mathematical proof. For linguists, language, and perhaps every mental aspect, is best comprehended in relation to discrete time and symbols (Port, 1996:495-496).

Theme two is about "phonetic theory as alphabet." In terms of the central point of motor, says Port, discrete phonetic notions can be perceived as supplying a ubiquitous alphabet comprising ascendant arrays. As such, this is the only phenomenon that is needed to indicate speech production and that in essence could be regulated by a language's grammar. Besides these elements, nothing else could be regulated by the grammar of the language. In terms of the central point of perception, it is believed that audition is accompanied by a "phoneticizer" that displays "categorical perception" and glosses continuous acoustic occurrences into discrete emblematic representations. Consequently, these two input and output instruments account for phonetic *Discreteness*; and, incomplete neutralization (in addition to numerous other data) makes the performance of the two instruments problematic (Port, 1996:497).

Theme three pertains to sound acquisition by adults versus children. According to Port, it is common knowledge that adults encounter significant difficulty when they try to obtain many sound elements in a non-native language. Thus, it is not unreasonable to say that adults may be unable to differentiate the new sounds of languages since they in some way no longer have much of their natural capacity to identify the sound variations of human speech. Evidently, for most lay speaker-hearers, only the phonetic sets employed in their native languages are customarily utilizable by adults in their perception of speech (Port, 1996:497-498). Absent from Port's discussion of the theme is the children's part.

Theme four relates to the "formal theories of linguistics." Beginning with symbolic theories, Port states that they basically propose that there exist a collection of emphatically observable linguistic symbols. Put differently, the grammar, or formal system, possesses discrete symbolic objects. The symbols have to be discrete if they are to be unerringly identifiable. They also have to exhibit solidity over unlimited periods. Next, formal models utilize the symbolic features to ensure that their procedures work and for data structures to remain reliably intact. Such reliability can be seen, for instance, in the performance of a computer program, which is a formal model, because of the engineering of the chip (Port, 1996:498).

In addition, Port talks about the standard theory of linguistic phonetics that had remained the same since the 1960s. The major shortcoming of this theory is its competence component which entails a precise divergence between *Symbols* and the expeditious *Transitions* among states. This representation of processing as being perfect (or better than reality) within competence treats time as if it does not exist. Real time, however, does not exist in a *Symbol* or in a *Transition*; instead, real time proceeds relentless and can always be observed over either much shorter or much longer graduated ranges of time (Port, 1996:499).

The fourth section entails "empirical arguments for phonetic *Discreteness.*" For this aspect, Port broaches the three major arguments he says come up frequently in discussions about why the phonetic *Discreteness* proposition and the *Identifiability* proposition are both legitimate. The first is the "limited perceptual resolution argument," which states that humans have restrictions on the quantity of variances practicable within any dimension, thereby substantiating discrete groupings. For a number of phonologists, only common sense is needed to ascertain the *Discreteness* of speech perception. Their belief is based on the premise that the variations of *Discreteness* an analyst can determine are finite at any particular level. Thus, they also opine that because of this finite limit, only a particular number of vowel differences are possible. Port is not convinced, however, about the validity of this notion, since discrete levels do not necessarily result from limits on resolution (Port, 1996:500).

The second argument Port identifies concerns "the categorical perception," which corroborates to and underpins the idea that discrete groupings with marked perceptual dividing lines are inherent to speech recognition. He informs us that this phenomenon has been common knowledge among linguists since the 1950s and that if an analyst were to introduce modifications to speech stimuli from one condition to another across an intricate continuous sequence of phonetic categories, the discernments of respondents will "jump" somewhat discretely. Nonetheless, Port maintains that this is not credible proof of a discrete "phoneticizer" in speech perception because the standard theory of linguistic phonetics calls for *all* phonetic differences to be discrete, whereas the "categorical perception effect" manifests more strongly for particular subcategories of sound differences, such as the point of articulation and voicing, than for others, such as vowels (Port, 1996:500-502).

The third argument delineated by Port pertains to the "quantal theory of speech," which postulates that for many phonetic groupings, *Discreteness* is an innate phenomenon. As he informs us, claims have been made that the acoustic reaction of the human vocal tract acts extremely nonlinearly for some alterations in articulation. As such, differences in articulatory precision will have little effect on acoustics given these nonlinearities for speech sounds in some positions. Thus, the "quantal properties" of speech indicate that some vowels are to a certain extent insensitive acoustically to variations in articulation. But as Port points out, these claims of the standard theory have not been backed by tenable empirical evidence demonstrating that naturally lodged in human perception is a discrete, consistently detectible inventory of sounds (Port, 1996:500, 502).

The fifth section tackles the issue concerning "what is lost" as "consequences for phonology." On this matter, as Port concedes, his is an incredulous perspective of the data collection procedure most frequently utilized in phonological inquiries. The procedure involves linguists and phonetic experts developing phonetic transcriptions that are said to have in symbolic representation all of the "linguistically relevant aspects" in speech recognition and making. These transcriptions are then utilized as the primary data for developing frameworks of data systems for a language. For Port, this approach has the following four shortcomings, since incomplete neutralization is only one of the many different aspects of human speech examined via experiments: (1) "many kinds of *subtle context effects* appear in speech articulation or speech acoustics"; (2) "many of these context effects are *manifested in the time domain*"; (3) "most of these effects are *specific to particular languages*, and cannot be phonetic 'universals'"; and (4) "essentially any of these subtle variables, under appropriate conditions, can be *employed by listeners in speech perception*" (Port, 1996:503).

The sixth section addresses the question about whether "phoneticians can fail where native speakers succeed." Using the German language as an example, Port ponders why phoneticians could mishear something that the native speakers of the language could hear quite effortlessly. The answer, he says, hinges on the fact that linguists and phoneticians are occupationally preoccupied with initiating and implementing a collection of phonetic elements that are tenable prospects for "the universal phonetic space." As such, they are not interested in results which reveal that there is not a statistically significant difference (say 70% accuracy) between categorical differences, a reflection of a "professionally motivated response bias." As such, with their interest geared toward large and consistent differences, these linguists and phoneticians disregard small dialect variations, speaker peculiarities, simple token-to-token differences, etc. (Port, 1996:506).

The seventh section titled "Toward phonological morphogenesis" is a call by Port to remove *Discreteness* from the analytical domain of phonology. His rationale is that since phonetics cannot elucidate the *Discreteness* of phonology, phoneticians are not in a position to provide an explanation for the clear-cut phonological aspects that they examine by relying on phonology. This is because languages appear to display "discretely different places of articulation, manners, vowel heights, etc." in the elements employed to record vocabulary items. Since the discipline of Linguistics is not distinctive in attempts to describe the nature of constructed phenomena, of characteristics that are constant under transformation, Port suggests that approaches in other disciplines be used to examine the biology of such features be considered (Port, 1996:507).

The eighth and final section entails the two conclusions drawn by Port based on what he determines to be the inferences of the incomplete neutralization principles. The first conclusion is that the absence of unabridged neutralization of syllable-final voicing in the German language is a reality. The second conclusion is that Ramer was correct in pointing out that the impact of "neutralization" is indistinguishable from cases of the "near-merger" of a difference, the restrained dilution of differences under "weakening" and "fast speech," and so on (Port, 1996:508).

Terminology Theory

One gleans from Pius ten Hacken's essay on Terminology Theory (i.e. the study of terms and how they are employed) that a question with which the theory has had to grapple is how "criteria terms" are differentiated from "general language words." He seeks to transcend the practical view on the question to offer linguistic implications based on the general linguistic framework of Ray Jackendoff (a United States linguist born in 1945) in his book titled *Foundations of Language: Brain, Meaning, Grammar, Evolution* (2002). To do so, Hacken divides his analysis into four major parts and a conclusion.

In the first part, Hacken proffers the axiom that terms are by necessity connected to a specific sphere. Thus, since there are numerous types of spheres, he suggests that the optimal method for delineating relevant features of the concept *term* would be to begin by taking into account a number of various spheres. For his analysis, Hacken chooses spheres encompassing examples from the legal, scientific and technical disciplines (Hacken, 2008:979).

To support his aforementioned axiom, Hacken discusses four examples pertaining to "domains and expressions." The first example is mathematical linguistics, which he says is a scientific discipline which yields text that is replete with terms. He provides the following theorems as proof: (a) "The set of context-free languages is not closed under complementation" and (b) "A context-free language is a language that can be generated by a context-free grammar" (Hacken, 2008:980). The second example is traffic law, which he describes as different from scientific disciplines and typical of the legal discipline. He states the following as evidence: (a) "Mopeds are allowed on dual-carriage ways but not on motorways," (b) "A moped is a motorcycle that has the following features...," and (c) "I saw three mopeds on my way home today" (Hacken, 2008:980). The third example is piano manufacturing, which he denotes as a specialized manufacturing sphere in which a significant number of terms are employed to talk about parts, procedures, and implements. The following is the statement he tenders as proof: "In an upright piano, the soft pedal moves the hammer rail closer to the strings" (Hacken, 2008:981). The fourth example is general experience, which he characterizes as different from the preceding example because it provides a general language sentence. For evidence, he provides this sentence: "Traveling by train is so much more pleasant than traveling by plane" (Hacken, 2008:981).

The second part explores the "classification criteria." As a means to arrange the preceding information according to an organized system, Hacken suggests three criteria. The first criterion deals with the "two criteria for termhood" which, according to Hacken, can be employed to differentiate statements in general language, and they also trigger various categorizations. The second criterion is "specialization" which, as he points out, is usually treated as the major characteristic differentiating terms from words, albeit its implementation can be quite complicated for a couple of reasons. To start with, specialization is a feature of impressionistic discernment. What one speaker perceives as "specialized" another speaker could perceive as "general." Next, specialization is measured in terms of degree. As such, asserts Hacken, the utilization of specialization as a criterion to decide what a *term* is makes the concept itself "prototype-based." The third criterion relates to "terms and prototypes" whose basic notion, Hacken says, is that classifications have systems. He cites the example of *bird* to explain the point as follows: "The prototype of a bird is often taken to be a robin. Eagles and falcons are less prototypical because they are big and carnivorous. This does not mean that they are less than 100% bird, but people are less likely to think of them when asked to name a bird" (Hacken, 2008:982-983).

In the third part, Hacken examines "term formation." He begins by explaining that from a parallel structural point of view, a word is a portion of a fact in the cerebral vocabulary of a speaker that allows him/her to connect a specific form—or sound image, a collection of syntactic restrictions regulating the span of viable places of the word, and a conceptual delineation. A new word is conceived in the cerebral vocabulary of a speaker. An existing cerebral vocabulary and outside precipitators influence the new word. The precipitators are of two types: (1) "general perceptual," such as visually discerning a new type of appliance; and (2) "linguistic," such as hearing the word someone else employs to call the appliance (Hacken, 2008:984).

Hacken continues by explicating that terms occupy a different position. He characterizes a term as having a concept, a domain, and a form; and contrasting with a word, a term is not an innate material thing and its central position of actuality is not the brain of a particular speaker. Conceptually, a term is ascertained by the collection of requisite and adequate states defining its borders. This collection of states is given priority over any speaker's capability (Hacken, 2008:984).

Also, Hacken states that there are two courses to the formation of a term. One course is the *onomasiological*, which he says is the "classical one" that is characteristic of the advanced sciences and typified in mathematical linguistics. This course involves a concept being described first and then assigned a name thereafter. The other course is the *semasiological*, whose formation he describes as beginning with "the form (word or multiple-word expression)" and the formation procedure entails ascertaining the borders of the concept for the form (Hacken, 2008:984).

The fourth part is about "medical terminology." About this aspect, Hacken tells us that the field of medicine is a copious spring of different kinds of terminology and the typical form of an applied science. An applied science is different from an empirical science in that the former is interested in not only defining and explaining occurrences, but also solving them. Thus, in the medical field, specialized vocabulary and terms in the restricted definition take place alongside one another, albeit their dispersal is not arbitrary (Hacken, 2008:985).

In his conclusion, Hacken asserts that a cogent idea of *term* cannot hinge upon the lack of correspondence between the level of the specialization of a discipline's discourse and the fact of a concept with precise borders, or both of the two characteristics. Accordingly, he suggests that we "call *specialized vocabulary* the expressions that signal a high degree of specialization of the discourse and to reserve *term* for expressions that designate a precisely defined concept" (Hacken, 2008:986). Consequently, he adds that "*terms* in this narrow sense cannot be described by referring to linguistic competence and language use alone. They also exist as non-mental, abstract entities" Hacken, 2008:986).

Continuous Mathematics

In her investigation of the *Continuity* and *Discreteness* of meaning representation, Fabienne Venant employs a geometric method within the structure of Continuous Mathematics. Her central thesis is that the rendering of denotations is a key idea in Natural Language Processing, especially for removing uncertainty of meaning from word sense or deriving information. Her focus is on the relevance of employing continuous models in semantics instead of the continuous or discrete tendency of vocabulary denotation. She uses *Continuity* as a mechanism to gain access to and maneuver vocabulary denotation, with the supposition that *Continuity* and *Discreteness* are features of descriptions of the distinctive attributes of theories upon occurrences instead of being features of occurrences. She also believes that *Continuity* and *Discreteness* can work together (Venant, 2010:3314). In order to demonstrate all this, she divides her exploration into four sections. Her treatise is summarized in the paragraphs that follow.

First, Venant broaches the "linguistic issues." Here, she addresses the question about why analysts should employ continuous semantic depiction. This is due to the fact that as soon as analysts attend to semantic depiction, they are faced with the issue of classification. A major vocabulary denotation (for example, as is exists in a dictionary) can be perceived as being equal to a category of fundamental semantic components. Based on the theoretical model utilized, these components can be exact semantic modulations, semantic attributes, contextual-usage denotations of terms, rational depictions, etc. Thus, a recurring issue that confronts analysts is how to denote benchmarks in order to decide whether a particular element is to be subsumed in a particular category (Venant, 2010:3314).

Second, Venant presents her "theoretical framework," which treats each linguistic element of an expression as connected to a semantic location where its various denotations are arranged in terms of semantic propinquity. The other elements of the expression denote a prospective activity that makes it possible for analysts to identify the area of the semantic location analogous to the denotation of the element examined within the expression. Consequently, the exact denotation of a "polysemous" element in a particular sentence is shaped by a dynamical structure on its semantic location. This "dynamic," according to Venant, "is parametrized by the other units present in the sentence," and "each attractor of the dynamics corresponds to s possible semantic value" (Venant, 2010:3314).

Third, Venant discusses "building semantic spaces" automatically by using an algorithm. She describes the appropriate algorithm as one that is suitable for examining a graph. The algorithm must show the structure of the vocabulary shaped by the graph in order for an automated system to get to the information contained in it. This technique allows analysts to establish local areas that indicate the semantics of a particular element, in addition to a universal area that indicates a vocabulary model in its entirety. The local areas can then be employed to calculate the denotation of the element being investigated when followed by a particular term in a sentence (Venant, 2010:3315).

Fourth and finally, Venant talks about "graduality and granularity." While she does not define them explicitly, one can glean from her discussion that *graduality* refers to delineating expressions that are capable of fulfilling a collection of benchmarks that can be treated as the prototypical features of a closely similar category, and *granularity* means making changes to certain elements so that they can be linked with other semantic features (Venant,

2010:3314, 3317). Venant states that her model is based on a continuous depiction of denotation, more specifically on the utilization of "cliques as the minimum semantic unit" for depicting meaning. Such a depiction, she argues, is quite useful when reckoning for faint semantic attributes or engaging in a deep analysis of the semantics of a particular vocabulary element. Using "co-occurrence data," she derives a self-activating technique to denote probable functions and to remove the uncertainty of the meaning of an item based on a preceding or subsequent word. She also maintains that such a task calls for denoting exactly what a contextual meaning is. As she further insists, analysts must not utilize "a clique as a semantic tag, because semantic distinctions between cliques are too subtle" and, therefore, analysts must "group cliques in order to obtain a more macroscopic tag set" (Venant, 2010:3316). Moreover, she recommends that analysts can marshal "cliques in sets corresponding more or less to dictionary definitions" (Venant, 2010:3316-3317).

African Linguistic Evidence

Presented in this section and in the chronological order in which they were published are summaries of findings delineated by some analysts who have studied manifestations of *Discreteness* in African languages. To commence with, Marina Savini examines earlier suggestions that had been made about the non-discreteness of the morphological derivations in the Afrikaans language of South Africa. She posits that the problem with the earlier perspectives "lies not with the criteria used to classify affixes, but rather with a fundamental assumption which all the level-ordering linguists and category distinction linguists made": i.e. "morphological categories are discrete entities" (Savini, 1991:206). Her systematic analysis of the Afrikaans data she had yielded the following findings: (a) there is a "tendency for certain derivational affixes to become more word-like and for the structures in which they occur to become more compound-like"; and (b) the tendency can be accounted for by assuming that there exists "a continuum of expression types and...the non-discreteness of expression types, but not...discrete morphological categories" (Savini, 1991:332).

Next, an alarm bell is rung by Carlos Gussenhoven when he states that should "floating" tones of the type that manifest in African languages to mark clause structure and which are placed into the tonal string at the beginning of the sub-clause are not to be included in intonational tones, then we must restrict these tones to those that either associate with metrically strong places ("central tones") or are used to mark off prosodic constituents ("boundary tones" or "peripheral tones"). To illustrate the point, he mentions the instances of some Nilo-Saharan languages that have breathy voice and laryngealized voice (Gussenhoven, 1999).

Also, having investigated tones in the Yoruba language numerously and over many years, Eunice Fajobi informs us that these linguistic features exhibit *Discreteness* when they are considered in isolation from their milieu for the method and practice of instruction, especially as an academic subject or theoretical concept. Thus, she recommends that the Yoruba tongue should be classified as a "register tone language" (Fajobi, 2005).

In addition, among the many linguistic features broached by Paul Kerswill in his inquiry about whether youth languages in Africa and Europe reflect "linguistic subversion or emerging vernaculars" is *Discreteness*, albeit very briefly treated. He raises the question about whether the sociolinguistic function of the youth language is discrete—i.e. completely distinct from other ways of speaking—or is in a continuum. His finding is that the function is "linguistically distinct from the multiethnolect (a language variety birthed from the contributions of several ethnic groups) even if the multiethnolect contains features from it, example the vowels" (Kerswill, 2010:30, 35).

Thereafter, we learn from Mirriam Nosiphiwo Ganiso that the sign language of South Africa is a completely developed linguistic system with all of the characteristics of a natural human language and, thus, *Discreteness*. According to her, this is the result of face-to-face interaction being a key requirement for communication in sign language (Ganiso, 2012).

Furthermore, as Sara Marzagora tells us, if we are to develop novel and useful evaluative viewpoints from a systematic analysis of words or phrases whose function is to link linguistic units together, cross-effects, and reciprocities among African language customs, between African literature written in African languages and African literature written in European languages, and between oral and written literature, we must "revise the most essentializing forms of nativism and nationalism" which are anchored on "a politicized idea of discrete and fixed identities" (Marzagora, 2015:4). This politicized idea, she asserts, prolongs a perception of Afrophone literatures as "narrow, peripheral, and constricted" (Marzagora, 2015:4).

Also critical of the most acute variants of deconstructionism, Marzagora disagrees with these philosophies' advocacy for a "de-essentialization" of African literary attributes, but at the same time rebuffs total "relativization." She therefore calls for a theoretical framework that would "conceive African literature as composed of multi-layered and intersecting literary systems—a perspective able to 'mediate between identity-as-essence and identity-as-construct, between the global connectivity and localized networks, between cosmopolitan dialogues and culturally defined concerns; between the continental, the national, and the ethnic'" (Marzagora, 2015:4). In addition, not buying into the euphoria over a novel "planetary system of literary circulation," she advocates for literary customs to have a "right to disconnectivity, or, in other words, the right to be disconnected from literary globalization 'without being labeled pre-modern or being considered unable, because of internal weaknesses or external constraints, to project themselves onto the global market'" (Marzagora, 2015:4).

Moreover, Mark Dingemanse alludes to the fact that since they were first identified in the languages of West Africa and systematically analyzed in the 1850s, ideophones, which are also referred to as "expressives" and "mimetics" and include onomatopoeia, have been characterized as "discrete linguistic systems" that serve a number of quite useful purposes because of their composite character as "gradient vocal gestures." These purposes include (a) facilitating the delineation and description of lexical categories of evocative sensory words; (b) using their African linguistics postulates to augment those of general linguistics and typology in order to explain similar ideophones across the globe; (c) their treatment by some Eurocentric linguists as an insignificant or peripheral concept notwithstanding, but still being employed by other linguists to influence "many areas of linguistics, from theories to phonological features to typologies of manner and motion, and from sound symbolism to sensory language"; and (d) making it possible to "reframe typological questions, reconsider the role of language ideology in linguistic scholarship, and rethink the margins of language" (Dingemanse, 2018:1).

The preceding presentation of Dingemanse's exposition on African ideophones would be incomplete without mentioning the definition of Bantu ideophones provided by Marion M. Johnson and quoted by Dingemanse as follows: "lexical items in which semantic representation of perceptual qualities are mapped directly onto phonological strings, without passing through the morphological component of the grammar" (Johnson, 1976:260; quoted by Dingemanse, 2018:12). The essence of Johnson's denotation hinges upon the fact that it allows us to envision nouns and verbs as, in the words of Dingemanse, "lexically discrete: i.e. change a single phoneme and you get an unrelated word or not a word at all" and, thus, "the iconicity and gradience of ideophones posed problems for models of language that rely on a strict dichotomy between meaning and phonological forms" (Dingemanse, 2018:12).

African Mathematical Evidence

First, we decipher the attribute of *Discreteness* from Ron Eglash's discussion of the spontaneous order in the board game commonly referred to as "*Owari*" and played across Africa in many varieties called "*ayo, bao, giuthi, lela, mancala, omweso, owari, tei,* and *songo,*" etc. He describes the way the game is played as follows:

> The game is played by scooping pebble or seed counters from one cup, and placing one of those counters into each cup, starting with the cup to the right of the scoop. The goal is to have the last counter land in a cup that has only one or two counters already in it, which allows the player to capture these counters. In the Ghanaian game of owari, players are known for utilizing a series of moves they call a "marching group." ...if the number of counters in a series of cups each decrease by one (e.g., 4-3-2-1), the entire pattern can be replicated with a right-shift by scooping from the largest cup, and...if the pattern is left uninterrupted it can propagate in this way as far as needed for a winning move (Eglash, 1999:101).

By way of illustration, Eglash posits the 13-iteration marching group strategy that follows (Eglash, 1999:106):

> 3421→523→43111→4222→3331→442→5311→42211→3322→433→4411→45 52→3321→4321

Eglash goes on to point out that in spite of its simplicity, the concept of "self-replicating pattern" underlies some sophisticated mathematical concepts. He adds that John von Neumann (1903—1957 AD), a major contributor to the advancement of the modern digital computer, also introduced the mathematical theory of "self-organizing systems," which was first based on self-reproducing physical robots (Eglash, 1999:101).

Second, Michael DeGiorgio, Mattias Jakobsson, and Noah A. Rosenberg provide us a notion of *Discreteness* when they mathematically modeled an "ancestral allele frequency spectrum" in their exploration of "migration outward Africa." They explain their procedure as follows:

> In computing the slope of the ancestral allele frequency spectrum as a function of allele frequency, we...evade a discreteness effect in which some frequency bins contain more markers than others due to more discrete frequencies being assigned to those specific bins. We used ancestral and derived allele assignments...for 407,001 autosomal markers in the data...For each population, for each locus, we computed ancestral allele frequency using 1,000 random draws from the empirical allele frequency distribution. Loci were binned by ancestral frequency into 20 bins, representing [0/20, 1/20), [1/20, 2/20), ..., [19/20, 20/20]. For each population, bin counts were normalized by the total number of loci. The slope of the linear regression of the normalized frequency spectrum on ancestral allele

> frequency was then computed using bins centered at 9/40 to 31/40 of frequencies 1/5 to 4/5 (DeGiorgio et al., 2009:16062).

In order to address the analogous computation from their simulation, DeGiorgio and his colleagues had to add another step to their procedure. They describe this step as follows:

> We used $n + 1$ bins, so that if a locus had the ancestral allele occurring i times, then the count for bin i/n was incremented. For a given simulation, for each population, counts were normalized by the number of segregating sites in that simulation. For each population, the slope of the linear regression of normalized frequency spectrum on ancestral allele frequency was computed using bins 10/50 through 40/50. In each population, we calculated an average slope across simulations, weighting the value for a simulation by the proportion of segregating sites observed in that simulated dataset relative to the total number of segregating sites from all simulated datasets (DeGiorgio et al., 2009:16062).

The preceding evidence prompted DeGiorgio and his colleagues to conclude that the increasing distance from Africa is its inclusion of an accumulating impact of genetic movement as humans settled in various parts of the globe (DeGiorgio et al., 2009:16057).

Third, in her study of "the interaction of mathematics teachers with learners' mathematical errors" in South Africa, Johanna Cornelia (Hanlie) Verwey broaches the notion of *Discreteness* when she discusses the case of a teacher who was instructing her students on how to generate mathematical graphs. She explains that the teacher used a prescribed mathematics textbook, followed the order of presentation in it, and basically referred to it. On a few occasions, the teacher would pose other problems to the learners that are not in the textbook. When the lesson is on financial mathematics, the teacher would bring to class currencies from a number of other countries to class, so that the learners can get a sense of the real value of the South African Rand. This approach also introduces a sense of contextualization. The teacher also introduced a context for direct proportion by referring to cellular phone contracts and purchasing groceries in bulk when teaching direct and inverse proportion. When it comes to illustrating inverse proportion using an example of a rectangle with a constant area, the teacher would stress the "algorithmic approach of constant product, inverse proportion; constant quotient, direct proportion." The result, Verwey's words, is that "while learners were graphically representing realistic situations such as paying postage for mail delivery of parcels, no reference to the realistic situation regarding discreteness of the graph or the constraints in terms of the domain or the range were raised" (Verwey, 2010:159).

Fourth, we detect an idea of *Discreteness* in One-Soon Her and Yun-Ru Chen's mathematical modeling of the "unification of numeral classifiers and plural markers" in which they mention that some African languages entail "cross-linguistic evidence of double-headedness, e.g., appositional compounds…the double-headed VV (verb verb) structures in Classical Chinese, and a double-headed verbal phrase structure for serial verb constructions" (Her and Chen, 2013:43). The upshot of this finding is that it contributes quality to the amalgamation of an obligatory numerical classifier (C) and a mandatory plural marking (PM) that ends with the implication of a general lexical count/mass distinction. Consequently, "C/PM's identical mathematical function [is] a multiplicand with the value *l* and the cognitive function of highlighting the discreteness, or the count nature, of the noun" (Her and Chen, 2013:44).

Conclusion

In the Ancient Kemetian language, *Discreteness* can be constituted in terms of putting together small units to make larger communicative units. Thus, while the words *khem, sbak, sbeq, sbek*, and *shnåu* all denote "small, little, slight, and tiny," it is the word *netch* that refers to "making something small." The words *au, āa* and *āai* all mean "to make something large" (for these translations, see Budge, 1978).

As we also gather from Henry G. Fischer, *Discreteness* or *Separateness* is one of the most conspicuous features of hieroglyphic writing. This is because, first, the phonetic and ideographic constituents are stated as individual signs. Second, when the elemental segments of the signs amalgamate contrasting features or perspectives, they are mostly joined together in a suitable manner, even when what results does not always depict the essence of what is depicted. Fischer adds: "In inscriptions dating to the classical period of the Old Kingdom their exceptions to the first rule bear out the second, but in other periods the second rule was also infringed to a greater or lesser degree" (Fischer, 1978:5).

Furthermore, as Andréas Stauder points out, *Discreteness* is manifest in the stronger properties of the Kemetian hieroglyphic system. This is due to the fact that the system "is not based on strokes or lines, but on intensive forms." This means that the forms can differ in "degrees of visual resolution and modes of formal actualization, or even morph," albeit minimally at the particular levels emphasized by the system (Stauder, 2018:367).

Chapter 10

Displacement

The work of Anouschka Bergman, Kathleen Curie Hall and Sharon Miriam Ross (2007) led me to state the following in my essay on the nexus between African languages and mathematics: "What is termed *Displacement* refers to the fact that speakers of a language can communicate about things, actions, ideas, etc. that are not spatially present while they are communicating." For instance, I add, "people can talk about Allah (SWT—*Subuhana Wa Ta'ala*, meaning in Arabic 'All Glory belongs to God; God is The Most Exalted and The Most High') without actually seeing Allah (SWT). They can talk about Prophet Muhammad (PBUH—i.e. Peace Be Upon Him) or the Day of Judgment" (Bangura, 2014:28).

The immediate question here is as follows: How has *Displacement* been conceptualized by linguists? It appears that more attention has been paid to *Displacement* as a concept compared to the other aspects of the phenomenon as evident in the next section. The perspectives of a handful of scholars who have written about the conceptual nature of the phenomenon are presented, although relatively briefly, in the rest of this section. While these scholars offer somewhat similar conceptualizations, they differ in terms of the aspects they highlight.

Starting with David Crystal, he states that *Displacement* is regarded as a denoting characteristic of human language viewed as different from other communication systems. It is a characteristic whereby language can be employed to mention milieus that are distant from the extant circumstance of a speaker: e.g., references to past time (like yesterday) and future time (like tomorrow). Contrastingly, the denotation of the signals of other animals is confined to the context in which they are employed. He cites the example of a hunger cry which denotes that hunger exists right "now," not in the past time or the future time (Crystal, 1992:107).

Next, for William O'Grady, Michael Dobrovolsky and Mark Aronoff, *Displacement* denotes speakers/writers and hearers/readers being capable of making references to occurrences that are spatially and temporally distant. They mention the following examples of *Displacement* or the lack thereof among other animal species: (a) *Displacement* is exhibited in bee dancing; (b) bird calls and songs have not been found to demonstrate *Displacement*; and (c) while baboons occasionally produce warning and combat noises long after a hostile battle, *Displacement* has not been identified as the reason, which leads O'Grady and his colleagues to suggest that the behavior could be due to a protracted unwinding of the baboons' emotions (O'Grady et al., 1989:407).

Also, while John Lyons likewise says that *Displacement* is a feature of human language utilized to talk about objects and occurrences that are spatially and temporally distant, he adds that the utilization is similarly distant from the discourse itself. He tracks down the concept to the behaviorist formulation of language and meaning, whereby the foremost task of words and utterances is to allude to attributes of the existent circumstances, with which they are linked as impetuses to answers, and the co-relationship between linguistic utterances with material things and occurrences external to the "situation-of-utterance" is a subordinate matter (Lyons, 1977:80). Lyons further notes that *Displacement* aids and relies on the advancement in the employment of the demonstratives (Lyons, 1977:657): i.e. words like *this* and *that* utilized to specify which things with distinct and independent existence are being talked about and to distinguish them from others.

In addition, Ronald B. Adler and Neil Towne provide a quite interesting conceptualization of *Displacement.* According to these scholars, the phenomenon transpires when we express truculent or antagonistic sentiments against people (or even things) that we view as less menacing than the individual or individuals who frightened us initially. For instance, people may be angry at their employers or supervisors, but instead of risk being terminated from their jobs if they were to direct their angers at the employers or supervisors, they would instead displace the anger by shouting at the people living with them at home. As Adler and Towne also succinctly put it, "*Displacement* almost always lets us preserve (at least to ourselves) the image that *we're patient*—that we're in control and can't be pushed around by forces beyond our control. The very act of displacing proves this is a lie, of course—but one the displacer fails to recognize" (Adler and Towne, 1993:339).

Furthermore, concerning *Displacement,* Asya Pereltsvaig writes with certitude that in contrast with other animals, humans can allude to bygone time, time to come, and alternative positions or points in space (Pereltsvaig, 2011:1). She deliberates over the essence of this design feature of human language as follows:

> I can tell you about my trip to Paris 20 years ago or about the future trip there that I am planning. We can even talk about non-existing and imaginary entities and locales, such as angels and fairies, Santa Claus and Superman, Heaven and Hell. Neither Greek mythology nor Viking legends would have been possible without this property of language. Nor would fiction be possible at all. Even the concept of 'would' is made possible only by displacement: it signifies the state of affairs in a

> different world, similar to our real world but also different from it in important respects. For example, the sentence *If I were a Rothschild, I would donate all my money to charity* means that in a slightly different version of the world (one where I am very rich), I give money to charity (Pereltsvaig, 2011:1).

Pereltsvaig adds that while the aforementioned aspects can be communicated via human language, they are impossible to communicate through other animals' communication systems because they are contrived solely to express the "here and now." They cannot be employed successfully to be causally connected to occurrences that are distant temporally and spatially. Pereltsvaig cites the example of a dog which can neither bark about the marrow bone that was given to it yesterday nor the cute little puppy residing north of town. In fact, she asserts, the dog is incapable of "managing" the notion of "north of town," and that there is neither a "doggie heaven" nor a "doggie hell" (Pereltsvaig, 2011:1).

Pereltsvaig concludes by challenging the notion advanced by some linguists that bee dancing demonstrates *Displacement* (see my earlier discussion of this claim by O'Grady and his colleagues; see also Nordquist [2018] for the same claim made by other linguists). She articulates her position on the issue so well that I will let her words speak for themselves as follows: "honeybees, however sophisticated their communication system is, cannot 'talk' about that delicious rose bush north of the town that they visited last weekend. They cannot do displacement in time and as it turns out they cannot do displacement in space either. All they can 'share' is geocentric position of the flowers with respect to the hive. However, remote the flowers are, they are indicated in relation to "here'" (Pereltsvaig, 2011:2).

Having provided the preceding setting, the rest of this chapter is geared toward demonstrating the nexus between the linguistic expositions on *Displacement* and the African linguistic and mathematical evidence pertaining to the language feature. These aspects are broached sequentially in the following three sections for the sake of coherence. But before doing all this, however, it would be fruitful to end this section with some lessons on *Displacement* in African thought as articulated by Jonathan O. Chimakonam in his very recent book titled *Ezumezu: A System of Logic for African Philosophy and Studies* (2019).

First, Chimakonam suggests what he calls "okwucentrism" (an African philosophy that advocates a "mode of conversations in an arena of unities") as a needed substitute, albeit not a *Displacement* narrative, for logocentricism as a way to salvage the latter (i.e. logocentricism) in a postmodernist mind mapping in what he refers to as "logomania." Second, he proposes that the ascription of validity to African content in order to liberate it from European "colonial" content would be a marginal *Displacement*, since not every strand of modern European intellectual agglomeration is colonial. Third, he proffers that since colonization was systematic, the *Displacement* of colonial content in Africa's educational systems and curricula need to be systematic as well. Fourth and finally, he tenders the notion of *Displacement* as a way to usher in African background logic to be used as a replacement for Western logic (Chimakonam, 2019:5, 17, 184, 186).

Linguistic Expositions on Displacement

As mentioned earlier, it appears that more observations have been made about *Displacement* as a concept than about its other features. This assertion is borne out by the fact that I could identify only the following seven items when I was conducting the research for this topic: (1) displaced modifiers and predication, (2) displaced speech, (3) displaced entities, (4) non-verbal systems, (5) phenomena included by Charles Francis Hockett, (6) evolution of language and cognition, and (7) communication. For lucidity, the perspectives of various scholars on these items are presented in the sequential order they are listed here.

Displaced Modifiers and Predication

We learn from Yuen Ren Chao that in the case of modifiers (i.e. words, especially adjectives or nouns, used attributively to restrict or add to the sense of a head noun—e.g., *good* and *family* in *a good family house*), *Displacement* can occasionally cause self-evident incongruities. He adduces as evidence the phrase "fills a much-needed void," whereby it is the term "need" that is a necessity and the term "needed" is to be eschewed. He adds that not so much customary, although very much used, is "displaced or dangling predication." He mentions as an example the road sign "Turn Right When Clear," to which a driver might query "You mean don't turn right when drunk?" For this statement, he posits, a "different logical subject" is presupposed after the word *when* (Chao, 1968:63).

So, a germane question that arises here is the following: Why do modifiers behave the way they do vis-à-vis *Displacement*? As Chao informs us, much more conventional, but less manifest, is the inconsistency between formal compositions and another uncommunicated composition which appears to be the real message. Representative of these inconsistencies are cases of displaced modifiers (Chao, 1968:63). Chao goes on to provide many examples that are worth restating here verbatim as follows:

> A sign in an American college cafeteria says "Bus [i.e. take back to the counter] your own dishes," where grammatically *your own* modifies *dishes*, but really it is *your own bussing* that is urged. Similarly, the library sign *Shelve your own books* does not imply that the library has an unusually acquisitive department of acquisition. When a girl "knows her men," there is a sense that the men are hers, but the expression usually emphasizes her knowing of the men. Likewise, when the Red Queen objected to Alice's saying "I have lost my way" because all the ways belonged to the Queen, it was not so much the way as the losing of the way that was Alice's (Chao, 1968:63).

Indeed, *Displacement* of a modifier can sometimes be quite paradoxical: i.e. certain statements or propositions that, despite sound (or apparently sound) reasoning from acceptable premises, lead to conclusions that seem senseless, logically unacceptable, or self-contradictory.

Displaced Speech

Displaced Speech, indicates Francis P. Dinneen, is characterized as people often uttering words like *peanut* when no *peanut* is present. Such a state of the utterer's body, which is an impenetrable challenge for a behaviorist analysis, is due to the fact that it is an important component of every situation, which is to account for *Displaced Speech*. Without this aspect, we would not be able to systematically analyze linguistic features such as irony, lying, narrative fiction, and poetry. An analyst's satisfaction with such an answer, as compared to the traditional mentalist explanation of the pre-linguistic procedures with those the mentalists refer to as "thought" and maintain that the traditional qualitative difference between "thought" and "material processes" is not one of kind but of complexity, can delineate differences. The mentalist perspective is therefore essentially pontifical, since it is based on a proposition that cannot be empirically validated (Dinneen, 1967:249-250).

Also, as Dinneen apprises us, it is disagreeable to assert that many forms of speech do not seem to achieve anything that is similarly difficult as that of *Displaced Speech* for two reasons. The first reason is that the wants, aims, cognition, etc. of utterers do play a part in the context of situation. The second reason is that a significant difference exists between literary and familiar utilizations of language in that the literary utilization is for a wider context, a particular task, and then to be forgotten; thus, it is closely connected with, and only completely comprehendible in, each context of situation (Dinneen, 1967:301). I must add here that the familiar utilization of language is the often encountered, experienced, or common use and, therefore, what is uttered is easily recognized.

In addition, in cases of *Displaced Speech* such as the one for the *peanut* I stated earlier, we learn from Geoffrey Sampson that for an utterer to use a word for which s/he is at that particular time of the utterance not being motivated by that word is for the utterer to reciprocate to some hidden internal impulses of that word which was connected at some point to his/her past with the impulses of the word. Sampson therefore argues that "anyone who is not prejudiced by attachment to the fallacious version of behaviorism will recognize that 'displaced speech' is the norm and cases like the Jack-in-Jill story are exceptional" (Sampson, 1980:68). To support his reasoning, Sampson mentions the example of an evening prattle around an English sitting-room fire, which he says "might concern anything from traditional Chinese architecture to the economics of the motor industry; if it restricts itself to the contents of the sitting room, it is likely to be a desperately dull conversation" (Sampson, 1980:68).

Displaced Entities

Concerning *Displaced Entities*, Gillian Brown and George Yule start off by telling us that these entities are almost solely recognized by examining the expressions of the form *the + property + noun*. Also mostly utilized to recognize *Displaced* (and *Current Entities*) is the simple definite noun phrase, *the + noun*. While *pronouns* do occur as expressions distinguishing *Current Entities*, they do not occur as expressions distinguishing *Displaced Entities*. Thus, the norms of distribution can differ from one genre to another. In addition, when implemented on copious data, the indexing analysis can be quite taxing (Brown and Yule, 1983:174).

Next, Brown and Yule make the observation that the indexing can turn out to be deficient as soon as an analyst investigates data involving change-of-state predicates. *Displaced Entities* are therefore frequently alluded to by complete lexical definite noun phrases. In limited data, an analyst could be able to determine which utterances initiate new entities into the discourse and which utterances are employed to initiate entities which have been expressed earlier. When this is done, the forms of those utterances become ready for linguistic analysis. While the description of the form of an utterance is of major interest to the sentence grammarian, the discourse analyst is more interested in analyzing the forms in terms of the context in which they are employed (Brown and Yule, 1983:176).

In addition, according to Brown and Yule, immediately after data become abundant and encompass change-of-state predicates, the certain difference between first and second mention becomes indistinct. In essence, an analyst cannot just count through the mentions in a text and state that an earlier mentioned entity is, automatically, "given" (Brown and Yule, 1983:176).

Non-verbal Systems

On the subject of *Non-verbal Systems* in terms of *Displacement*, Lyons begins by referring us to Charles Francis Hockett's postulate that a message is displaced when the major attributes in its precursors and effects are temporally and spatially removed from its transference. Lyons then mentions Hockett's assertion that the earliest language-utterances of children, reminiscent of their *Non-verbal Systems*, are not displaced and that empirical evidence on language acquisition by children has been provided by researchers indicating that displacement happens later for children, instead of earlier, in the acquisition process. Lyons adds that the determination of the displacement of any non-human signaling-system or any non-human *Non-verbal System* hinges upon the characterization of "spatiotemporal remoteness and reference" (Lyons, 1977:80-81).

Also, the idea of displacement in an utterance such as *The partition of the African continent was a process of imperialism that encompassed invasion, occupation, colonization and annexation of the continent's territory by European powers between 1881 and World War I in 1914*, a bird's alarm call in response to a predator that is still a distance away, or the dance of the honey bee in reference to a faraway source of nectar would belong to the same purview, according to Lyons. Nonetheless, as he mentions, contemporary research on chimpanzees has shown that these animals have the capacity to make and translate signals to things with distinct and independent existence that are far away from these animals' close surroundings (Lyons, 1977:81).

Phenomena Included by Charles Francis Hockett

For the discussion here, I cull again from Lyons' observation. He informs us that the phenomena included by Hockett in his disquisition on *Displacement* are those that are subsumed within the ambit of *Prevarication*: i.e. the likelihood of employing a semiotic system to mislead or give false or inaccurate information. Lyons mentions Hockett's example about a child crying from hunger only when s/he is actually hungry, whereas the linguistic signal *I am hungry* can be said whether or not the person saying it is actually hungry. Based on this example, Lyons asserts that *Displacement* seems apprehensively like a behaviorist concept that is injected in order to give an account of a typical attribute of language-behavior which is more precisely designated as "freedom from stimulus-control." This notion of "freedom from stimulus-control," he maintains, is doubtlessly a prerequisite of untruthfulness and deceit in the stringent meanings of these words; thus, the notion constitutes a prerequisite of *Prevarication*. He therefore also argues that *Prevarication* should not be treated as an aspect of the semiotic system, but instead as an aspect of those employing the system (Lyons, 1977:84).

So, the immediate question that should be addressed here is the following: Why is *Prevarication* essential in terms of *Displacement*? As Lyons points out, many linguistic researchers have determined that *Prevarication* is the ingredient which, with *Reflexivity* (i.e. the capacity of a natural language to refer to, or describe, itself), most evidently differentiates human language as a semiotic system from all other signaling systems. He notes, however, that some linguistic researchers have indicated that *Prevarication* appears in behaviors of some birds and non-human mammals. This purported evidence leads him to ask about whether one species cajoling another into a trap by sending out heartening signals constitutes deceit or not, and whether the behavior of a bird which fakes an injury and entices a predator to chase it away from the nest rationalizes the utilization of the concept *Prevarication*. His response to these questions is that the very fact that these questions could even emerge is evidence that in the inquiry of non-human animal signaling, linguistic analysts must provide an outside explanation of *Prevarication* in relation to the monitored behavior of the sender and that of the receiver. He adds that, in the inquiry of human signaling, however, it would be benumbing for linguistic analysts to foist upon themselves such unwarranted and extraneous constraints and to forgo, on rectitude, any entreaty to viewpoints and aims (Lyons, 1977:83-84).

Evolution of Language and Cognition

In broaching the development of language and cognition apropos to *Displaced Reference*, Marieke Schouwstra chunks her analysis into five sections. The first section is about how the phenomenon has been addressed in the debate on the gradual development of language and the mental action or process of acquiring knowledge and understanding through thought, experience, and the senses. She starts off by stating that given that no direct proof exists concerning the advent of language is a significant point in the exchange of views on the issue. She then points out that while many researchers from various backgrounds have introduced a number of techniques for collecting indirect proofs to paint a portrait about how human language emerged, her interest is in *Displaced Reference* because of its appurtenance for many particulars of the gradual development of language (Schouwstra, 2007:1).

Next, similar to the definition of *Displacement* tendered by the other scholars mentioned earlier, Schouwstra defines *Displaced Reference* as "reference to things that are not present *here* and *now*. In other words, to objects or events that are remote in space or in time" (Schouwstra, 2007:1). Thereafter, she offers the following characteristics of *Displaced Reference*: (a) an interesting perspective of the gradual development of language since the

1960s; (b) present in human communication but appears absent in other animals' communication systems, although some bits of it might emerge in plain non-human animal call systems; and (c) exists in languages and also identified in other spheres such as the simple "lugging of a stick or a stone around" (Schouwstra, 2007:1).

The second section deals with how the comparative approach is used to investigate *Displaced Reference.* On this topic, Schouwstra begins by describing the comparative approach as one which employs methods that have been developed in the field of comparative biology to analyze language and perception by investigating other species so that inferences can be reached about human language. A major axiom of this approach is that if a behavioral or physical characteristic exists in some species that developmentally came before another, then that characteristic exists in the species that follows, except if it has not been selected. A major objective of the approach when used to investigate the gradual development of language is to determine the extent to which humans are distinctive and what the major differences between humans and other animals are (Schouwstra, 2007:2).

Also, Schouwstra tells us that while the comparative literature dealing with the evolution of language has focused only on the *communication* systems of humans and other animals, her focus is on the *cognition* of these species because it allows her to shift her focus away from *Displaced Reference,* since it is inappropriate to discuss "reference" when the matter of interest is about cognitive capabilities. Nonetheless, she says that it is possible to perceive displaced entities, a phenomenon referred to in the comparative literature as "object permanence." Researchers writing about this phenomenon have indicated that when an animal has attained "object permanence," it is conscious of the continuous existence of an object, and also when no sensory information is available on that object (Schouwstra, 2007:3).

In addition, as Schouwstra informs us, James R. Hurford (a British linguist born in 1941) has developed two types of test to determine whether an animal has gained "object permanence." The first is the *Visible Displacement* test, whereby "an object is visibly (to the animal subject) moved behind an obstruction, and the animal is tested to see whether it searches for the object behind the obstruction" (quoted by Schouwstra, 2007:3). The second is the *Invisible Displacement* test, whereby "an object is, in sight of the subject, placed inside a box. Then the box is moved behind an obstruction and the object is removed from the box, without the animal seeing this. Then the subject is shown the empty box, and the experimenter observes whether the animal looks behind the obstruction" (quoted by Schouwstra, 2007:3). The overall results from these tests, according to Schouwtra, revealed that every animal passed the first test and some animals passed the second test, which indicate that "in these animals, the representation of an object is controlled by knowledge (or memory), and not merely perception." She adds that "moreover, the fact that primates that are closest related to humans perform best at the displacement task suggests that displacement is a key capacity in the development of animals towards a linguistic capacity" (Schouwstra, 2007:3-4).

The third section pertains to what Schouwstra refers to as "L2-acquisition as a window on language evolution" (L2 refers to non-native, target, or second or more language(s), whereby its counterpart L1refers to native or first language). For starters, she talks about the first stage of untaught second language acquisition among adults being an instance of a limited linguistic system. The reason for this situation, she says, is the fact that at some point in the acquisition process, second language learners develop a somewhat solid genre of the target language—one, she notes, that has the following features: (a) "it appears to be organized by a small number of organizational principles"; (b) "it is to a great deal independent of both the source language and the target language of the speaker"; (c) "it is simple but rather successful for many communicative purposes"; and (d) "approximately one third of the learners fossilizes at this stage" (Schouwstra, 2007:4).

Next, Schouwstra apprises us that this first stage of second language acquisition, which is called the "Basic Variety" (BV), is designated as a self-contained linguistic system. Before this stage, learners go through a more rudimentary one referred to as the "Pre-basic Variety" which, she says contrary to what is popularly believed, harbors very little remnants of the source language. After the Basic Variety stage comes the "Post-Basic Variety" stage during which more complex grammatical forms are acquired, albeit not all learners get to this stage. As Schouwstra also informs us, even though the demarcations among the three language acquisition stages are not always quite transparent, each of them, nonetheless, has its own grammatical system (Schouwstra, 2007:4-5). She describes the grammatical systems of the three stages as follows:

> The Pre-basic Variety is based on nominal structure. The different elements of an utterance are related on the basis of pragmatic principles. Usually, utterances consist of two or three constituents: mostly nouns, but also some adjectives and adverbs; almost no verbs. The speaker, in the Pre-Basic Variety, relies heavily on context for his utterances to be interpreted correctly by the hearer. In the BV, (non-inflected) verbs appear, and the utterance is generally structured around the verb. In the Post-basic Variety, verb inflection appears and the grammar becomes more like that of the target language (Schouwstra, 2007:5).

Schouwstra adds that since the Pre-Basic Variety and the Basic Variety stages are particularly somewhat self-sufficient in relation to both the source and target languages, the two varieties can offer an opening on the gradual development of language (Schouwstra, 2007:5).

The fourth section is in reference to how the windows approach is employed to analyze *Displaced Reference*. Here, Schouwstra commences by defining the windows approach as a technique employed to investigate the gradual development of language via a linguistic framework. The units of analysis used by the approach, which comprise limited linguistic systems, include initial stages of untaught second language acquisition by adults, home sign systems, and pidgin languages. These language forms tend to emerge in abnormal sets of circumstances, since the resources that are usually available for learning a first language in normal circumstances are lacking for learning the second language. By investigating the system of the language forms that emerge in these circumstances, researchers may discover some aspects of the reasoning procedures on which language elaborates and, subsequently, add to the discourse on the evolution of language (Schouwstra, 2007:2).

Thereupon, Schouwstra culls data from the European Science Foundation (ESF) Second Language Database, which is a compilation of data on second language acquisition from various universities across Europe, including longitudinal data on adult learners of various languages, to tease out findings from studies focusing on temporality in order to learn more about *Displaced Reference*. She insists that temporality and *Displaced Reference* are not synonymous. As she explains, on the one hand, temporality is *less* compared to *Displaced Reference*, since the latter always entails allusions to entities that are spatially distant; on the other hand, temporality is *more* compared to *Displaced Reference*, since the manifestation of temporality deals with *reference* to entities that are temporally distant and says much more. Nonetheless, as she also points out, during the initial phases of second language acquisition, the manifestation of temporality is at a quite unsophisticated level that is not much different from *Displaced Reference*. She further mentions that mechanisms exist for conveying temporality in languages that are well-developed (Schouwstra, 2007:5-6).

In addition, focusing on the Pre-basic Variety acquisition phase, Schouwstra edifies us about how we would look for this stage to supply us with thought-provoking information pertaining to temporality. Unfortunately, in this phase, as she indicates, neither verbs nor even inflected verbs are utilized, and quite limited resources are available to the learner. This assessment, she also notes, hinges upon the normal linguistic anticipation that tense forms are used to communicate temporal reference. In practice, however, as she further points out, this is not the case, as learners employ intelligent techniques to communicate about past and future even when they are not acquainted with the language (Schouwstra, 2007:5-6).

Also, as Schouwstra iterates, only nouns and some adjectives and adverbs can be obtained by a learner during the Pre-basic Variety phase. Therefore, the learner during this phase can only utilize noun-like expressions or adverbials such as "calendar noun phrases" to convey temporality. To mark temporal connections, the learner can employ context. First, the learner can mimic the temporal reference started by a speaker ("or the contextual now") and thereafter initiate an original reference by placing an adverbial. Additionally, the learner utilizes reference to time span, or tense, before s/he even starts to utilize an aspectual notion on circumstances; the aspectual notion is only initiated during the Basic Variety phase. The most significant idea here is that the learner can communicate about circumstances that do not take place at the moment. Thus, *Displaced Reference* is envisaged as an essential attribute of language because it takes places in phases of language acquisition during which the linguistic means are quite limited (Schouwstra, 2007:6-7).

Furthermore, Schouwstra ends her discussion of the windows approach by positing that since *Displaced Reference* manifests itself quite early, even before enough resources are accessible to the language learner to communicate it in the habitual manner, it is quite evident that the necessity to communicate about entities that are not currently present temporally and spatially is quite urgent. Obviously, she argues, even during the initial phases of acquiring a language, it is not sufficient to just speak it and not have the means to communicate about distant entities. Therefore, she maintains, being a paramount feature of human language, *Displaced Reference* ought to have been in existence during the very early history of language—i.e. "protolanguage" (Schouwstra, 2007:7).

The fifth and final section has to do with what Schouwstra calls "an evolutionary path." She begins by restating the fact that since the capability for object permanence has played a significant role in the gradual development of language, during the phase before a language even existed, it is not farfetched to assert that a developmental path that begins in object permanence in animal intelligence yields the feature of *Displaced Reference* in human language. The windows approach buttresses the perspective that if *Displaced Reference* was already manifest in the earliest phases of the gradual development of language, then its roots must have first fully developed in intelligence. Evidence on *Displaced Reference* reveals that it is possible to link the comparative and windows approaches to paint a fuller portrait of the emergence of language. The amalgamation of the two approaches also gives credence to the notion that many levels of human intelligence had come into existence way before humans began to speak (Schouwstra, 2007:8).

Despite all of the foregoing evidence, however, Schouwstra insists that it is not conclusive. She argues that while the evidence provided by the comparative and windows approaches underpins each other's perspectives, it is nonetheless inconclusive. A question that remains unanswered, she says, concerns the circumstances that must be consummated in order for an animal to begin utilizing *Displaced Reference* to convey information. In order to provide the appropriate answer to this question, she contends, more indirect evidence is necessary. When such indirect evidence is obtainable, she adds, the chances of combining a variety of results from a variety of investigations become feasible (Schouwstra, 2007:8).

Communication

Kaori Tamura and Takashi Hashimoto in their exploration of the phenomenon of *Displacement* in communication begin by introducing what *Displacement* is. Thereafter, they discuss four facets of the phenomenon and end with a summary and a conclusion. The rest of this subsection encompasses synopses of their delineations in the order they are presented for clarity.

First, in introducing *Displacement*, Tamura and Hashimoto commence by defining the concept, similarly to the other definitions stated earlier, as an element of human language with the capacity to mention entities that are spatially or temporally distant, or both, from the point of communication. In light of this attribute, humans can talk about entities that exist in the present and the past or the future as well as imagined ones. Differing strikingly, most of the other animals' symbolic communication systems, such as the alarm calls of vervet monkeys to warn about predators, communicate only occurrences taking place at the moment but not those in the past or future (Tamura and Hashimoto, 2012:352).

Next, according to Tamura and Hashimoto, *Displacement* is an idea that denotes a contrast between humans and other animals, and comprehending the idea adds to the examination of the emergence and gradual development of human language. Nonetheless, the classical denotation of the idea includes honeybee dances as instances of *Displacement*. In light of this classical definition, the determination of which characteristic of *Displacement* is distinctive to humans becomes quite opaque. By focusing on the *Displacement* that occurs between a speaker and a hearer, there is a greater chance of comprehending the phenomenon as something distinctive to human language (Tamura and Hashimoto, 2012:352-353).

Second, Tamura and Hashimoto broach the notion of *Displacement* "on the learner side and…not based on direct experience." For starters, they mention that in the cognitive and language development disciplines, *Displaced*

Reference "or" *Displaced Speech* has been investigated for *Displacement.* (Actually, *Displaced Reference* and *Displaced Speech* have been treated separately—the former by investigators interested in cognition, especially in terms of language acquisition, and the latter by those interested in the historical evolution of language.) While the focal point of some researchers has been on the capacity of human speech to refer to entities that are not present, it has been quite a challenge to distinguish the *Displacement* that is reported to take place in bee dancing from that of humans, so long as the center of attention remains on *Displaced Reference,* which is about *Displacement* as it pertains to the speaker (Tamura and Hashimoto, 2012:353).

Also, as Tamura and Hashimoto remind us, human language is frequently employed to express goals and ideas to others so that they could be commonly shared. This aspect of communication makes it possible for a hearer to get an idea that is in the mind of a speaker. Thus, due attention must be paid to the *Displacement* that takes place on the hearer's end in order to comprehend the *Displaced References* on the speaker's end. Also required is the investigation of *Displacement* that occurs between speaker and hearer reciprocities. As such, *Displacement* on the hearer's end is considered as "displaced understanding," making it possible to treat it as communication that incorporates both *Displaced Reference* and *Displaced Understanding* (Tamura and Hashimoto, 2012:353).

In addition, Tamura and Hashimoto point out that a shortcoming of previous investigations on *Displacement* is that they do not separate issues pertaining to the phenomenon from those pertaining to memory (Tamura and Hashimoto, 2012:353). To illustrate the point, they cite the research on the pointing gestures of pre-linguistic infants and chimpanzees to absent objects conducted by Ulf Liszkpwski, Mark Schafer, Malinda Carpenter, and Michael Tomasello (2009) and state the following:

> They demonstrate that prelinguistic infants are capable of engaging in such pointing, while chimpanzees are scarcely able to do so. However, these researchers do not clearly indicate if this divergence in behavior with absent objects stems from a difference in memory or in displacement ability. If the latter, it appears that some ability associated with displacement exists separate from memory. To specify this distinction, we regard communication about absent objects based on what the speaker and/or listener have experienced in advance as "displacement based on memory" (Tamura and Hashimoto, 2012:353-354).

Third, Tamura and Hashimoto deliberate over "the classification of displacement." They start by categorizing *Displacement* in human communication in order to make the phenomenon as it pertains to humans transparent. To do so, they set up two axes: (1) a horizontal axis entailing speaker and hearer distinctions and (2) a vertical axis encompassing the distinctions of *Displacement* that hinges upon memory and that which does *not* hinge on unmediated encounter. The columns indicate the differences among three conditions: (1) where the entity being alluded to is currently present, (2) where the entity being alluded to is not currently present, and (3) where the entity being alluded to is neither currently present and nor in memory (Tamura and Hashimoto, 2012:354). The results from Tamura and Hashimoto's categorization can be summarized as follows (Tamura and Hashimoto, 2012:354-356):

a) Cases exist whereby speakers make allusions to entities they had previously encountered. Ingenuity is significantly evident in these cases, as speakers produce referents that are similar to their circumstances.

b) Cases exist whereby what are referred to exist for both the speakers and the hearers and, therefore, they are devoid of *Displacement*.

c) Cases exist whereby the states in which what are referred to exist for the hearers but not the speakers, who know these ahead of time. Evident in these cases is *Displaced Reference* but not *Displaced Understanding*, since the hearers can comprehend what the speakers are referring to given the entities that are present.

d) Some cases exist whereby *Displacement* is unique to human linguistic communication and other cases exist whereby the phenomenon is present in other animals' symbolic communication.

e) Cases exist whereby what are being referred to exist for the speakers but not for the hearers, who comprehend what the speakers refer to because of previous memory.

f) Cases exist whereby what are referred to do not exist for either the speakers or the hearers; their references and comprehensions hinge upon their memories. These cases reflect the classical denotation of *Displacement*.

The aforementioned results prompt Tamura and Hashimoto to assert that the principal element for determining *Displacement* as being distinctive to humans is displaced comprehension that does *not* hinge upon personal encounters. They add that it is beneficial to consider statements made by speakers that make it practicable for hearers to grasp *Displacement* (Tamura and Hashimoto, 2012:356).

Fourth, Tamura and Hashimoto propose an "experimental framework to investigate displacement in communication." They initiate the discussion by pointing out that in their effort to examine *Displacement* in communication (particularly *Displaced Understanding* that does not hinge on firsthand encounter), they had to contrive an experimental model. They state their goal as involving grasping the mental procedures of *Displacement* that are distinctive to humans by teasing out those features entailed in communicative *Displacement* (Tamura and Hashimoto, 2012:356).

Next, as Tamura and Hashimoto explain, humans employ natural language as a means to discern *Displacement* normally and effortlessly. Therefore, they posit, natural language is not appropriate for teasing out attributes of *Displacement* for systematic analysis. To address this limitation, they choose to employ an experimental model that limits the communication media. While the artificial symbol model is usually utilized as a communication medium, they believe that its span of expression is too restricted to allow a speaker to participate in unhampered *Displaced Reference*. As such, they examine how a speaker makes *Displaced Understanding* likely. Accordingly, they utilize drawing as a symbolic medium, since it is not a customary technique for communication and needs certain instruments to realize *Displacement* that can be analyzed (Tamura and Hshimoto, 2012:356-357).

Also, Tamura and Hashimoto extracting from the work of Nicolas Fay, Simon Garrod, John Lee, and Jon Oberlander (2003) in which they employ drawing as a medium in their experimental model to communication, they (i.e. Tamura and Hashimoto) investigate *Displacement* in communication with a focus on *Displaced Understanding* by reorganizing the model of Fay and his colleagues. Before stating how they reorganized the Fay et al. model, they first describe it as presupposing two roles: (1) the speaker plays the role of a "director" who draws objects to communicate them and (2) the hearer plays the role of a "matcher" who identifies the objects drawn by the speaker/director. The research participants are paired; they draw pictures of the same objects and alternate the two roles. The overall findings by Fay and his colleagues lead them to assert that it is through the interchange between the research participants that the change from icon to symbols takes place in graphical expressions (Tamura and Hashimoto, 2012:357).

In addition, as mentioned earlier, before utilizing the Fay et al. model, Tamura and Hashimoto had to reorganize it in two ways. First, the drawing tasks given to the research participants by Tamura and Hashimoto comprised of objects the speaker and hearer knew beforehand. In essence, these tasks are predicated on *Displacement* via memory, although Tamura and Hashimoto altered the tasks to include entities a hearer has not heard or seen ahead of time. Second, they allowed a hearer to point out without restrictions what a speaker means, as opposed to producing finite options from which to choose. They did this based on the rationale that by giving a hearer finite options, the requirement that the entity mentioned by a speaker should not be encountered before an occurrence is violated (Tamura and Hashimoto, 2012:357).

Furthermore, Tamura and Hashimoto looked at how a speaker plans his/her own utterances that would allow a hearer to extrapolate a thing with distinct and independent existence so that they could study how a speaker frames utterances that make *Displaced Understanding* a reality for a hearer. Thus, they shared the answer provided by a hearer with a speaker in order for the speaker to re-sketch the picture to make the subsequent answer as error-free as possible. They go on to describe the procedure they utilized as follows: "Subjects communicate using electronic drawing pads and computer displays in separate rooms and not through the usual communicative media, such as verbal exchange or eye contact. They can use only black lines for drawing. The use of linguistic characters, such as alphabet or kana, and the use of meaningful symbols, such as algebraic signs, is prohibited" (Tamura and Hashimoto, 2012:357-358). These modifications, Tamura and Hashimoto say, allowed them to investigate the mental procedures that are necessary for *Displacement* in communication (Tamura and Hashimoto, 2012:358).

Fifth, Tamura and Hashimoto talk about "feature extraction as a cognitive mechanism for displacement." They kick off their discussion by informing us that based on the experiment they conducted using their modified version of the Fay et al. model, they discovered that research participants made sketches in which two or more graphical representations were put together to represent entities that were not present. Teasing out a fragmentary attribute of each sketch allows for the depiction of an attribute of an entity. Tamura and Hashimoto report the following occurrence as an example of this aspect: "A subject expressed the task 'sour boot' by drawing pictures of 'lemon' and 'boot' in combination. In this example, the subject, who drew 'lemon,' inferred that 'boot' had the feature 'sour,' the same possessed by 'lemon'" (Tamura and Hashimoto, 2012:358).

Also, when a speaker identifies what a hearer has not encountered personally, the speaker may depict the attributes of the entity by using another entity that entails the same attributes. Tamura and Hashimoto refer to this phenomenon as "feature extraction" and hypothesize that "feature extraction permits displacement in communication." Along with this procedure, research participants sketched pictures that depict the occurrences where the attributes a speaker wishes to communicate are frequently encountered. Nonetheless, these occurrences also cause confusions, as hearers tend to comprehend speakers' sketches according to their own encounters. When a speaker communicates what a hearer has not encountered, this situation tends to cause confusion (Tamura and Hashimoto, 2012:358-359).

In addition, according to Tamura and Hashimoto, in order to comprehend a speaker's utterance, a hearer must conjecture the attributes the speaker seeks to communicate. This is the only way Tamura and Hashimoto are able to show the habits of human feature extraction and elucidate one of the mental procedures that allows a hearer to grasp *Displaced Understanding*. Moreover, like feature extraction, analogical inferences also appear to operate as procedures that manifest *Displacement*, in addition to feature extraction. Tamura and Hashimoto propose two kinds of analogical inferences. One of them is *cognitive analogy*, which is an analogical inference that depends on resemblance. It emerges when the hearer comprehends the entity being referred to that s/he has not encountered personally via contrasting it with an entity that possesses parallel attributes. As such, *cognitive analogy* is perceived to depend on the "cognitive common ground" of a speaker and a hearer. The other kind is *linguistic analogy*, which is based on construction rules that allow for the amalgamation of forms that stand for entities and assist in the communication of what is referred to by joining entities that are difficult to amalgamate from a cognitive perspective. All this leads Tamura and Hashimoto to proffer the hypothesis that the "three mechanisms, feature extraction, cognitive analogy, and linguistic analogy, enable displacement in human communication" (Tamura and Hasimoto, 2012:359).

Finally, Tamura and Hashimoto end their analysis by summarizing key findings and drawing an attendant conclusion. First, they state that for us to comprehend the particulars of *Displacement* that are peculiar to humans, we must analytically appraise *Displacement* in the milieu of communication and not only *Displaced Reference* or *Displaced Speech*. Second, they say that to get a better understanding of the features of *Displacement*, we must investigate them in terms of (a) the dissimilitude between speaker and hearer and (b) the characteristic of *Displacement* in terms of cognition and *not* personal encounter. Third, they maintain that in order to deduce the *Displacement* that is common to humans, we must *not* base our comprehension of the phenomenon on personal encounters. Fourth, they suggest that an appropriate experimental model is needed to systematically investigate the mental procedures that discern *Displacement* in communication. Fifth and finally, they conclude by restating the hypothesis that feature extraction acts as a procedure that allows *Displacement* in communication (Tamura and Hashimoto, 2012:360).

African Linguistic Evidence

I present in the ensuing paragraphs synopses of the findings on *Displacement* as it relates to African languages proffered by researchers who have studied the subject in the chronological order in which their works were published for the sake of simplicity. Starting with Russel Schuh, in his investigation of paradigmatic *Displacement* in West Chadic languages, he identifies the following two aspects as being the most obvious characteristics of the feature in these languages: (1) "analogical leveling of original distinctions on the model of one or more members of the paradigm" and (2) "to level variation at its original position while shifting it to some other point" (Schuh, 1980:355). To support his proposition, he mentions the following two examples: (1) Ngamo subject pronoun tone and (2) Bele object pronoun vowels. Nonetheless, he contends that two elements are needed for paradigmatic *Displacement* to manifest itself. The first element is that there should be a site to which the *Displacement* distinction is to be moved. The second element is that the displaced distinction should not disrupt the original distinctions to the point where communicative efficiency is hampered (Schuh, 1980:355-356).

Next, as we can glean from Jennifer Margulis and Peter Nowakoski's discussion of the theoretical and scholarly exchanges concerning language, the notion of *Displacement* is evident in the way African writers deal with an authoritative colonial language. They state that African writers do so in the following two-step procedure: (1) they displace the colonial or standard language by denoting it with a lower-case letter (e.g., "e" in "English"); and (2) replace the colonial or standard language with an indigenous African language "that does not have the perceived stain of being somehow sub-standard, but rather reflects a distinct cultural outlook through local usage" (Margulis and Nowakoski, 1996:5). Margulis and Nowakoski add that some African writers have named these two procedures as "abrogation" and "accommodation/appropriation." The former procedure "is a refusal of the categories of the imperial culture, its aesthetics, its illusory standard or 'correct' usage, and its assumption of a traditional and fixed meaning 'inscribed' in the words" (Margulis and Nowakoski, 1996:5). The latter procedure "is the process by which the language is made to 'bear a burden' of one's own cultural experience...Language is adopted as a tool and utilized to express widely differing cultural experiences" (Margulis and Nowakoski, 1996:5).

Also, according to Gillian Sankoff, in contemporary times, certain languages, including those in Africa, have made inroads into domains where they were not spoken in earlier times. In postcolonial contexts, in particular, speakers of local African languages are making inroads by speaking the languages of wider communication. He cites the example of "trilingual Kenyans who are transferring vowel harmony, stress and pitch patters from their native Kelenjin into Kiswahili and English" (Sankoff, 2001:646). He adds that from a sociolinguistic perspective, such cases reflect local groups bilingual in languages imported from outside. In essence, "a latter-arriving group, rather than assimilating to the language already spoken in the area, imports a new language that is subsequently spoken by those already living there" (Sankoff, 2001:646). He further points out that "insofar as ensuing bilingualism (and possibly, language shift) is often a very slow process, substratum influence may flourish under such conditions" (Sankoff, 2001:646).

In addition, Maarten Mous (2003) apprises us about two significant aspects pertaining to *Displacement.* One aspect is that displaced Bantu languages such as Ngoni from South Africa to Tanzania and the Bantu Mushungulu from Tanzania to Somalia reveal insights into general language adaptation, and that the mixed Swahili varieties along the Mozambique coast also show a loss of a special contact situation. The other significant aspect has to do with the fact that because of the prolonged war in their region and the fact that none of them had more than a few thousand speakers, Koman, a group of several languages in the Sudan-Ethiopia border, encountered *Displacement.*

Thereafter, one learns from Lawrence Wright that *Displacement* has been a central issue in the debate over official language policy dealing with "English versus African language education, multilingualism and the choice of languages of learning and teaching" in South Africa. On one side of the debate are the proponents of "the mother tongue principle." This group argues that the *Displacement* or neglect by English and/or principally English approaches inflict a handicap on approximately 83% of the socioeconomically poor learners in South Africa. On the other side of the debate are the defenders of the principally English education. This group argues that "the mother tongue principle" is "unfair" and amounts to "riding roughshod over parental views—that would be utterly undemocratic" (Wright, 2004:176). Given the two divergent positions on the issue, Wright opines the following:

> Language policy debate is often obscured by two factors: failure to acknowledge different time-frames attending contrasting positions, and failure to recognise that ordinary people are motivated by their perceived best interests in the present...the key to more general public acceptance of linguistic ecological diversity in South Africa is to shift

> the emphasis from policy development to practical language cultivation issues. Provide the requisite cultivation support, and acceptance of a revitalised future for African languages becomes more assured. It should also be understood that the modernisation of African languages in South Africa has a political dimension concerning which South African language commentators are strangely silent. This political thrust may not be entirely congruent with the concerns of those whose brief for African languages is primarily cultural or ecological—if, indeed, they are even aware of it. Finally, it needs to be recognised that language development under conditions of controlled influence, as in the civil service or schooling, is potentially achievable (with whatever difficulty), but that this must be complemented by authentic contemporary intellectual work published in African languages if the linguistic dimension of the African Renaissance is to take off (Wright, 2004:175).

Subsequently, the *Displacement* concept is utilized by Oliver Bourderionnet to analyze the musical productions and circumstances of Corneille and Tiken Jah Fakoly, two African francophone singers and songwriters residing in France. His aim is to offer answers to inquiries about selfhood, language, and the social function of the African pop artist in France and other francophone countries. He discovers that the two artists "embody the mediation between publics from two continents and the global music industry in a particular geopolitical context. They also belong to a generation of artists whose productions signal a shift in French popular music representations of Africa and Africans. Discussing these artists' choice to sing in French will also allow us to reflect on the position of the French language African artist in the English-dominated world of pop music" (Bourderionnet, 2008:14).

Thereupon, Samantha Ross Hepworth (2008) investigates the idea of *Displacement* in terms of the link between Kiswahili and minority languages in Tanzania by paying special attention to one district of the country in order to demonstrate that while Kiswahili has made a major domain shift as it leaks from the public or formal domain into the vernacular, minority languages have been able to hold their own in at least the area of plant identification, knowledge and practices. She says that the motive for the minority languages to maintain this ethnobotanical knowledge is based on the fact that local wild plants are basic sources to the speakers' livelihoods, as the plants are a source of food and medicine.

Furthermore, as Danny Hieber explicates about the nature of the debate over language *Displacement* in South Africa, should people be apprised about the false idea of the phenomenon, they would not be so quick to stop speaking their languages to their children. He concedes, however, that parents have the right to make the decision. In South Africa, as he also points out, people consciously decide to move from areas where they have access to a system of outcomes and advantages for speaking their mother tongues to areas which they perceive will provide them with an assortment of economic opportunities, but where the social advantages for speaking the languages of the new residences are lacking (Hieber, 2012:4). He adds:

> Far too many well-intentioned individuals are too quick to substitute their valuations for those of the last speakers of indigenous languages this way. Were it up to them, these speakers would be resigned to misery and poverty and deprived of participation in the world's advanced economies in order that their language might be passed on. To be sure, these speakers themselves often fall victim to the mistaken ideology that one language necessarily displaces or interferes with another. Although the South African Department of Education is trying to develop teaching materials in the local African languages, for example, many parents are pushing back; they want their children taught only in English (Hieber, 2012:4).

Moreover, Salikoko S. Mufwene edifies us about three aspects pertaining to *Displacement* vis-à-vis the knowledge African languages can add to evolutionary linguistics. The first aspect deals with *Displacement* leading to all of the Pygmy languages disappearing and only a few of the Khoisan languages remaining due to contacts among groups speaking different languages since the huge migration out of East Africa. Mufwene posits that "whether or not Proto-Bantu was homogenous and did not vary internally, contact with the Pygmy and Khoisan languages must have contributed to its speciation, notwithstanding later contacts between, say, earlier western and eastern offspring of Proto-Bantu which would have contributed to more differentiation" (Mufwene, 2013:58).

The second aspect relates to the *Displacement* of some languages and the several movements and quantity of language contacts that followed the phenomenon presenting a significant challenge to initiatives geared toward reconstructing the proto-versions of those languages. The African languages that prevailed over the ones that faltered and the new ones that emerged pose a challenge in determining when and from where the not so common linguistic attributes emanated. Mufwene suggests that we must even question the oft-stated claim that the Khoisan populations are the oldest residents of Africa and, thus, their languages are the oldest, since their clicks as African phonemes constitute a later episode in the diversification of human speech (Mufwene, 2013:59).

The third aspect concerns the *Displacement* triggered by colonization styles explaining why many indigenous languages of North Africa (survived by Aamazigh and Tuareg, for example) have been displaced by the Arabic language. The case of South Africa is similar to those of the other countries in Africa South of the Sahara in that the Afrikaners did not establish an assimilationist settlement, most probably because they were also later colonized by the British in an exploitative manner. In fact, the colonization of most of Africa was based on an exploitative, non-assimilationist approach, whereby European languages were treated as "elite lingua francas" (*linguae francae*) and served formal or official roles in educational and governmental systems (Mufwene, 2013:63-64).

African Mathematical Evidence

Beginning with Cheikh Anta Diop's disquisition titled "Square Root, So-Called Pythagoreum Theorem, and Irrational Numbers" in his famous book, *Civilization or Barbarism*, we learn about how Ancient Kemetians/Egyptians rigorously extracted the square root, even of the most complicated whole or fractional numbers. He describes and states the equation succinctly as follows (Diop, 1981/1991:258):

> The term that served to designate the square root in the Pharaonic language is significant in that respect: the right angle of a square, *knbt*; "to make the angle" = to extract the square root. Now the Egyptians defined a fundamental unit of length called "double remen," which is equal to the diagonal of a square of little side a = one cubit (royal); in other words, if d is that diagonal, then one necessarily has, by definition of the length itself, "double remen,"
>
> $d = a\sqrt{2} = (\sqrt{2} \text{ X } 20.6) = 29.1325$ inches.
> The royal cubit = 20.6 inches
> The remen $= \frac{d}{2} = \frac{\sqrt{2}}{2} a = 14.6$ inches

Diop also points out that since the Ancient Kemetians knew how to extract the surface of a triangle, they therefore wrote the following equation, followed by the extraction of the square root (Diop, 1981/1991:260):

$$S = \frac{1}{2} a^2 = \frac{1}{4} d^2 \rightarrow a^2 = \frac{1}{2} d^2 \rightarrow 2a^2 = d^2$$

whence: $a\sqrt{2} = d = a$ double remen.

Concomitantly, the following attestation is made by Diop:

> This definition of the "double remen" by itself, and its mathematical implications, clearly show that Pythagoras was neither the inventor of irrational numbers (incommensurability of the diagonal and of the side of the square) nor of the theorem that bears his name: he took all these elements from Egypt where he had been, as reported by his biographers (cf. Jamblichus), a pupil of the priests for twenty-two years (Diop, 1981/1991:260).

What the foregoing evidence clearly demonstrates is that for thousands of years, Africans have engaged in the exchange ideas, things, etc. that are not spatially present while they are communicating.

Next, as Gregg S. Gonsalves, Edward H. Kaplan, and A. David Paltiel elucidate with their mathematical prototype for ameliorating sexual violence by providing more toilets to the residents of Khayelitsha in South Africa, spatial *Displacement* is mitigated for location-specific opportunities that attract violent offenders (who, of course, use linguistic and extra-linguistic cues to perpetrate their crimes). To arrive at this conclusion, the co-researchers start with a model that operationalizes "sexual assaults as a constant rate per unit of exposure time." The attendant axiom is that "sexual violence [is] an opportunistic behavior that can be reduced by expanding the number of available toilets and thereby decreasing the number of exposure opportunities" (Gonsalves et al., 2015:10). Thereafter, they postulate that "if assault were more additive in nature, then reducing the exposure time might induce assailants to modify their predatory behavior, driving them towards other locations, shifting the geography of women's vulnerability and risk, and not affecting the overall number of sexual assaults" (Gonsalves et al., 2015:10).

In addition, we get a notion of *Displacement* from how Elsa M. Ordway, Gregory P. Asner and Eric F. Lambin describe "deforestation risk" precipitated by "commodity crop expansion" in Africa South of the Sahara and predict that distant markets will affect future land-use alterations in the region. Their prognostication is based on a number of verities they tendered: (1) large-scale deforestation resulted in South America and Southeast Asia in recent decades due to the rapid integration of agricultural markets across the world and the cropland displacement that ensued; (2) pressure is increasing on tropical forests as commodity crops are expanding in Africa South of the Sahara, particularly in four Congo Basin countries, Côte d'Ivoire, Liberia, and Sierra Leone; (3) these countries' average for the highest percent forest cover is 58% ± 17.93 and the lowest average of possible cropland that is not in forest areas is 1% ± 0.89; (4) foreign investment is mostly in oil palm production, which is 81% and a median investment area of 41,582 thousand ha.; and (4) cocoa represented 57% of global expansion between 2000 and 2013 at a rate of 132 thousand ha yr^{-1} (Ordway et al., 2017:1).

Also, an idea of the *Displacement* feature can be discerned in a study of the rainfall in South Africa's Kruger National Park done by Sandra MacFadyen, Nick Zambalis, Astrid J. A. Van Teeffelen, and Cang Hu. The feature can be perceived because the researchers use a long-term mathematical model of "local rainfall measurements as a function of global rainfall surfaces, elevation and distance to the ocean using a generalized additive mixed-effects model to produce fine-scale (1 km^2) monthly rainfall surfaces from July 1981 to June 2015" (MacFadyen et al., 2018:2506). The results generated after their systematic data analysis lead the researchers to suggest significant short- and long-term changes in spatio-temporal rainfall sequences at Kruger Park. They also say that the rainfall will continue its sequence of rising from north to south and from east to west as altitudes rise closer to the escarpment in Drakensburg (MacFadyen et al., 2018:2507).

Conclusion

In the case of Ancient Kemetians/Egyptians, the notion of *Displacement* in terms of communicating about things that are not spatially present is best represented by how the term "absent" is conceptualized. The term *uai* denotes "that which is remote, away from a person or place, and being far off" (for this translation, refer to Budge, 1978).

Accordingly, as Scott Ramsey recounts, in order to initiate and perpetuate the hegemony of the Memphite, *Displacement* discourse of Heliopolitan cosmogony was deemed imperative by Kemetian officials. Ramsey retells the reason for this strategy as follows: "in 3100 BC, Upper and Lower Egypt were joined and the capital became Memphis. This began a new theory of creation. Ptah, the high god of Memphis was deemed creator. At some point, Ptah was even declared to be Nu (thus placed above Alum, high god of Heliopolis). The Ennead of Heliopolis was said to be merely a manifestation of Ptah" (Ramsey, 2015:4).

In addition, Barbara Egedi explicates that there was a *Displacement* of the affix *–w* in the Ancient Kemetian language in order to develop a novel dimension of the meanings of words. As she puts it, "in this case the intrusive grapheme is /.w/ and it is probably due to a displacement of accent…[and also] probably due to another later shift which is also popularized in the so-called broken plural and also took over for other suffixoid word forms, for example in the *nisbe*). Another intrusive /.w/ is hypothesized in forms where the feminine ending /.t/ was not pronounced but a shift in stress to the suffix occurred, i.e., dpt - boat ~ dpwt-f - his boat" (Egedi, 2012:117).

Chapter 11

Productivity

On the basis of the discussion about *Productivity* by Anouschka Bergman, Kathleen Curie Hall and Sharon Miriam Ross (2007), I state the following in my article referenced at the end of the following extract:

> *Productivity,* which augments *Discreteness,* is about a language's capability to develop fresh meanings from discrete units. *Productivity* contrasts with *Discreteness* in that the latter feature only requires re-combinable units in a fixed set of ways in which they could be combined while the former has no fixed set of ways in which units can be combined. Language *Productivity* therefore makes it possible for humans to develop and comprehend any number of innovative statements that they may have never heard/read before, allowing them to express postulates that may never have been proffered before (Bangura, 2014:30).

Given the preceding excerpt, a profitable question here is the following: How has *Productivity* been conceptualized by linguists? In the rest of this section, I summarize the views of a number of linguists pertaining to the question.

To begin with, David Crystal provides a conceptual definition of *Productivity* as being occasionally referred to as *Creativity* or a defining feature of human language and denoting the ability of the users of language to develop and comprehend an indefinitely humongous quantity of sentences. He adds that in a limited tenor, an arrangement is said to be *productive* when it is employed time and again to introduce additional occurrences of the same kind—an example being the utilization of *–s* to make plural nouns in the English language. Contrastingly, as he also mentions, *nonproductive* or *unproductive* designs do not possess the ability to do the same—an example being the plural form for *mouse* in the English language is *mice.* He further points out the other contrasting *semiproductive* forms that have limited or desultory *Productivity*—an example being the employment of *un-* to make lexical contrasts like *unhappy, unqualified, uninformed,* etc., albeit not *unsad,* in the English language (Crystal, 1992:314).

Next, for William O'Grady, Michael Dobrovolsky and Mark Aronoff, *Productivity* concerns new communication on any subject being constructed at any moment. They indicate that while *Productivity* is decidedly a key feature of human language, it is limited in bee communication, and it is nonexistent in bird calls. They add that evidence of "recombination" (i.e. the process of merging entities) exists for birdsongs, such as those of the laughing gulls and some calls by certain monkeys like the macaques, although the question remains unanswered about whether the "recombination" produces new communication (O'Grady et al., 1989:407-408).

In addition, John Lyons discusses a number of aspects in his conceptualization of *Productivity.* First, he informs us that he utilizes the idea to denote a feature of a language-system a speaker of the language employs to put together and comprehend an unlimitedly huge quantity of expressions, including those that s/he had never previously experienced. The essentiality of this idea is particularly emphasized in the study of how children acquire language. The fact that children at a quite young age are capable of making utterances they have never heard previously is adequate evidence that language cannot be acquired only by memorizing and mimicking entire stretches of discourse. By earlier studies on the roots and acquisition of language, focusing more on how the meanings of isolated words and utterances are developed, they paid little attention to *Productivity.* This issue is significant because no matter what answer is provided by the earlier investigations, it would not sufficiently explain how language emerged and how children acquire it (Lyons, 1977:76).

Second, as Lyons argues, it is imperative that when assessing semiotic systems as they relate to the feature of *Productivity* that an evaluator pays attention to the conventional intricacy of the postulates through which novel expressions may be derived. This intricacy may not automatically be evident in the quantity of newly composed expressions vis-à-vis the quantity of those expressions that had already been experienced. Lyons cites as misleading the example of bee dancing which is believed to have the element of *Productivity,* since a bee can make unlimited and numerous contrasting signals to communicate the length and path to a place where there is honey by methodically changing its position in terms of the sun and the vigor of the motions of its body during the dance. As he insists, the systematic plainness of the postulate which establishes the *Productivity* of bee-dancing is quite different from the intricacy and diversity of the systematic postulates of syntax which govern language *Productivity* (Lyons, 1977:76-77).

Third, Lyons apprises us that when *Productivity* is explicated according to a language's grammatical structure, it is to a greater extent typified by the attribute of *Arbitrariness.* To elucidate this proposition, he mentions the cases of the simple attributive adjective coming before the noun that it modifies in the English and German languages, but coming after the noun in the French language; and the verb generally coming at the beginning of a sentence in the

Irish language, but coming at the end of a sentence in the Turkish language. In these cases, word-order is arbitrary, as it is in most grammatical systems of other languages. Lyons cautions us, however, not assume that everything in grammar is arbitrary. He cites the following case to buttresses the cautionary note: "the fact that in *John came in and he sat down*, but not in *He came in and John sat down*, 'he' can refer to John is not just an arbitrary and inexplicable fact of English" (Lyons, 1977:77). For more on this point, see the chapter on *Arbitrariness* in the present book.

Fourth, according to Lyons, there exist expressions whose newness is not based only in the fact they had not manifested themselves in the previous experience of a communicator, but in being acknowledged for their style. It is for this type of newness or inventiveness that the concept "creativity" is applicable. What remains debatable, however, is whether creativity should be treated as an attribute of languages or an attribute of how languages are utilized by certain communicators in certain instances. No matter one's position on the issue, what is obvious is that it is through the semantic configuration of the language-system that creativity emerges in the metaphorical utilization of lexemes (i.e. a *lexeme* is a basic lexical unit of a language, consisting of one word or several words, considered as an abstract unit, and applied to a family of words related by form or meaning.) or their atypical, but stylistically successful, amalgamation. This aspect has therefore been of considerable interest to semanticists (Lyons, 1977:77-78).

Fifth and finally, Lyons elucidates that the design feature of *Productivity* allows us to realize that the perception of a hyponym (i.e. a word of more specific meaning than a general or superordinate term applicable to it; for example, *spoon* is a hyponym of *cutlery*) can usually be investigated as a result of the perception of its superordinate and of certain syntagmatic (i.e. denoting the relationship between two or more linguistic units used sequentially to make well-formed structures) reconstruction of the superordinate. Through clear syntagmatic reconstructions (e.g., 'African continent,' 'huge African continent,' 'huge rich African continent,' and so on), languages provide us with the resources to put together an unlimitedly huge quantity of hyponymous utterances. They also allow language to assist us in being accurate when giving accounts about events, persons, things, etc. (Lyons, 1977:309-310).

Also, Henry Brice starts off by telling us that in linguistics, *Productivity* is a procedure that is employed *ad hoc*. He mentions the instance of derivation, whereby one can employ it copiously to coin novel terms. He adds that with a few exceptions, inflection almost always employs *Productivity*, since it is utilized paradigmatically on an entire category of terms. Thus, according to him, *Productivity* underlies the majority of linguistic procedures. He cites the following examples to support his claim: "in the English [language] the plural derivation is something like [*add the suffix 's'*] and this is a productive process—if someone invents a new noun, then [s/he] can pluralise it in this manner. One of the earliest and best-known psycho-linguistic experiments involved the nonce-word *wug*—even if you ask a small child, [the child knows] that three of them would be called *wugs*" (Brice, 2017:1). It behooves me to add his clarification about the first example here: "Without going into any sort of formal notation, it's obviously actually much more complex than this—there are phonological process going on that turn this into /s/ (as in *cats*), /z/ (as in *dogs*) or /iz/ (as in *horses*). But to go into the complexities of exactly how the plural derivation works is a whole other question" (Brice, 2017:1).

Next, Brice points out that there exist different levels of *Productivity*, ranging from "fully productive" to "completely unproductive." He notes that in the English language, for example, pluralization nearly invariably displays the feature of *Productivity*, with a small number of exceptions such as *women*, *geese*, *oxen*, etc. These exceptions comprise a closed category, which one just has to know. They therefore can be subsumed into the unproductive category, just as verb derivations such as *be* with its present form *is* and its past form *was* (Brice, 2017:1). Moreover, as Brice also mentions,

> ...there are also semi-productive processes, such as the zero-derivation of denominal verbs. You can take most nouns, and use them as verbs, without changing anything other than adding verbal morphology where needed. So not only do you have examples like *a hammer—to hammer*, you can also make them up—*I belt my trousers* may sound a little unusual, but it works perfectly fine. The neologism *to google* has become so common it's in some dictionaries. But you can't necessarily do this always, so it isn't quite fully productive. If you choose a noun that doesn't have any sort of prototypical action associated with it, it may work less well-what does *I sofa the cushion* mean? What about *I yoghurted the soup*? You might guess that this means that you added yoghurt to the soup, but you'd need quite a bit of context in order to make that certain (Brice, 2017:1).

Furthermore, one gleans from Richard Nordquist's essay on *Productivity* that the phenomenon is a language feature that is employed to utter novel entities. He states that from a broader perspective, *Productivity* is also recognized as "open-endedness" or "creativity." He adds that from a narrower perspective, *Productivity* is used to characterize certain "forms or constructions (such as affixes) that can be used to produce new instances of the same type. In this sense, productivity is most commonly discussed in connection with word formation" (Nordquist, 2019b:1).

With the preceding views on the general conceptualization of *Productivity*, in order to probe the link between it and its existence in African languages and mathematics, the remainder of this chapter is segmented into three sections. The first section encompasses linguistic expositions on the phenomenon, and the second and third sections entail evidence of its existence in African languages and mathematics, respectively.

Linguistic Expositions on Displacement

After enjoying a great deal of attention from the 1960s to the late 1980s, precipitated by the 1960 work of Charles Francis Hockett on the design features of language, *Productivity* received very little coverage until more recent times when interest in the subject was rekindled by relatively younger linguists exploring methodological approaches to analyze lexical, morphological, and syntactic features of *Productivity*. I have placed the themes I discovered during my research on the subject into the following 14 classifications to be discussed in the order they are listed for lucidity: (1) production strategies, (2) semiotic system, (3) syntagms, (4) grammar, (5) morphological measures, (6) quantitative and qualitative aspects, (7) psycholinguistic aspects, (8) productivity restrictions, (9) motion constructions, (10) rules and productivity, (11) computational model, (12) the Tolerance Principle, (13) characteristics of verbs in spontaneous production, and (14) distributional semantics.

Production Strategies

O'Grady and his fellow scholars posit that *Production Strategies* are utilized by second language learners when they endeavor to employ in a discourse the linguistic knowledge they have gained. These approaches may incorporate the preorganization or application of expressions in addition to the rectification of expressions (O'Grady et al., 1989:313).

In addition, O'Grady and his colleagues state that preorganization in *Production Strategies* necessitates the selection of parts of an expression (adjectives, adverbs, affixes, inflections and nouns, and their attendant phonetic attainments) with a communicative goal in mind. Rectification and observation procedures are a component of the application attribute of *Production Strategies*. Learners of second languages utilize observational procedures when they rectify what they intend to say ahead of saying it or quickly thereafter (O'Grady et al., 1989:313).

Semiotic System

We are told by Lyons that a *Semiotic System* has been observed to operate to a certain degree in the syntactic system and *Productivity* of chimpanzees.

However, the determination of whether the ability of the chimpanzees to utilize these features is similar to that of humans, he says, hinges upon how "language" is denoted. What is obvious is that the two features the chimpanzees have learned are far less grammatically complex compared to the language-systems adult humans employ. Nonetheless, the two features learned by the chimpanzees do not vary significantly from the language-systems utilized by young children or the American Sign Language used by capable adults in terms of formal complexity (Lyons, 1977:92).

Lyons' inference is based mostly on the empirical findings of two female chimpanzees named Washoe and Sarah that were taught to use human language systems. Washoe was trained to use the gestural system of the American Sign Language, which is not situated on finger spelling, and the concepts which determine the fusion of the manual gestures of which expressions are arranged vary from those which determine the grammatical arrangement of words in the English language. Almost three years later, Washoe had amassed a lexicon of 87 signs and was able to utilize these productively in approximately 300 two-sign amalgams, many of which, it appeared, Washoe had not encountered before. She also composed some longer expressions, and a subsequent generation of chimpanzees did even more (Lyons, 1977:91).

Sarah was taught to learn the lexicon of more than 100 pieces of plastic—which varied in terms of color, shape, and size—in order to connect them with certain denotations. The associated meanings included some adjectives, adverbs, common nouns, proper nouns, and verbs. The plastic word-forms were to be arranged linearly in terms of the syntactic order of the system delineated by the researchers. Just like Washoe, Sarah produced accurate amalgamations she had not encountered before (Lyons, 1977:91-92).

Syntagms

Syntagms (a *Syntagm* is a linguistic unit consisting of a set of linguistic forms—phonemes, words, or phrases—that are in a sequential relationship to one another), Lyons asserts, require us as speakers of a language to know and for descriptive linguists to find out that for many lexemes there is basically a collection of *Syntagms* in which they can be utilized and another collection in which they cannot. He also recommends that we must be aware that a native speaker of a language has the capacity to utilize most lexemes of syntagms that s/he has not encountered before and will be assessed by other speakers of the language about whether s/he has employed them accurately. Consequently, he posits that *Syntagms* constitute a feature of *Productivity* instead of *Creativity*. Nevertheless, he warns us not to go from the radical position in asserting that the collocation of a lexeme is based on its denotation or denotations (where a denotation is perceived to be independent of syntagmatic deliberations) to the other radical position of denoting a lexeme as more than the collection of its collocations (Lyons, 1977:265).

In addition, Lyons explicates that there has been an implicit agreement among linguists that all lexemes are initially applied, "phylogenetically and ontogenetically," in quite definite and fixed circumstances and that they are in turn confined syntagmatically. Therefore, the most linguistic analysts can do is to theorize about the furthest roots of language, and diachronic semantics can only take analysts but so far in chronicling the history of a language or a family of languages. Nonetheless, the available diachronic findings indicate that for every epoch, semantic change has followed generalization and abstraction as well as specialization. This is evident in all examples cited in the classic works of diachronic semantics and etymological dictionaries (Lyons, 1977:265-266).

Grammar

In his work on Theoretical Linguistics, Lyons begins by indicating that each grammar presupposes a *lexicon* (i.e. vocabulary) in which the words of its language are categorized in terms of their belonging in the distributional categories mentioned in the rules of that grammar. The grammar and the lexicon can be analyzed from two perspectives: (1) *Recognition*, whereby the linguist is interested in the *analysis* of the body of expressions; or (2) *Productivity*, whereby the linguist is interested in the *synthesis* of the grammatical sentences. While practical considerations may influence a linguist to arrange the grammar and the lexicon in a relatively different manner, in terms of whether they are being utilized for *Recognition* or *Productivity*, it is imperative for the linguist to know that these categories are themselves impartial in terms of the difference between the grammar and the lexicon. Thus, a body of confirmed expressions can only be analyzed adequately as a "sample" of the sentences generated from the grammar. As such, no polarity exists between descriptive and generative grammar (Lyons, 1971:158).

Also, as Lyons argues, the fact that grammar is in general impartial in terms of analysis and synthesis does not imply that the promotion of one perspective instead of the other would be devoid of practical implications. When the grammar is employed for synthesis, the lexicon should be arranged in such a manner that certain word-class symbols, such as *N* for "noun," can easily be subsumed in the relevant category for the application of the lexical-substitution rules. A clear way of doing this is to arrange the lexicon as a collection of lists. Contrastingly, when a particular text is to be analyzed, it would be less taxing to work with a master list in which the words are arranged in terms of some principle, such as alphabetical order, which would then allow an analyst to identify immediately any discrete words evident in the sentences being examined and find their grammatical features (Lyons, 1971:158).

In addition, Anthony Arlotto presents the concept of *Productivity* as a contrast to *Nonproductivity* of grammatical forms and procedures referring to a language in existence at a certain moment in time. On the one hand, certain affixes of the language are productive in that they can be attached to *any* term of the appropriate grammatical category so that novel terms or inflections can be formed. On the other hand, some affixes are non-productive and, therefore, exist in a relatively quite small number of forms and cannot be comprehended when extended to other forms. By way of explanation, Arlotto mentions the case of the plural marker *–s* (as in *Zimbabwean-s*) in contrast to the plural *–en* (as in *festschrift-en*) in the English language. Evidently, the *–s* is productive and is comprehended when tagged to any English noun, including new concoctions such as *Klingon* (plural *Klingons*). But if we are to attach the un-productive suffix *–en* to *Klingon* in order to pluralize it as *Klingon-en*, it would appear strange and, most probably, not comprehended by hearers (Arlotto, 1972:138-139).

Furthermore, according to Arlotto, during the course of time, the *Productivity* of any form is bound to metamorphose. He cites the example of the suffix *–wise*, which he says was employed some years ago to derive adverbs from nouns, was quite seldom and appeared only a quite small number of forms such as *clockwise*. But because of the business activities on Madison Avenue in New York City, the suffix has become quite productive in nouns such as *moneywise*, *clotheswise*, etc. This, Arlotto says, is a case of a productive relationship in what is needed for analogy to work (Arlotto, 1972:139).

Morphological Measures

Anke Lüdeling and Stefan Evert demonstrate the link between qualitative and quantitative analyses of *Morphological Productivity* (which they define as the capacity with which an existing word-formation procedure is used to generate novel terms) by discussing a number of aspects. First, they argue that *Morphological Productivity* is important because while a qualitative analysis by itself does not parallel the instinctive idea of *Productivity*, a quantitative analysis which must augment it hinges decisively upon a comprehensive intensional (i.e., providing the denotation of a word by indicating the needed and sufficient circumstances for when the word should be employed) comprehension of the morphological procedure and sub-procedures (Lüdeling and Evert, 2003:475).

Second, Lüdeling and Evert broach the two underlying qualitative and quantitative models. One is the *Morphological Model*, which the co-researchers say is undergirded by two postulates that are not axiomatically common among morphologists. In Lüdeling and Evert's model, "stem changes such as the insertion of 'linking' elements, the fronting of certain vowels (umlauting), or the elision of some elements are considered lexical specifications that are listed as 'word-formation stem forms' together with the lexical entry of each morphological element" (Lüdeling and Evert, 2003:476).

Also, for their *Morphological Model*, Lüdeling and Evert employ what is referred to as the "compounding theory of derivation," which denotes that derivation in the German language is thought of as bona fide series and that suffixes constitute morphological heads with part-of-speech- classifications. The result is that no classification contrast exists between affixes and stems and also none between derivation and compounding. Rules can be indicated at the highest level, meaning that each morphological component can be analyzed on all linguistic strata. While certain rules entail variables that are intensionally denoted, others lack such variables. Thus, in this model, any rule that has variables is productive at the highest level, meaning that every component that meets the intensional denotation can be inserted (Lüdeling and Evert, 2003:476).

The other model proffered by Lüdeling and Evert is the *Productivity Model*, which they say takes into consideration the verity that in addition to the lexicalized variety, a language's lexicon includes all kinds of "mental lexicon" that can be productively formed by a speaker. Accordingly, a mathematically productive rule is commonly regarded as an enormous or unlimited lexicon size, as each productive rule yields a significant quantity of low-probability kinds in the lexicon. Referred to as the distributions for "large number of rare events (LNREs)," these distributions of type probabilities tend to get unlimitedly bigger as more and more text is engendered by the model and are surveyed in an output. Contrastingly, the "growth curve will flatten out" for an unproductive process until the lexicon is used up (Lüdeling and Evert, 2003:477).

Third, Lüdeling and Evert explore the "consequences for the treatment of productivity." As they enlighten us, their exploration of the issue was based on their desire to learn more about *Productivity* as it relates to the word-formation procedure that yields *lich*-adjectives. Their major interest is in gaining knowledge about why *lich*-words do not appear to adhere to a single model since they are not derived by a single homogenous procedure but by several, with each of them playing a different role in generating the ultimate curve. They do this by making two distinctions. The first is about what Lüdeling and Evert characterize as the quite evident distinction between various parts of speech of the base: i.e., *lich* affixes to adjectives, nouns, and verbs. Within each sub-pattern, -*lich* (an adverbial and adjectival suffix in the German language that is a cognate with -*ly* in the English language) possesses disparate limitations and also disparate semantic consequences. Generally, analysts using the quantitative technique on *–lich*-derivations do not contrast these sub-patterns, as their focus is on calculating the probability of a novel –*lich* variety. This state of affairs leads Lüdeling and Evert to hypothesize that speakers contrasts sub-patterns, thereby suggesting that measures of *Productivity* must be computed for every sub-pattern (Lüdeling and Evert, 2003:479). I must mention that Lüdeling and Evert say nothing about the qualitative method here.

The second distinction made by Lüdeling and Evert has to do with what they label as "subpatterns." As they point out, the variations in *Productivity* evident for the various base categories do not dispute the validity of their model. Nonetheless, they add, a closer examination of the *–lich* derivations from nouns which are not well articulated in their statistical model will lead one to notice that there appears to exist "subpatterns" dealing with nouns that cannot be differentiated unambiguously but reveals more kinds than would probably be anticipated (Lüdeling and Evert, 2003:481). For illustration, Lüdeling and Evert provide the following evidence:

> One example: we find 71 types of *lich*-adjectives that are derived from complex nouns which have the head noun *Geschichte* 'history': *naturgeschichtlich* from *Naturgeschichte* 'natural history', *literaturgeschichtlich* from *Literaturgeschichte* 'literary history', *architekturgeschichtlich* from *Architekturgeschichte* 'history of architecture', or *religionsgeschichtlich* from *Religionsgeschichte* 'history of religion' are just some examples. Similar 'clusters' can be found with a number of other head nouns such as *Recht* 'law', *Wissenschaft* 'science' or *Wirtschaft* 'economy' (Lüdeling and Evert, 2003:481).

Finally, Lüdeling and Evert conclude that the mathematical features of a quantitative investigation of the formation procedure of a word's *Productivity* can be an indicator for a qualitative investigation. They add that in cases where statistical models fall short of connecting the empirical data satisfactorily, it is imperative for a researcher to rectify his/her qualitative inquiry (Lüdeling and Evert, 2003:482).

For Amir Zeldes, *Productivity* can be perceived as a linguistic feature that has been dealt with mainly as the *binary ability* of a word-formation procedure to derive novel words or as the *scalar degree* of how effortlessly novel elements emerge in the formation. He goes on to make the other following observations: (a) "some definitions concentrate on spontaneous generation of items not *encountered before* by the speaker through *regular* combination of a compatible base with a word-formation process to produce a *transparent* item whose meaning can be inferred from the base and the formation"; (b) "the degree of productivity is often associated with *type frequency* of the formation—i.e. how many adjectives with a" suffix there are—"which can be ne measured in a *corpus*, and the proportion of productive cases therein"; (c) "it is difficult to determine for all items whether the speaker was familiar with them and whether they are transparent"; and (d) the Baayen method "uses *hapax legomena*—i.e. words appearing only *once* in a corpus—to estimate productivity," the rationale being that "neologisms form a subset of these, though words appearing two or three times may also be relevant through repetition of a neologism" (Zeldes, 2009:1).

Quantitative and Qualitative Aspects

In his review of the quantitative and qualitative aspects of *Productivity*, Ingo Plag starts by stating the oft-repeated observation that a speaker of a language can generate novel words by using word-forming constituents to add on existing words. The cases of the suffix *–ness* to turn the adjective *cute* to the noun *cuteness*, for example, and the combination of existing words such as *train* and *connection* to form *train connection* are an outcome of *Productivity*. A careful examination of such word-formation procedures indicate that what is taking place in this arena is dictated by rules, as there are formulaic connections of form denotations among closely-related words that are morphologically multiplex. Thus, axiomatically, it can be stated that in the English language, "adjectives regularly can take the suffix *–ness* and that *–ness* derivatives regularly express a meaning that can be paraphrased as 'the property of being X,' with 'X' standing for the meaning of the base" (Plag, 2004:2).

Plag goes on to make the observation that the preceding morphological rules, designs or procedures through which multiplex words are generated allow an analyst to effortlessly notice that certain rules, or affixes, are frequently utilized to develop novel words while other rules are rarely or never utilized for the practice. As an instance of this phenomenon, he adduces the cases of no novel verb can be discerned in the contemporary English language with the assistance of the prefix *en-* (as in *entail*, *engender*, *ensure*, and so on), whereas the verbal suffix *–ize* is continuously added to adjectives and nouns to generate novel verbs (an example would be the verb *Mandelarize*, which was first used in 2018 by the South African newspaper *Daily Sun* to describe the practice of the African National Congress to name every landmark in the country after the country's great leader Madiba Nelson Rolihlahla Tatamkhulu Mandela). Therefore, as Plag suggests, certain morphological rules can be considered as *Productive* and other rules as *Unproductive* or less *Productive*. This reality leads to the emergence of a couple of issues: (1) the elements that render a particular rule *Productive* or *Unproductive*, and (2) the calculation of the *Productivity* of a particular rule and the elements that lead to the lack of consistence in the *Productivity* of morphological procedures (Pkag, 2004:2-3).

Plag continues by noting another significant theoretical issue concerning whether *Productivity* should be treated as a "theoretical primitive" (i.e. a non-derivable attribute of word formation principles) or "epiphenomenon" (i.e. an attribute that is derived from other attributes of the relevant rule or a procedure that has not yet been identified). For instance, the *Productivity* of a rule is never unlimited since any particular word may act as its base. Essentially, possible bases, or the derivatives, can possess morphological, phonological, semantic and syntactic constraints that may hamper the *Productivity* procedure. Add to this the idea of *Productivity* being also pertinent for the quotidian difference between inflection and derivation. It is generally held that while inflectional procedures are completely productive, derivational procedures are marked by different degrees of *Productivity*, with most of them not being completely productive (Plag, 2004:3).

Next, Plag talks about the actual aspects of qualitative and quantitative characteristics of *Productivity*. At issue here is whether the idea of *Productivity* is qualitative or quantitative. Thus, the argument is that if, on the one hand, the idea is qualitative, then a procedure or affix could be characterized as either having this feature, or not having it. If, on the other hand, the idea as often characterized is a gradual occurrence, then either the morphological procedure is approximately productive than others and, thus, a fully unproductive or a fully productive procedure only indicates the end-positions of a scale (Plag, 2004:4-5).

About the qualitative approaches, to which Plag pays relatively little attention compared to the quantitative approaches, after presenting many definitions proffered by other scholars and him for these techniques with their strengths and limitations, he informs us that implicit in these denotations is that *Productivity* is a feature of morphological procedures with no middle ground. After assessing the various propositions tendered for these definitions, he concludes that three kinds of procedures are evident in a qualitative approach to *Productivity*: (1) the distinctly unproductive, (2) the distinctly productive, and (3) the not easy to categorize as either productive or unproductive. In light of the challenges that qualitative approaches entail, he tells us that a number of researchers have given up on the notion of a qualitative idea of *Productivity* and have instead directed their attention to the precise calculation of "profitability": i.e. the degree to which a morphological procedure can be utilized to generate novel relevant forms (Plag, 2004:6-7).

Concerning the quantitative approaches, to which Plag pays much greater attention compared to the qualitative approaches, after also presenting many definitions provided by other scholars and him for these methods with their strengths and shortcomings, he makes the following observations as the differences between the methods:

a) "Using a text corpus or a large dictionary, productivity can be measured by counting the number of attested different words with a particular affix (i.e. the type-frequency *V*)" (Plag, 2004:14).

b) "Productivity can be measured by counting the number of neologisms in a given period, using, for instance, a large historical dictionary. The greater the number of neologisms in that period, the higher the productivity of a given affix in that period" (Plag, 2004:14),

c) "Productivity can be measured by counting the number of hapaxes with a given affix (n_1) in a large corpus. The higher the number of hapaxes, the greater the productivity" (Plag, 2004:14).

d) "...by dividing the number of hapaxes with a given affix by the number of tokens with that affix, we arrive at *P*, which indicates the probability of finding new words among all the tokens of a particular morphological category" (Plag, 2004:14).

Thereafter, Plag illustrates and discusses the preceding methods of *Productivity* by looking at some of the suffixes for which the methods are easily accessible. These suffixes include "*–ion, -ist, -ity, -ish, -less, -ness,* and *wise.*" After providing a detailed analysis to determine which suffix is the most productive, he summarizes his findings by pointing out that researchers have access to many different methods they can use to gauge the *Productivity* of word-formation procedures. Each of these methods focuses on distinct features of *Productivity* and entails its own methodological shortcomings in terms of data sampling and analysis. Accordingly, distinct methods must be considered by researchers in order to provide sound analyses and draw logical conclusions concerning the *Productivity* of a particular affix (Plag, 2004:18).

Psycholinguistic Aspects

Drawing again from the work of Plag, one learns about the answers to questions dealing with (a) how a speaker can know that a particular affix can be employed to generate novel words, (b) the features that productive procedures share but unproductive procedures do not, and (c) the productive attributes of affixes that lead to the emergence of various extents of *Productivity*. To address these questions, he alludes to several key findings from recent psycholinguistic investigations.

First, Plag tells us that the fundamental rationale for utilizing frequency to calculate *Productivity* is based on the verity that the frequency of multiplex words greatly dictates how we operationalize and preserve them. Concomitantly, most contemporary frameworks of morphological computing deal with access to morphologically multiplex words in cognitive vocabulary via two routes: (1) the *whole word route*—i.e. directly examining the representation of the entire word; and (2) the *decomposition route*—i.e. examining the disintegrated constituents of the word. This implies that any emerging multiplex word is concurrently computed in the preceding two ways, with one of the ways achieving the desired result. The *whole word route* allows us to look up a word as a single element in the cognitive vocabulary, while the *decomposition route* allows us to look up each constituent of the word independently (Plag, 2004:19).

Second, according to Plag, for most complex words, the base appears more often compared to the derived word; thus, the reciprocal rate of occurrence is lesser than unity. In relation to psycholinguistics, the base has a more intense delineation, or higher "resting activation," in the cognitive vocabulary compared to the derived word. The upshot for the decomposed rule is that it becomes dominant because of its high "resting activation," which allows the base to be retrieved each time the derivative comes into the system. Contrastingly, when the derived word is more recurrent than the base, there is a whole word partiality in analyzing a string or text into logical syntactic components, typically in order to test conformability to a logical grammar., since the "resting activation" of the base is lower compared to that of the derivative. Based on these findings, Plag hypothesize that "In general, the higher the frequency of the derived word in relation to the base word, the less likely is decomposition; alternatively, the lower the frequency of the derived word in relation to the base word, the more likely is decomposition" (Plag, 2004:21).

Third, as Plag apprises us, relative frequency also recurs with other features of morphological groupings. Therefore, a correlation exists between low frequency and high *Productivity* and also a correlation exists between low "relative frequency" and high "semantic transparency." In essence, combined with the preceding finding on analyzing a string or text into logical syntactic components, the correlations here suggest that a strong relationship exists between relative frequency on the one hand and parsing in discernment and morphological *Productivity* on the other hand (Plag, 2004:22).

In essence, following Plag, from a psycholinguistic perspective, *Productivity* is explicable in terms of characteristic combinations of parsing, phonological and semantic transparency, and relative frequency serving as major ingredients. Vis-à-vis whether *Productivity* is a derived concept or a "theoretical primitive," it is apparent that the *Productivity* of an affix yields a multiplex element given the aforementioned procedures. Of these procedures, phonological and semantic transparency are both psycholinguistically and structurally ascertained, since it is "the semantic and phonological structure of affixes and their derivations that co-determine processing and storage of these forms," which are not "exceptionless principles but strong probabilistic tendencies which are sometimes overruled by structural restrictions" (Plag, 2004:24).

Productivity Restrictions

Culling once more from the work of Plag, we glean that a major *Productivity* limitation is the functionality of a neologism for the speakers of a language. Preconceived functionality is a necessary condition for productively formed derivatives to be birthed, no matter which role a specific derivative plays in a given circumstance. Nonetheless, some limitations operate, as not every potentially functional word is generated and utilized. Consequently, a distinction must be made between the potentiality to use a word-formation rule to coin novel words and the chances of employing such neologisms in speech. The two aspects lead to two contrasting restrictions: (1) *pragmatic restrictions* and (2) *structural restrictions* (Plag, 2004:24-25).

Beginning with *pragmatic restrictions*, which Plag discusses somewhat briefly, he states that "fashion" is one of the most apparent usage-based aspects that affect *Productivity*. He mentions the incline and decline of the uses of affixes such as "*mega- giga-*, *mini-*, or *–nik*" comprising instances of the outcomes of extra-linguistic occurrences in society which render particular words or morphological constituents preferable for utilization and, thus, productive. An additional pragmatic prerequisite that novel lexemes must fulfill is that they must define an entity that can be given a name. While this requirement is not adequately denoted, it, nonetheless, indicates a notable perception: i.e. "the concepts encoded by derivational categories tend to be rather simple and general (e.g., adjectival *un-* 'not X,' nonverbal *–en* 'make X,' etc.) and may not be highly specific or complex. This does not mean, however, that more complex notions cannot be encoded by affixes, but that this requirement seems to be language-specific and is a mere tendency" (Plag, 2004:25-26).

Also, Plag mentions that the shortcoming of pragmatic restrictions is that in the face of an apparently unfeasible novel formation, it is not obvious whether it is ruled out because of either structural or pragmatic cogitations. Thus, before an analyst declares that a particular form is unfeasible because of pragmatic restrictions, it is imperative for him/her to first engage in a very careful examination of the structural restrictions that frequently show that a form is unfeasible due to its contravention of relevant morphological, phonological, semantic, or syntactic restrictions (Plag, 2004:26).

Next, concerning *structural restrictions,* which Plag broaches relatively extensively, he begins with the observation that these restrictions in word-formation may deal with the conventional levels of linguistic study: i.e. morphology, phonology, semantics, and syntax. Thus, a common issue that emerges from the study of the restrictions concerns which of them should be thought about as being distinctive to the specific word-formation rule under consideration and which restrictions are more common that work on all or at least some categories of morphological procedures (Plag, 2004:26).

In addition, Plag alludes to the fact that rule-specific restrictions may deal with the attributes of the base or of the derived word. For instance, phonological restrictions can refer to independent sounds or to prosodic occurrences such as stress or syllable structure. A specific example provided by Plag is the following: "suffixation of verbal *–en,* as in *blacken,* is subject to the segmental restriction that it only attaches to base-final obstruents (e.g., *blacken* vs. **finen*) and to prosodic restrictions that it does not take bases that have more than one syllable" (Plag, 2004:26-27).

Thereafter, Plag intimates that besides its sensitivity to phonological restrictions, affixation may hinge upon the morphological system of the relevant base. The suffix amalgamation *–ize-ation* is a good example of this phenomenon. Thus, any word with the suffix *–ize* at the terminal position can be turned into a noun by simply adding *–ation.* Systematically eliminated from such a procedure are "conceivable deverbal nominal suffixes" like *–age,* -*al,* -*ment,* and so on. Semantic restriction is exemplified by the suffix *–ee,* as in *committee,* which must define sentient elements, as demonstrated, for instance, by the unfeasibility to employ "*amputee* to refer to an amputated limb." In addition, restrictions of *Productivity* can introduce syntactic features. An oft-cited example is the restriction of word-formation rules to a collection of a particular syntactic grouping, as in the case of the adjectival suffix *–able* which generally fastens to verbs like *electable* or the adjectival suffix *–al* which fastens to nouns like *controversial* (Plag, 2004:27).

Finally, Plag shifts his focus to "blocking," which he says is a more principled type of *Productivity* restriction. He tells us that the concept *blocking* is employed in different ways, albeit he restricts his discussion to two varieties of *synonym blocking.* One variety is *token-blocking,* which he says concerns the blocking of a prospective normal form by a synonymous word that is already in existence. Examples of this type of blocking are **arrivement* blocked by *arrival* and **stealer* blocked by *thief.* The other variety is *type-blocking,* which involves the blocking of the implementation of a rule by another rule. An example of this type of blocking is *–ness* and *–ity* suffixation (Plag, 2004:27-28).

Motion Constructions

In his thesis on this topic, Joel Olofsson explores the *Productivity* of Swedish *Motion Constructions* (or *Motion Event Expressions*) from the viewpoint of a "usage-based construction grammar." His major attention is on the *double adverbial constructions*, each of which is comprised of a verb proceeded by a directional adverb and a propositional phrase. The data for his analysis were gleaned from four of his own articles on the topic and analyzed by employing experimental methods (Olofsson, 2018:1-2). From the perspective of grammar, *Motion Construction* can be generally defined as the arrangement of words according to syntactical rules to depict the action or process of moving or being moved.

For starters, Olofsson briefs us that a key component of the investigation of motion expressions adopts a typological point of view. In the case of the Swedish language, its motion expressions have been compared to those of the French, Icelandic, Spanish, Thai, and other languages (Olofsson, 2018:3).

Next, as Olofsson notes, there exists a difference between "subject-centered motion," whereby the grammatical subject is usually the moving unit, and "object-centered notion," whereby the grammatical object is usually the moving unit. Also, in the Swedish language, a morphological difference exists between directional and locative adverbs, in which the directional adverbs are usually comprised of one morpheme—e.g., *ut* "out," *upp* "up," *ner* "down," *hem* "home," etc.; and the locative adverbs are usually comprised of derivatives entailing a directional adverb and the suffix *–e* or *–a*—e.g., *ute* "outside," *inne* "inside," *uppe* "up," *nere* "down," *hemma* "at home," etc. (Olofsson, 2018:3-4).

In addition, Olofsson mentions that Swedish is appraised as a "satellite language" in that its path in a move usually exists in a complement to the verb—an example being an adverb or propositional phrase. This is different from the Spanish language in which the path in a move is usually homogenized in the verb. "Directional adverbs," asserts Olofsson, "are satellites typically used to describe *translocative* scenes, where the moving entity shifts from one point to another, including boundary-crossing scenes, as opposed to locative scenes" (Olofsson, 2018:4).

Also, according to Olofsson, construction grammar concerns the study of linguistic designs as constructions, which are traditionalized form-meaning arrays via which form deals with morphological, semantic and syntactic features, and meaning deals with functional and semantic features. Based on this viewpoint, language is considered as a huge collection of constructions with different levels of complexity. A key principle in the usage-based methods of construction grammar is that encounters with languages influence language structures, as language encounters affect the mental representation of language. This perspective therefore challenges the strict division between competence and performance. It also represents a language as not an organized modular system, but the result of the interconnections among various mental capacities—for example, "domain-general processes such as categorization, cross-modal association, chunking and analogy" (Olofsson, 2018:5).

Thereupon, we learn from Olofsson that *categorization* (i.e. putting novel grammatical constituents into a grouping) is a core procedure that allows general linguistic concepts to arise. In other words, constructions are generated by the grouping of occurrences. An example noted by Olofsson is "high token frequency," which he says is suggested to yield "entrenchment," which is the strengthening of a string of tokens that eventually yields automation. Contrarily, "high type frequency" is perceived to enlarge the productive utilization of constructions. Basically, constructions may entail both open (i.e. "more general/schematic") and/or specific (i.e. "lexically filled") places. The schematic places comprise collections of elements that have been systematically classified into groupings. All this propels Olofsson to make the following observations: (a) "a specific instance of a construction leads to an abstraction through recurrence—i.e. it becomes more entrenched in the minds of the speakers (*lexical strength*)"; (b) "entrenchment is mainly discussed as a token frequency—i.e. highly frequent instances not only serve as prototypical instances of a construction, but are also proposed to be more accessible and, thus, easier to process in language production"; and (c) "categorizing…is rather a matter of type frequency—i.e. when speakers hear utterances, they classify them into categories" (Olofsson, 2018:5-6).

Moreover, Olofsson posits that *Productivity* constitutes a core concept in construction grammar and deals with the potentiality of employing a particular construction with novel lexical elements or an existing element with a novel denotation. Two determinants are frequently suggested to contribute to *Productivity* in usage-based construction grammar. The first determinant is *type frequency*, which denotes the quantity of the distinct elements (such as the position of the verb or that of the proposition phrase) that take place in a schematic spot in a construction. This determinant is perceived to add to *Productivity* in that there is a correlation between the higher level of *type frequency* and the higher probability of the elements being employed by speakers in the spot. This determinant also varies from *token frequency*, which is the quantity of times a particular element takes place in a particular spot. The second determinant is *semantic variability*, which refers to the semantic span of the kinds that take place in a schematic spot in a construction. Normally, speakers are hidebound and tend to utilize constructions productively when the new occurrence is akin to instances that are already known. Consequently, variability adds to *Productivity* by the same rationality as *type frequency*: i.e. there is a correlation between the larger the span of semantic types and the higher likelihood of speakers utilizing a construction productively, since more semantic types will exist in corresponding usage (Olofsson, 2018:6-7).

Rules and Productivity

From Charles Yang, we learn about *Rules and Productivity* as they concern three aspects. The first aspect is the English past tense. To begin with, Yang elucidates that the most fruitful consensus reached in the debate over the past tense is that regular and irregular verbs are decidedly different. Skipping the lengthy debate on the issue, he agrees with the basic postulate (not the details) of the "Words and Rules" model that whereas regular verbs are inflected by the utilization of the productive rule "add –d," irregular verbs are anomalies and must therefore be memorized (Yang, 2005:268).

Next, as Yang evinces, even irregular verbs are inflicted by *morpholexical* rules which, unlike the "add –d" rule, are exerted on an established collection of lexical elements and are not spread widely to cover new tokens. The irregular inflected forms must still be memorized, albeit not directly, as a speaker need only memorize the particular list of words to which each morphological rule is relevant. Appropriately, "frequency-dependent associative memory" continues to perform a function in morphological calculation, since interconnection is manifested between words and rules instead of between words and words as in the Words and Rules model. This alternative framework, referred to as the "Rules over Words" model makes it possible for a speaker to ensnare the well-acquainted frequency effects in learning the past tense and the greater attention to detail needed when expounding the collected data (Yang, 2005:268).

Also, Yang apprises us that whichever of the morphological models is utilized for the purpose of learning, it is still incumbent on the learner to find the answer to the *Productivity* issue. The learner must first realize that a particular verb is regular or irregular for the Words and Rules model to operate in order for the verb to be placed into the proper "module" (i.e. "memory or rule"). As such, this procedure calls for the learner to formulate the "add –d" rule based on the input and to note that this rule is really a "productive/default" variety. For the Rules over Words model to operate, the learner must formulate the "add –d" and morpholexical rules for the irregular verbs based on the input and note that the "add –d" rule is productive, and that the irregular rules are unproductive by the virtue of their being morpholexical (Yang, 2005:268).

In addition, according to Yang, empirical evidence shows that children are quite proficient at the issue of *Productivity*. Almost every mistake a child makes in past tense learning is due to "over-regularization": e.g., "*hold-holded.*" Children's keen perception of *Productivity* in the past tense augurs well with the universal findings on morphological learning. Cross-linguistic evidence indicates that the morphological mistakes children consistently make are either ones of over-generalization or omission of default forms, while mistakes on wrong morphological forms, or substitutes, are rare. In sum, children do very well at learning the correct utilization of rules they identify and generalize productive rules when they learn by heart the regulated utilization of unproductive rules (Yang, 2005:269).

The second aspect is in regards to learning and *Productivity*. On this facet, Yang starts off by recounting that only after children learn the rules of *Productivity* can they evaluate them. He adds that there exist so many morphological learning models which have been proffered by various scholars over the past several decades that their review would be beyond the scope of his essay, especially when his focus is on the *Productivity* of rules as opposed to models like "connectionist networks" which dismiss the utilization of rules on a theoretical basis (Yang, 2005:269).

Next, Yang informs us that for models such as Words and Rules and Rules over Words, the issue of rule learning always emerges. For the default "add –d" rule, it is asserted that it can be learned by paying attention to the procedures used for most word kinds. This notion, however, encounters problems in instances where the default category is not statistically ascendant. Nonetheless, some models such as Artificial Intelligence that have been developed by computer scientists harbor the potential for addressing the issue of rule learning. But as Yang is also quick to point out, while the learning model yields the *form* of rules, it does not address their *Productivity* (Yang, 2005:271).

The third aspect concerns some terminologies. Yang names the terms "default," "productive," and "(ir)regular," which are used as substitutes for one another by many authors as needing distinctive and precise denotations when discussing *Rules and Productivity*. The following are synopses of the definitions he tendered for the three terms: (1) *default* denotes a "backup option, or when all fails—i.e. when more specific rules fail to apply, the default is used"; (2) *productive* denotes "predictable and generalizable—a rule is productive if it automatically applies to a set of lexical items characterized by certain properties, producing predictable derives forms, and can extend to others, including novel items that have the same properties"; and (3) *(ir)regular* alludes to words and not rules (Yang, 2005:272-273).

Computational Model

Again from Yang, we become aware that first, as it pertains to a computational model of *Productivity*, there is a cost for exceptions to rules. He enjoins us to consider a child learning the rules of the past tense in the English language as an example. When dealing with irregular forms, the child must spend a considerable amount of time and energy to memorize them in order to be able to contend with lexically marked exceptions, which are unlike the regular forms for which the *Productivity* rules can be easily utilized. Based on this perspective, a rule's *Productivity* depends on the assessment process that addresses the appropriate type of cost-benefit evaluation that depends on the elements that conform well and those that do not for a productive rule (Yang, 2005:274-275).

Also, as Yang briefs us, the widespread interest in a computational model of *Productivity* is nothing novel. This interest is underscored by the significant amount of the literature in the fields of Artificial Intelligence and Psychology addressing rule *Productivity* as it relates to categorization, decision making involving inconsistent data, inductive generalization, and a number of other approaches. Nonetheless, according to Yang, the selection of an assessment technique for a linguistic theory is not determined deductively; instead, it is the result of an empirical hypothesis about the essence of language (Yang, 2005:275).

In addition, Yang suffuses that two measures are commonly employed to compute complexity: (1) space and (2) time. Both measurements are basically logical and their empirical standings dictate which one to be utilized for a particular case. However, induction of linguistic structures within the framework of Minimum Description Length (MDL) has gained the interest of some linguistic analysts. This approach is based on the notion that grammar is a "data compression device" that is used to limit the structural interpretations of pertinent linguistic data. Yang argues that to some extent, this approach appears veracious, since a major notion of generative linguistics is to get rid of unneeded entities in order to reach abstemious delineations of a grammar. He adds that given the level of accuracy of the Rules over Words model, the human learner seems to make space-saving rules instead of direct storage of inflected forms, even when the saving gained for the approximately 150 irregular verbs is somewhat small. Still, the issue of applying the MDL framework to grammar which calls for autonomous stimuli remain unresolved (Yang, 2005:275).

Second, Yang explicates morphological knowledge and use. He starts with the following three procedures for computing words that are generated morphologically: (1) *word search*, which involves the "lexical lookup of the stem (e.g., *walk*)"; (2) *rule selection*, which involves the "search for the appropriate morphological rule(s) for the stem (e.g., "add –d" for past tense)"; and (3) *rule application*, which involves the utilization of "the rule to generate the inflected form (e.g., *walked*)" (Yang, 2005:276). He then asserts that these three procedures must be carried out separately and in a logical order for them to be grammatically efficacious (Yang, 2005:276).

Next, Yang alludes to the fact that in general, morphological processing seems to be directed toward "real-time processing efficiency." Strong evidence of this can be seen in the frequency effects. An early and quite profitable finding, as mentioned previously in this chapter, is that lexical processing of high-frequency words are noted more quickly and consistently compared to low-frequency words. More recent findings point to gradient results such as prevalence in the vocabulary as being quite knotty for the abstract depictions of linguistic systems pertaining to generative grammar (Yang, 2005:276-277).

Third, Yang points to several evidential characteristics of the Elsewhere Condition Serial Search (ECSS) model. One characteristic has to do with the model predicting that there is a correlation between lexical access time and word ranking—i.e. how they are positioned on the search list. While rank is linked to frequency, it is a somewhat comparative idea instead of an absolute one. The two ideas are not easily differentiable, although recent research findings indicate that rank is a better match to reaction time data in lexical decision compared to frequency. Nonetheless, Yang describes the upshot to the problem as follows: "although ranking a list by frequency in a strictly decreasing, and thus optimal, order appears to be computationally taxing, there are a number of so-called randomized algorithms that are comparatively trivial yet guarantee near optimality of the sorted list" (Yang, 2005:278-279).

Another characteristic of the ECSS model concerns the forecast of frequency consequences in the inflection of the exceptional forms. An adequate evaluation of the forecast must maintain the stem-cluster frequencies immutable in order to counteract the impact from the frequency consequences in the word search procedure. Frequency consequences amid the exceptions, such as the irregular verbs in the English language, are among the consistent results in morphological processing. This evidence has been generally noted as supporting the perspective of the frequency-sensitive minding fetching procedure being accountable for the storage of irregular forms (Yang, 2005:279).

The other characteristic of the ECSS model relates to the notion that during the rule search process, certain words will not be retrieved until every rule-defying exception has been evaluated and jettisoned. In essence, non-words cannot be identified until every actual word in the search has been assessed and discarded. This prediction, as Yang acknowledges, presents challenges for empirical evaluation. What is obvious, however, is that the inflection of a word calls for the following three procedures for computing words that are generated morphologically discussed earlier: i.e. (1) *word search*, (2) *rule selection*, and (3) *rule application* (Yang, 2005:279).

The Tolerance Principle

One gleans once more from Yang that the *Tolerance Principle* is an optimization mechanism used to process words online by taking into consideration token and type frequencies. The optimization of token frequency is evident in the ECSS model that sorts clauses in terms of their frequency. Nonetheless, the *Tolerance Principle* itself is illustrated as a function of frequencies (Yang, 2005:282).

After providing a lengthy technical description of the principle, a repetition of which is beyond the scope of this section, Yang goes on to discuss "tolerance principle in action" by focusing on three aspects. The first aspect deals with tolerance and learning. For starters, Yang inculcates that the low tolerance of exceptions can propel an analyst to prognosticate that there is a connection between child language learning in tenuous connections and the distributions of word frequency. By the time a learner realizes a productive rule, s/he must have accumulated a significant amount of rule-following elements than exceptions. Yang cites as an example the case of learning the past tense in the English language as follows: taking the initial occurrence of over-regularization as confirmation for the effective attainment of the regular rule, at that particular point there must be substantially more regular verbs than irregulars in the child's lexicon as forecasted by the Tolerance Principle (Yang, 2005:284).

Also, an affiliated conjecture Yang discusses deals with the standing of the U-shaped curve (i.e. a bimodal distribution whereby frequencies steadily fall and then steadily rise again) in language learning. Among some children, the acquisition of the past tense begins with a phase of unerring inflections of both the regular and irregular forms. Thereafter, over-regularization ensues as the regular rule manifests itself. In the English language, this system is intrinsic under the Tolerance Principle in light of the particular word frequency distributions. This is mainly due to the fact that since irregular verbs make up the majority of token frequencies in the English language, they are more likely to be first acquired by children. Thus, few regular verbs will be in the children's learning repertoire. Despite this reality, a few regular verbs could propel the acquisition of the "add –d" rule (Yang, 2005:285).

The second aspect of the "tolerance principle in action" broached by Yang concerns tolerance and *Productivity*. According to him, three distinct prognoses about morphological *Productivity* are projected by the Tolerance Principle. One prognosis is apropos to delineating artificial language experiments whereby the distributions of regular and unpredictable irregular forms are deliberately controlled in an experiment. A researcher can then look at how children make generalizations when they encounter variable occurrences. Another prognosis is in regards to the prospect to test the principle based on the abundant data that exist on historical evolution. For instance, words have shifted morphological categories over time. The other prognosis relates to corpus analysis of synchronic morphological systems that have shown relevant sets of regular and irregular forms. The relative ease of this prognosis allows for a robust analysis of *Productivity* and facilitates the evaluation of traditional postulates on morphological *Productivity* such as the affix problem in the English language (Yang, 2005:286-287).

The third aspect of the "tolerance principle in action" explored by Yang is comprised of regular "irregular" nouns. Focusing on irregular nouns in the German language, he maintains that their "horrors" are quite real as there is an enormous number of them to be placed in their various irregular categories. Despite this enormity in the number of irregular nouns in the German language, at least one of the four "irregular" categories of nouns—i.e. the feminine nouns—is in fact pluralized via a productive add –(e)n rule. As such, a learner need not memorize a majority of these feminine nouns, except for a small number that do not add –(e)n and must therefore be treated as exceptions (Yang, 2005:288).

Characteristics of Verbs in Spontaneous Production

In their study of the "lexical, morphological, and syntactic characteristics of verbs in the spontaneous production of Italian children," Laura D'Odorico, Mirco Fasolo, Rosalinda Cassibba, and Alessandro Costantini start by considering why verbs are not the first linguistic utterances of children acquiring a language and reference to actions or status is quite frequently done by using adverbial phrases. They say that this phenomenon is due to that fact that verbs manifest themselves in the vocabulary of children when the size of their lexicon is significantly large, as the tools with which verbs are acquired are more complex and opaque compared to those used to acquire nouns. But verbs, nonetheless, are imperative in the syntactic development of children because children need them to start talking about who does something when, where, and to whom. The question that emerges concerning this phenomenon has to do with the knowledge children have when they start utilizing verbs. The discourse on this question has pitted various schools of thought. The nativist school argues that children's linguistic knowledge is comprised of a large collection of abstract classifications and that the linguistic elements to which syntactic rules are applied are more abstract than those which underlie the ultimate expressions. Opposing this perspective are those theories which maintain that in the early stage of language development, children's syntactic knowledge is verb, or item, specific and, thus, the knowledge of one verb cannot be relayed to that of another verb and the knowledge of a linguistic structure cannot be relayed to that of another linguistic structure. This latter perspective essentially postulates that the abstract classification "verb" manifests itself through a procedure of generalizations and abstraction after many different particular verbs and the occurrence of their argument features have been learnt (D'Odorico et al., 2011:1).

Next, D'Odorico and her colleagues question the validity of the nativist theory on the grounds that the syntactic system of initial multiword expressions entailing verbs is quite undeveloped among children across the globe. Consequently, some linguists have suggested that children at initial phase of language acquisition express less than what is in their minds. This same argument has been promoted by other linguists studying the single-word expressions of children, adding that these single-word expressions are more complex than the naming of an object and can be identified via their extralinguistic milieus. On the one hand, proponents of the "performance limitation hypothesis" argue that even when the linguistic competence of children is complex during the initial stages of their lives, the length of the expressions they can make is constrained by limited processing capabilities. On the other hand, proponents of the "competence limitation hypothesis" advance the view that the initial verbal utterances of children are a reflection of their knowledge at that phase. This hypothesis undergirds the theories of language acquisition that hinge on the constructivist framework of language development (D'Odorico et al., 2011:1-2).

After that, D'Odorico and her coauthors broach "subject production." They indicate that the utilization of subject noun phrases by children is especially useful for retrieving information concerning the inherent essential characteristics of the items that are left out in initial sentences, especially when contrasting languages such as English that need a conspicuous subject in tensed clauses with languages such as Italian that allow a null subject. They mention the following examples: the English language will allow the statement "I am a good kid," but will not allow "Am a good kid"; the Italian language will allow both *Io sono bravo tato* (I am a good kid) and *Sono bravo tato* ([I] am a good kid). Unfortunately, they point out, there are very few studies that have investigated the use of the subject in null-subject languages. After reviewing the few studies that have been conducted on the phenomenon as it operates in the

Italian language, D'Odorico and her colleagues conclude that the results from the studies are inconclusive. The reason for their claim is that the studies employed small sample sizes and non-comparable sentences across investigations even from a cross-linguistic perspective since the characteristics of the groups were dissimilar (D'Odorico, 2011:2).

Subsequently, D'Odorico and her fellow scholars treat the notion of "direct object production." Following their scrutiny of some of the works on the notion, they compare the utilization of the obligatory direct object of Italian versus English children. Their finding is that while it behooves English children to employ subject and direct object in their constructions of transitive verbs in order to form grammatically accurate sentences, Italian children can only express direct objects with transitive verbs, so long as the use of subjects is not imperative. When the Italian children are capable of producing only two-word combinations, a la the performance limitation hypothesis, they would choose to employ direct objects as opposed to optional subjects with transitive verbs. Regrettably, as D'Odorico and her colleagues add, data do not exist on the expression of indirect objects by Italian children. An investigation of the use of indirect objects would yield useful information about the knowledge children possess concerning the essence of transitive and intransitive verbs (D'Odorico et al., 2011:3-4).

Furthermore, D'Odorico and her colleagues explore the "acquisition of verb morphology." They compare how Italian and English children acquire verb morphology in order to determine the validity of either the "parameter setting" notion or the nativist theory of language evolution. After reviewing a number of studies that have been done on the topic, D'Odorico and her companions conclude that none of the studies addressed the utilization of contrasting morphological inflections with the same verbs which, they believe, is a quite necessary requirement for pinpointing an actual procedure of generalization via an early "parameter setting." Contrarily, they add, "if Italian children, at the early precocious stages of language development, use a verb only in one or two morphological inflections, while they are able to use other possible inflections with other verbs, this may constitute evidence of an item-by-item learning" (D'Odorico et al., 2011:11).

Distributional Semantics

Florent Perek instructs us about "using distributional semantics to study syntactic productivity in diachrony" by using a fully data-driven framework. Based on his thesis that "a distributional approach to word meaning not only is adequate for the study of syntactic productivity, but also presents the advantage of allowing the use of visualization techniques and statistical analysis," Perek divides his analysis into four major sections and a conclusion (Perek, 2019:2).

The first section deals with "determinants of syntactic productivity." Here, we learn from Perek that the *Productivity* design feature of language has a long history in the area of morphology, wherein it is referred to as the attribute of a word construction procedure employed by speakers and writers to invent neologisms. He notes that it is only recently that this idea has been used in the area of syntax, which has for a very long time been controlled by the idea of grammar as a system comprising abstract rules divorced from the vocabulary that allows speakers and writers to construct an unlimited number of sentences, including novel ones. Based on this perspective, vocabulary units can be freely amalgamated with syntactic systems as long as both attributes are parallel. The amalgamation of words and syntactic constructions is not a trivial undertaking since the positions of the constructions are not equally likely to be filled by any vocabulary unit, even in cases where an amalgamation would make sense. It frequently happens that amalgamations that could be anticipated either do not manifest themselves; or if they do, they are marginal; or can be even rejected (Perek, 2019:2).

Also, Perek alludes to the fact that syntactic *Productivity* can be used to delineate different types of linguistic behavior. He describes these as follows:

> In language acquisition, it accounts for the children's ability to generalize beyond their inherently limited input by producing combinations of words and constructions they might not have witnessed in the speech of their caregivers. Of course, children are not the only productive speakers, as adults, too, do at times use the resources of their language creatively to produce unconventional expressions. By the same token, the concept of syntactic productivity also applies in synchrony when new words enter the language, in that there may be several constructions in competition for using a novel word in a sentence. Finally, in language change, syntactic productivity may refer to the property of the slots of a syntactic construction to attract new lexical fillers over time, thereby forming novel combinations of words and constructions (Perek, 2019:3).

After discussing some studies that have examined the preceding types of linguistic behavior which can be described by using the notion of syntactic *Productivity*, Perek goes on to assert that the consensus on the phenomenon is one that is significantly connected to the previous utilization of constructions. In essence, speakers and writers are more likely to use a construction in the same manner they had seen or heard it before and would only abandon it if the distribution structure compels them to do so. Concomitantly, early studies of the phenomenon suggest a significant semantic component whereby new utilizations must be semantically parallel to previous utilizations. The essence of semantics for syntactic *Productivity* therefore suggests that the denotation of vocabulary units has to be taken into consideration when analyzing the distributions of constructions (Perek, 2019:3-5).

The second section is about "factoring in semantics in studies of syntactic productivity." Piggybacking to his earlier observation that existing findings on syntactic *Productivity* are based significantly on semantics, Perek maintains that any assessment of the validity of the models via empirical data calls for a measurement of the semantic features of *Productivity*. More precisely, it calls for an evaluation of such features of denotation because of the lack of consistency in the collection of units and similarities among units. Operating from the premise that linguistic denotation cannot be viewed directly as that of phonetic or morphosyntactic features (particularly for corpus investigations, with corpus only entailing signifiers or forms), Perek says that gleaning the denotation of these features needs a human translator. This is because, according to him, thoroughly scrutinizing a corpus, unlike gathering behavioral data from native speakers and writers, is the only manner to espy the earlier phases of a language. Thus, the question of factoring in semantics is imperative for the analysis of syntactic *Productivity* from a diachronic viewpoint (Perek, 2019:5-6).

After perusing a number of works on semantic factoring in the case of syntactic *Productivity* and their shortcomings, Perek suggests another potential framework for evaluating semantic similarity which, he says, had been mentioned by some scholars but not been investigated for examining syntactic *Productivity*. This framework is "automatic automation" which, he claims, refrains from the necessity for hand-checking by human annotators—a quite expensive and time-consuming process. Similarly, says Perek, the vector-space model he uses is quite profitable for demonstrating "distributional semantics and its key computational linguistic implementation" because it can be employed to do the following: "augment corpus data with information about lexical meaning in an automatic, data-driven way that, to a large extent, dispenses with the need for human semantic intuitions" (Perek, 2019:7).

The third section elaborates on the "distributional semantics and vector-space models" briefly introduced by Perek at the end of the second section of his essay. Here, he begins by pointing out that distributional semantics has been and remains the most utilized and successful framework for semantic analysis in the field of Computational Linguistics (particularly for practical applications in "word-sense disambiguation," "automatic thesaurus generation," and "information contraction"). The framework allows an analyst to cull from the maxim that words appearing in similar milieus are inclined to share affiliated denotations. Accordingly, a technique for retrieving the denotations of words is via their distributions. In brief, as Perek puts it, "in distributional semantics, the semantic similarity between two words is related to the number of their shared frequent collocates in a vast corpus of naturally occurring texts. Conversely, differences in the distributional profile of two words are expected to correspond to differences in their meanings" (Perek, 2019:8).

The fourth section is in regards to a case study of "the *hell*-construction." To summarize the lengthy discussion on this topic is unnecessary since Perek is generous enough to provide free access to his article via the Internet. Only a few points will be presented here. Perek begins by stating the syntactic form "V *the hell out of* NP" as they appear in the following sentences culled from the Corpus of Contemporary American English (COCA): "(a) *Snakes just scare the hell out of me*; (b) *It surprised the hell out of me when I heard what he's been accused of*; (c) *Damn this man loved the hell out of this woman*; (d) *Me and Jeff want to beat the hell out of each other*; and (e) *You might kick the hell out of me like you did that doctor*" (Perek, 2019:11). This construction, he adds, is generally employed with two-participant verbs and fundamentally manifest in two-argument constructions whereby the post-verbal argument comes after the phrase *the hell out of*. He also notes that unlike a regular transitive construction, the *hell*-construction typically communicates an intensifying function, albeit in a broader sense (Perek, 2019:11).

Finally, after his lengthy deliberations on the nature of the corpus data from the Corpus of Historical American English (COHA), the vector-space model, evaluation of the vector-space model, and statistical analysis, Perek concludes that a distributional technique for analyzing meaning yields a fitting operationalization of semantic similarity and allows an analyst to employ approaches like data visualization and statistical analysis when quantitative explanations are necessary. He adds that the technique also engenders the possibility of evaluating the effect of contrasting features of the semantic space—beyond density—on *Productivity* (for details, see Perek, 2019:12-27).

African Linguistic Evidence

Beginning with Russel G. Schuh (1989), who in his work on "low tone raising" (LTR) in the Hausa language broaches the phenomenon of linguistic *Productivity*, he tells us that words in which the vowels are now long should be subject to LTR since the productive rule is applicable in such a case. He also says that a significant proportion of these words were borrowed long before lengthening and determiners stopped being routine and productive surface occurrences.

Next, Schuh in his other work titled "Palatalization in West Chadic" (2002a), demonstrates that a kind of morphological palatalization that is systemic and productive is operative in the Biu-Mandara languages. He also shows that Duwai and Ngizim languages possess a productive rule that changes alveolar consonants other than lateral fricatives, *l* and *r*, to their palatal counterparts when followed by an *–i*. He additionally shows that the procedure for the palatalization of alveolars and consonants with palatal coarticulation is virtually 100% productive in the Hausa language.

Also, in his "The Locus of Pluractional Reduplication in West Chadic" (2002b), Schuh alludes to the fact that in the Hausa and some of the other Chadic tongues, "reduplication of a root initial syllable" acts as the "productive method of forming pluractional verbs." In addition, he demonstrates that the productive pattern of verbal pluractional formation in the Hausa language is the doubling of the fist consonant / vowel / consonant (CVC) of a verb stem, with some phonological procedures that commonly modify the final consonant of the initial syllable and/or its vowel. He further illustrates that the only completely productive kind of reduplication that exists in the Bole language is the initial consonant / vowel (CV) of the verb.

In addition, Gideon Omachonu and David Abraham apprise us that in the Igala language, a West Benue-Congo tongue used in north-central Nigeria, compounding is a highly productive word-formation process in terms of varieties/forms and functions. These scholars also note that semantic criteria above phonological and syntactic considerations are favorable when defining "compoundhood" and distinguishing compound words in Igala. They go on to explain this phenomenon as follows:

> Apart from noun + noun compounds, other compound types such as synthetic and verbal compounds which could equally be accounted for using semantic criteria have been attested in Igala. Even though endocentric nominal compounds in Igala generally obey Left-Hand head rule (N1 as head), synthetic compounds in Igala, it has been observed, exhibit the possibility for either Left-Hand head position or Right-Hand head position. However, the Right-Hand head rule...may not be very productive in the language. It is adjudged as an exception rather than the rule. Lastly, in addition to the general function of lexical expansion through creation of new lexical categories or lexemes, compounding has been used copiously in naming concepts, particularly foreign institutions, ideas, items or objects and concepts that were hitherto non-existent in Igala (Omachonu and Abraham, 2012:201; also quoted in Bangura, 2014:31).

Thereafter, we can learn from Gerrit J. Dimmendaal about a quite profitable aspect of *Productivity* of Ngiti (or South Lendu, and also called Ndruna by its speakers), a Nilo-Saharan or Central Sudanic language spoken in the Democratic Republic of the Congo. As he points out, when morphemes enter into a collocation, they affect one another both formally and semantically. This semantic specialization, he notes, accumulates as the frequency of utilization of the *Productivity* of a morpheme increases. The outcome may be the loss of semantic transparency. Another factor that leads to the loss, or even obsolescence, of *Productivity* he adds is the formal neutralization of the fundamental and derived forms. As such, the derived causative involving tonal modification in the Ngiti language is no longer formally different from the root form from which it is derived (Dimmendaal, 2000:174).

Dimmendaal also tells us that the application of the productive rules of word formation is just one of the several techniques used for the extension of lexicon. To illustrate this point, he cites the case of the word *kòòmàà hááyá* (literally "return back" describing a "reactionary person or ideology") in the Hausa language (a Chadic tongue with approximately 44 million speakers, spoken mainly in northern Nigeria and Niger, and also in Benin, Burkina Faso, Cameroon, the Central African Republic, Chad, Congo, Eritrea, Germany, Ghana, The Sudan, and Togo). He also explains that the word is "an exocentric compound, consisting of a verb plus (adverbial) complement, which was created by analogy with an existing and highly common patterns of exocentric compound formation in the language" (Dimmendaal, 2000:181).

Thereupon, one can also glean from my writing in the *Culpability of the Tran-Atlantic Slave Trade* (2017) that Llaila Olela Afrika enlightens us about the following aspects of African linguistic *Productivity* in American English:

a) The first concept of "wonderland" is from the African concept of "Gondwanland" which denotes the notion of "paradise on earth" (Afrika, 1989:7; Bangura, 2017:128).

b) Sassafras Mountain, which is the highest point in South Carolina, was given the name derived from the African language: i.e. Sassa (Afrika, 1989:21; Bangura, 2017:128).

c) The gospel song "Swing Low Sweet Chariot" was used as a cryptic code when Whites were around from "Swing Low Sweet Harriet," indicating that Harriet Tubman was around to take enslaved Africans to freedom through the Underground Railroad to Canada (Afrika, 1989:22; Bangura, 2017:128).

d) African words that became American slang words became popular (Afrika, 1989:22; Bangura, 2017:129).

e) "Tall tales" or African myths were published by David Crockett. These myths have been traced to the book titled *Coming Forth by Day and by Night*, or the so-called *Egyptian Book of the Dead*, written before 1500 BC (Afrika, 1989:22; Bangura, 2017:129).

Furthermore, Katherine E. Rohrer edifies us that enslaved Africans in the United States employed the linguistic features they retained from their African tongues to develop novel Black language varieties in order to express distinctiveness from their enslavers. This phenomenon, she says, was motivated by the fact that the enslaved Africans were compelled by their human desire and capability to maintain significant elements of their past while adjusting to the realities of their new milieu in their new environment. Concomitantly, very early in their arrival to North America, enslaved Africans started building the linguistic system that would evolve into what came to be known as Black English. And, just like any other living language, Black English is biotic and is in a state of perpetual transformation (Rohrer, 2008:197).

Moreover, as Anne Njoki Hinga reveals, African linguistic *Productivity* as it relates to the teaching of the German language in Kenyan schools is a quite challenging undertaking because the endeavor is heavily affected by the multilingual nature of the existing linguistic context. She describes the linguistic context as one which entails approximately 40 indigenous languages, with English and Kiswahili as official languages and the learners are fluent in one or more indigenous languages. The prestigious role enjoyed by the English language is due to it being (a) the main medium of instruction from upper primary school onward; (b) a mandatory subject in primary and high schools; (c) the medium of communication is business, government, industry, and judiciary; and (d) perceived as the language of the Kenyan elite. German therefore serves as the second foreign language after English for learners who select to study it as a subject in secondary school. Therefore, according to Hinga, the indigenous and English languages affect the manner students acquire the German language. She adds that "these learners are mainly either bilingual in English and Kiswahili or in some cases trilingual in the indigenous language, English and Kiswahili, and it is very likely that they already bring with them not just the knowledge of other languages but also learning experiences and strategies as they start to learn German" (Hinga, 2010:1).

Consequently, as it concerns *Productivity*, as Hinga observes, a majority of the learners of the German language, who express themselves in spoken but mostly in written form, use their competence in the English language to acquire German (Hinga, 2010:2). She explains the reason and provides an example for this *Productivity* strategy as follows:

> This dominance of the English language tends to aid communication since English and German are similar in wide portions of the basic vocabulary as well as grammar. Moreover, German and English share a writing system and literary tradition since they are both western Germanic languages. For example, English spelling is very similar to German spelling. In this case, English acts as a true friend. This is especially so for cognates. A good example here is *der Park* / *'park.'* One should on the other hand beware of false friends. For example, *das Gift* / is best translated as '*poison*' and has nothing to do with *a present*. However, in some cases, there occurs negative transfer and this leads to errors. A good example is word order. Word order is dissimilar in both languages; for example, in subordinate clauses, gender marking of nouns that is present in German is non-existent in English. English, for the students of German, is in this regard more like a foe (Hinga, 2010:2).

Finally, Nicholas Wade inculcates that since words evolve expeditiously, many linguists believe that the origins of language cannot be tracked down to a very distant past. The Indo-European family language tree, the oldest so far reorganized, dates back to approximately 9,000 years. Thus, it was confounding for many linguists when Quentin D. Atkinson, a biologist at the University of Auckland in New Zealand, when analyzing the sounds of languages spoken across the globe recognized an ancient signal indicating that the origin of modern human language is in southwestern Africa. The finding parallels the findings from fossil skull and DNA (i.e. deoxyribonucleic acid, a self-replicating material which is present in nearly all living organisms as the main constituent of chromosomes; it is the carrier of genetic information.) that Africa is the original home of modern humans. African linguistic *Productivity* is therefore manifest in how phonemes (consonants, vowels, and tones which constitute the simplest features of language) are found by Atkinson to exist in some 500 languages spoken across the globe, with a language environment using fewer phonemes the farther that early humans had to travel from Africa to get that environment. For example, according to Wade, Atkinson found that "some of the click-using languages of Africa have more than 100 phonemes, whereas Hawaiian, toward the far end of the human migration route out of Africa, has only 14. English has about 45 phonemes" (Wade, 2011:1). This configuration of diminishing diversity based on distance from Africa, Wade adds, mirrors the inveterate diminishing genetic diversity also based on distance from the continent (Wade, 2011:1).

African Mathematical Evidence

To start with, Cheikh Anta Diop provides us with a notion of African mathematical *Productivity* when he discussed Problem 79 in the *Ahmes Papyrus* which deals with a geometric progression of ratio seven and 3,500 years later was appropriated for the Mother Goose rhyme in English. He interprets Problem 79, called the problem on "the inventory of *goods* contained in a house," from hieroglyphics as follows: "there are seven houses; in each house there are seven cats, each cat kills seven mice; each mouse had eaten seven grains; each grain would have produced seven *hekat*. What is the sum of all the enumerated elements? (What is the total of all these things?)" Diop adds that the reasoning of the scribe yields the same numerical result as the application of the formula for modern algebra that yields the following sum of a geometric progression (Diop, 1981/1991:270):

$$S = a\frac{r^n-1}{r-1} = 7 \text{ X } \frac{16{,}807-1}{7-1} = 7 \text{ X } \frac{16{,}806}{6} = 7 \text{ X } 2{,}801 = 19{,}607$$ [45]

Beatrice Lumpkin informs us that almost three and a half millennia later, the preceding geometric progression was appropriated and named the "Mother Goose" rhyme in English as follows (Lumpkin, 1986:103):

> "As I was going to St. Tues,
> I met a man with seven wives.
> Each wife had seven cats..."

Next, while Ayo Bamgbose's treatise on the topic is not based on a mathematical model, it is his discussion of the subject of Mathematics itself (especially being a leading linguist) that is quite instructive in espying a notion of *Productivity* in his essay. We get a sense of *creativity* as African learners must study mathematics in an imported official language which forces them to develop "additive" instead of "subtractive" bilingualism in order to master the mathematical concepts. Unfortunately, as Bamgbose also points out, in addition to the practice propagating a negative perception of African languages, since the imported languages are privileged in the learning process, the education systems are not set up to nurture "additive" bilingualism in learning mathematics (Bamgbose, 2011:7). Bamgbose therefore proposes the following suitable solution:

> Ideally, a properly planned mother tongue-based multilingual education should make provision for the first language to be used as a medium of instruction at least for the entire duration of basic education, while the imported official language is taught as a subject.

> The advantage of this model is that children will develop enough confidence in their own languages before they make any transition into another language. They will not lose mastery of their language in trying to transit into another language, rather they will add to what they already have (Bamgbose, 2011:7).

Also, we can decipher an idea of *Productivity* from the investigation of the effects of climate change on the production of crops in Africa by Jerry Knox, Tim Hess, Andre Daccache, and Tim Wheeler. The idea is evident in these scholars' conclusion that African national governments and regional decision-makers have developed a comprehension of the infinite linguistic and extra-linguistic evidence that has been provided to explain climate change impact on crop productivity to continue in at least the short- and medium-term (Know et al., 2012:7). Before making this determination, the scholars had first determined, through their mathematical model, the following: (a) areas which already lack food security are significantly vulnerable to climate change effects on food productivity, an asseveration that is supported by a meta-analysis of 52 original studies systematically selected from a collection of 1,144 studies; (b) by the 2050s, the project mean change in the yield of all crops is -8%; (c) the estimated mean yield changes for wheat is -17%, maize is -5%, sorghum is -11%, and millet is -10%; and (d) for rice, no mean change in yield was identified (Knox et al., 2012:1).

In addition, in the article by Mustapha D. Ibrahim, Sahand Daneshvar, Mevhibe B. Hocaoğlu, and Olasehinde-Williams G. Oluseye, *creativity* is manifested vis-à-vis how policy makers across Africa South of the Sahara were able to logically weigh efficiency and productivity in their healthcare systems. As the coauthors point out, via their mathematical model, the policy makers determined that government plans of action (i.e., "the rule of law and government efficacy") would affect the efficiency of healthcare systems more compared to public spending on health and, thus, decided that increased spending on healthcare systems is less important than managing the systems' resources more efficiently (Ibrahim et al., 2019:371).

Furthermore, *creativity* is evident when Alcardo Alex Barakabitze and his seven coauthors, employing a variety of information communications technology (ICT)-based mathematical models of education and a transformational framework, point out that through rhetoric African policy makers have been able to address fundamental questions pertaining to the utilization of ICT in several areas of education and student outcomes in the STEM (science, technology, engineering, and mathematics) fields. These areas include "retention and learning achievement, affordances that directly change the nature of teaching and learning, delivery mechanisms for teaching and learning, and the foundational ICT-based pedagogy that matters to current and future students" (Barakabitze, 2019:2).

Conclusion

The *Productivity* feature in the Ancient Kemetian/Egyptian language is evident in two concepts. One concept is *uati*, which denotes "production" and "creation." The other concept is *per-t seshu*, meaning "literary production," "Precepts of Amenemhat" (also referred to as the "Teaching of King Ammenemes I to His Son Sesostris" is a short Ancient Kemetian poem of the sebayt genre written during the early Middle Kingdom; it is a highly dramatic monologue delivered by the ghost of the murdered 12th Dynasty pharaoh Amenemhat I to his son Senusret I, describing the conspiracy that killed Amenemhat and warning his son to trust anyone), and "the produce of the scribe" (for these translations, refer to Budge, 1978). In fact, it is no exaggeration to assert that to this day, Ancient Kemetian scribes are revered for their ingenuity in developing fresh meanings from discrete units in every field.

Also, Antonio Loprieno alludes to a much higher degree of *Productivity* demonstrated by the "nexal negation of sentences with verbal forms of adjectival verbs" in the Ancient Kemetian language. He adds that the rules for negating verbal sentences remain constant for sentences with *nj nfr.n=f* negation of the unmarked present state and the *nj nfr=f* employed for negating past quality (Loprieno, 1995:126).

In addition, according to Ruth Kramer, of the three numbers in the Ancient Kemetian language—i.e. (1) the singular, (2) the dual, and (3) the plural—it is the dual that encountered limited *Productivity* by the time of Middle Kemet (i.e., from around 2050 BC to around 1710 BC). The dual was restricted to "the number two, paired body parts (e.g., lips) and fixed expressions (e.g., *ṯꜣ-wj* land-DU 'the two lands' = Egypt)"/Kemet (Kramer, 2009:5). She further mentions that "The dual was realized by a suffix, *-wj* for masculine nouns and *-tj* for feminine nouns. The plural was also realized by a suffix, *-w* for masculine nouns, and *-wt* for feminine nouns" (Kramer, 2009:5).

Chapter 12

The Role for Mathematicians in the African Renaissance

It behooves me to mention here that the discussion in this chapter is culled from the section with the same title in my essay, "Domesticating Mathematics in the African Mother Tongue" (2014) which, as I said earlier, was the impetus for this book. While the copyright of the article, like all others in the *Journal of Pan African Studies*, rests with me as the author, as there is no copyright transfer protocol, I did, however, out of courtesy, seek permission from the editor of the journal to reuse the section in this book. Unfortunately, the many correspondences via several media yielded no response. I later learned that the editor had passed away on June 25, 2019, after a long illness.

As stated earlier, mathematicians did play a significant role in the development of African societies during antiquity and that they are sorely needed now to help fulfill the African Renaissance. In this section, I suggest two ways mathematicians can work with experts in other disciplines to do so: (1) Mathematics of Sustainability and (2) Mathematics of Tipping Point. These are discussed in the following two sections.

Mathematics of Sustainability

To assert that African countries, like those on other continents, are quite ambitious when it comes to launching development projects is hardly a matter of dispute. The constant challenge, however, has been to ensure the Sustainability of many of these projects. The challenge prompted, to the best of my knowledge, two high-level meetings and two scholarly books to address it. The first meeting was convened by the United Nations on November 21, 2011, at which the organization's Deputy Secretary-General Asha-Rose Migiro emphasized the point that sustainable development is critical for addressing the economic, social and environmental challenges in Africa (United Nations News Centre, 2011). The second meeting was held in Cape Town, South Africa and brought together scientists and lawmakers from across Arica to discuss ways to promote sustainable development in Africa South of the Sahara (Voice of America, 2012). The first book titled *Sustainable Development in Africa* (2005) comprises papers from various contributors suggesting tools from various disciplines to examine factors limiting sustainable development in Africa and making recommendations. The second book, *Chemistry for Sustainable Development in Africa* (2013) entails a series of papers dealing with current chemical research in Africa focusing on environmental chemistry, renewable energies, health and human well-being, food and nutrition, and bioprospecting and commercial development. The major shortcoming with all of these efforts from an African-centered perspective is that they offer no encompassing or indigenous African scientific approach to address the challenge of Sustainability in Africa. An answer to the challenge is Mathematics of Sustainability. Before discussing this aspect, it makes sense to first answer the following poignant question: What is all this talk about Sustainability?

Each year, we use more and more natural resources to maintain our way of life. Some of these resources are renewable and will never run out while others are fading quickly. Natural gas is slowly fading and efforts are being made to harness other energy sources such as solar, wind and, of course, electricity. Houses are being powered by solar panels, cars by electricity, and businesses by wind. These efforts help to keep our environment safe while also giving us the energy we need to survive. These new sources of energy do not only help us personally and environmentally, but also economically. These new ventures generate new businesses, and they can be quite profitable as well. Governments, different organizations, and numerous corporations are making an effort to provide a sustainable way of life, making the concept of *Sustainability* quite significant right now in science. But, what exactly is Sustainability? How do we use it? What purpose does it serve? According to the United States Environmental Protection Agency (USEPA),

> Sustainability is based on a simple principle: Everything that we need for our survival and well-being depends, either directly or indirectly, on our natural environment. Sustainability creates and maintains the conditions under which humans and nature can exist in productive harmony, that permit fulfilling the social, economic, and other requirements of present and future generations (USEPA, 2013).

For North Carolina State University (NCSU), the "concept of sustainability centers on a balance of society, economy and environment for current and future health. Responsible resource management in all three areas ensures that future generations will have the resources to survive and thrive" (NCSU, 2013). The NCSU definition adds the economy, while the EPA definition just deals with society and nature living in harmony. Thus, the NCSU definition is more comprehensive by providing us with a nexus among society, nature, and economy.

With the broader definition, the fields in which the concept of Sustainability is being used can now be discussed. Just looking at the definition, it would seem that Sustainability can be used in almost any field, since it deals with society, nature, and the economy. But for the sake of this essay, let us look at the available research. Steve Cohen, director of the Master of Public Administration program in Environmental Studies at Columbia University, has found that there has been much growth in the field of Environmental Studies. As he notes,

> One of the most encouraging trends I've seen in recent years has been the growth of environmental studies programs in many American Universities. For the past decade I've directed an MPA program in Environmental Science and Policy at Columbia University's School of International and Public Affairs and Earth Institute. Today over 500 graduates of that program are working as environmental professionals all over the world (Cohen, 2012).

Cohen would go on to discuss how the school later launched a program in Sustainability Management and also how many other schools are developing Sustainability programs.

Kate Galbraith wrote an article for the *New York Times* titled "Sustainability Field Blooms on Campus" (2009), describing how schools are ramping up their Sustainability programs to meet the demand in society. Galbraith also discussed how many older individuals are going back to school for jobs in Sustainability fields. As she put it,

> Mr. Gressens's trajectory will sound familiar at educational institutions across the country, whose continuing education arms have seen influx of students interested in the relatively new field of sustainability. At Harvard's extension school, enrollment in environmental courses has soared by more than 70 percent in two years, according to the university, which has responded with new offerings in fast-changing fields like carbon neutrality and environmental economics (Galbraith, 2009).

Obviously, these institutions realize that this is a fast-growing market and are boosting their programs to meet the needs of society.

But still, what industries are using Sustainability the most? This question leads to an article by Helen Coster in the *Forbes* magazine which listed the top 100 most sustainable companies. The list, which includes companies such as General Electric, H&M, Nokia, and Vodafone, shows that there exists a wide range of industries that use Sustainability. With this concept's popularity and the growing threat of damage to our world, more industries and academic institutions are taking Sustainability more seriously and really putting their thoughts and money behind how to become more sustainable. As Cohen states,

> The challenges of global sustainability are also starting to influence higher education. The liberal arts have begun to reemphasize science, including earth sciences, ecology and environmental biology. The physics and finance of energy production and consumption are of obvious and increasing importance. My own academic field of organizational management is beginning to study the growing importance of the earth's resources in determining organizational effectiveness (Cohen, 2012).

So Sustainability is being used in almost every field now and everyone has a feeling of responsibility to society and the earth.
Along with the numerous fields in which Sustainability is being used, it is also being employed to study different aspects of life. In essence, Sustainability has an almost limitless potential for studying anything: the way we live, the way we do business, the way we go to school, and even the way we deal with other nations. Sustainability is used to study how we impact society, the environment, and the economy, which is mostly how we live in general. In short, Sustainability allows us to study the aspects of the way we live our lives.

A number of mathematical models have been used or suggested for measuring Sustainability. The following is an inventory of the works in the chronological order in which they were published:

a) Amir Abbas Rassafi and Manouchehr Vaziri (2003) use pairwise correlation to examine Sustainability in 52 African countries.

b) The Global Community WebNet (2004) suggests the use of the Gross Environmental Sustainable Development Index (GESDI), which quantitatively describes quality indicators rather than merely measuring different variables by focusing on the interactions between four major quality systems: (1) people, (2) economic development, (3) environment, and (4) availability of resources.

c) Marion Hersh (2006) employs state space system representations, mental models, systems methodologies, optimization, mathematical decision making, multi-criteria problems, multi-criteria decision support methods, and fuzzy set operations to develop models for sustainable development.

d) System Analysis Decisions (2009) utilizes a gauging matrix mathematical model to measure sustainable development, which is characterized by two main constituents: (1) security of population (Isec) and (2) quality of their life (Iql). The generalized sustainable development measure is then presented by a quaternary (Q) with an imaginary scalar part j(Isec), which describes the security of people, and a real vector part (Iqf), which describes quality of life in the space with three dimensions: (1) economic (Iec), (2) ecological (Ie), and (3) social-institutional (Is).

e) Jason Phillips (2010) applies ideas of coupled environment-human systems to develop a mathematical model of sustainable development.

f) Kalu A. Ugwa and A. Agwu (2012) tap difference equations, ordinary differential equations, partial differential equations, optimization modeling, simulation modeling, and function fitting data modeling to develop a mathematical tool for sustainable development in Nigeria.

g) George Assaf et al. (2013) suggest life cycle analysis (LCA) that is useful for comparing the impacts of different activities in order to make hard choices for creating a climate-friendly sustainable energy future.

h) Colin W. Clark (2013) implements the "logistic" model of population dynamics to measure sustainable resource management.

i) Charles R. Hadlock (2013) uses differential equations, probability statistics, simulation, evolutionary game theory, and network theory to show how human society can be sustained in a rapidly changing world.

j) Mark Alan Hughes and Elise Harrington (2013) show how the mathematics of Greenworks is comprised of baselines, absolute versus relative measurements, and interpolation.

k) David Kung (2013) employs numbers, statistics, and functions to demonstrate the connection between social justice and Sustainability.

l) Fred S. Roberts (2013) calls for the utilization of statistical and algorithmic methods of data analysis, advanced computational tools, and new cryptographic tools to aid us in making management and policy decisions about the electric power grid.

m) The Sheward Partnership, LLC (2013) applies the LEED Rating System, which is the internationally-accepted benchmark for high performance green buildings, to calculate return on investment in sustainable design.

The problem with these models from an African-centered perspective is that a majority of them are not indigenous to the continent. The few that have their origins in Africa are not so identified, probably because the authors are oblivious to the fact.

Indeed, the notion of Sustainability was quite present in the minds of classic Egyptians. In Egyptian Hieroglyphic, the word *sānkhu* represented the word "peasant," "sustainer," or "vivifier" (Budge, 1978:645), meaning to give or bring life to, animate, to make more lively, intense, or striking enliven (http://www.thefreedictionary.com). In fact, in the funerary texts of later periods (c. 300 BC), Isis, the goddess of motherhood, magic and fertility (380-362 BC), due to her great power, was considered the mourner, protector and sustainer of the deceased in the afterlife. Isis' "protective and sustaining roles were extended to nobles and commoners and her power and appeal grew to the point that she eventually eclipsed Osiris (her brother and husband) himself and was venerated by virtually every ancient Egyptian" (http://egyptian-gods.99k.org/isis.html).

From Joyce Tyldesley, we also learn that Pharaoh Hatchepsut/Hatshepsut (throne name Maatkare, meaning Foremost of Noble Ladies, 1508-1458 BC), in her effort to ensure Sustainability in ancient Egypt, launched a number of bold initiatives. First, Hatchepsut increased the demand for scribes which led to an expansion of the education system. Second, the Pharaoh poured a lot of money into the building of monuments to ensure that the past is resuscitated and her name lives on forever, which in turn benefited a lot of artists and sculptors (Tyldesley, 1996:39-40). Third, she launched theological and technological advances to make sure that "New Kingdom Egypt remained tied to Middle and Old Kingdom Egypt by an unparalleled continuity of language, religion and artistic/architectural convention, and by the idiosyncratic Egyptian view of the world, and the position of Egypt, her people and her gods within the world, which had remained basically unchanged for over a thousand years" (Tyldesley, 1996:5-6). Fourth, she dreaded uncontrolled chaos, which was by definition a *maat*-less period, and therefore did everything in her power to avoid it by engaging in many international trade and diplomatic ventures (Tyldesley, 1996:8-10). Fifth, she advocated for a large number of offspring, especially for the royal family which was plagued by the dearth of children, with sons particularly being in short supply and single daughters being the norm (Tyldesley, 1996:73-74).

In terms of the Mathematics of Sustainability in Africa, Diop teaches us plenty. First, the geometric progression of ratio seven, commonly called the problem on "the inventory of *goods* contained in a house," and the classical formula for the arithmetic progression, which I mentioned earlier, were invented in ancient Egypt. The mathematical series also invented included the following polygonal numbers (Diop, 1981/1991:270-272):

a) trigonal or triangular numbers whose ratio, or difference of terms, is equal to 1

b) tetragonal or square numbers such as 1, 4, 9, 16, 25, etc. whose difference in terms is 2, meaning that of the odd numbers 1, 3, 5, 7, 9, etc.

c) *gnomons*, or successive rectangular belts, that allow one to obtain all squares from one unit square from the arithmetic series of odd numbers

d) summing up of arithmetic progression: $S = \frac{N}{2}[2a + (n-1)d]$[49]

In sum, these are all the elements that led to the "discoveries" of Pythagoras' theorem. This fact is echoed by Corinna Rossi when she points out that the first unambiguous evidence of the use of the theorem is a Demotic papyrus that dates back to the third century BC on which the numbers involved correspond to three triplets which might have been used as early as the Old Kingdom to construct some pyramids. Rossi observes that this truth is reflected in the insistence of the late Greek sources to link the 3-4-5 triangle to ancient Egypt (Rossi, 2007:64). She goes on to also cite the following most quoted passage from Plutarch's *De hide et Osiride* ("His and Osiris") to prove the point:

> ...the better and more divine nature consists of three elements—what is spiritually intelligible, the material and the element derived from these, which the Greeks call the cosmos. Plato is wont to call what is spiritually intelligible the form and the pattern of the father; and the

material he calls the mother, the nurse, and the seat and place of creation, while the fruit of both he calls the offspring and creation. One might suppose that the Egyptians liken the nature of the universe especially to this supremely beautiful of the triangles which Plato also in the Republic seems to have used in devising his wedding figure. That triangle has a vertical of three units of length, a base of four, and an hypotenuse of five, which is equal, when squared, to the squares of the other two sides. The vertical should thus be likened to the male, the base to the female, and the hypotenuse to their offspring; and one should similarly view Osiris as the origin, Isis as the receptive element, and Horus as the perfect achievement. The number three is the first and perfect odd number; four is the square of the even number two; five is analogous partly to the father and partly to the mother, being made up of a triad and a dyad (Rossi, 2007:64).

Second, the following simple equations were invented by classic Egyptians (Diop, 1981/1991:272-274):

a) the abstract and symbolic notation of the unknown quantity by assimilating the privileged number 1 for X

b) the method of false supposition:
A quantity (any) plus $^1/_7$ of it = 19; in modern algebraic form,
A quantity X plus $^1/_7$ of it = 19; find X

c) the simple equation of a more complex form:
$X + \frac{2x}{3} - \frac{1}{3}(x + \frac{2x}{3}) = 10$

d) the simple complex equation of $^2/_3$, plus $^1/_{10}$ of a number = 10; immediately the equation is written as follows:
$(\frac{2}{3} + \frac{1}{10})X = 10 \quad X = 13\frac{1}{23}$

Third, classic Egyptians invented the following quadratic equations (Diop, 1981/1991:273-274):

I. $\begin{cases} X^2 + Y^2 = 100 \\ 4X - 3Y = \ \ 0 \end{cases}$

II. $\begin{cases} X^2 + Y^2 = 400 \\ 4X - 3Y = \ \ 0 \end{cases}$

The explicit wording of this system of simultaneous equations for the quadratic problem is the following: How to divide 100 into pairs, so that the square root of one of them is $^3/_4$ that of the other. Stated in modern symbols, the equation reads as follows:

$$X^2 + Y^2 = 100 \rightarrow Y = \frac{3}{4}X \rightarrow X^2 + \frac{9}{16}X^2 = 100$$

Fourth, the balance of quantities (*pesou*) was the invention of classic Egyptians. An example is that if the *pesou* of a loaf of bread is 12, then it means that this loaf of bread contains $^1/_{12}$ of a bushel. Also, the sacred right-angled triangle shows that some mathematical proportions had a divine essence. Furthermore, a table found from 2000 BC to 600 AD has a division of the number 2 by the odd numbers from 3 to 101. The table $2/n$ has no error. Even more, the Egyptian notation of fractions of 2200 BC was as follows (Diop, 1981/1991:274):

$$\frac{1}{17} \text{ of a silver talent} = 352 + \frac{1}{2} + \frac{1}{17} + \frac{1}{34} + \frac{1}{52} \text{ drachmas}$$

Finally, in the area of arithmetic, the following three observations by Diop are poignant (1981/1991: 276:276):

1) The originality of Egyptian arithmetic is that it does not require any effort of memory. Multiplication and division are reduced to addition after a series of duplication. Only the multiplication by 2 needs to be known in order to easily carry out the most complex calculations.

2) Operations on the fractions generally deal with fractions whose numerators equal a unit; however, the Egyptians knew and used also the following complimentary fractions: $^2/_3$ (frequently used); $^3/_4$, $^4/_5$, $^5/_6$ (less frequently used). Owing to this fact, the Egyptians had made a table of factorization of the fractions of type $2/n$, including the fractions from $^2/_5$ to $^2/_{101}$.

3) By the third millennium onward, the Egyptians had already invented decimal notation and discovered or portended the zero. The proportional divisions were known. They knew how to rigorously extract the square root of any number, even fractional ones. All these are evident in the examples dealt with by the Egyptian scribe Ahmes in the *Berlin Papyrus.*

And even more interesting is that, as Diop demonstrates, many Egyptian mathematical terms have survived in the Wolof language (1981/1991:226-228).

It should also be noted here that other scientific areas utilized to promote Sustainability in ancient Egypt included (a) Astronomy, encompassing the phases of the moon, the calendar, the orientation of the monuments, and the decans; (b) Medicine, (c) Chemistry, including the etymology of the word chemistry itself and the metallurgy of iron; and (d) Architecture, comprising the mathematical bases and the aesthetic canon of Egyptian art (for more on this, see Diop 1981/1991:278-307).

Mathematics of Tipping Point

As Brett Cherry (2011), the Rensselaer Polytechnic Institute (2011) and Christian Kuehn et al. (2013) assert, the recent political revolutions that took place in Tunisia and Egypt are characteristic of Tipping Points, as they could result in vast social changes throughout North Africa and the Middle East, and affecting the rest of the world in the process. This conjecture raises at least two related questions: (1) Why are some African political regimes suddenly changed, with no obvious trigger other than a slowly changing environment, while others that have the attributes of vulnerability thrive? (2) How can this phenomenon be predicted, so that mechanisms can be put in place to facilitate a political system's stability? A solution to this dilemma is Mathematics of Tipping Point.

When thinking of society and everything that happens on a day-to-day basis, when do things break or change? It seems as if life moves at a constant pace, consistently keeping us in a loop. At some point, it feels like one small event pushes us over the edge and leads us to a major change that breaks us from this loop. This happens every day in our personal lives, societies, businesses, schools and, especially, our environment. What I am hinting at here is *Tipping Point.* This relatively new scientific concept seems to be popping up everywhere. But, what exactly is it? In my attempts to keep things simple, I will use a definition from dictionary.reference.com, which states that Tipping Point is "the point at which an issue, idea, product, etc. crosses a certain threshold and gains significant momentum, triggered by some minor factor or change; the point in a situation at which a minor development precipitates a crisis" (dictionary.reference.com). There is another definition of Tipping Point from the Mathematical Association of America (MAA), which states that

> The term "tipping point" describes the moment when a system suddenly changes state, with no obvious trigger other than a slowly changing environment. Tipping points are difficult to predict and difficult to

> reverse. Examples range from capsizing boats to fishery collapse; they include financial market crashes, the poverty trap, melting polar ice caps, shifts in ecosystems, and mood changes (MAA, 2013).

Like the example I used earlier, Tipping Point is when a situation like your life has a major change from a minor event in it. Since I used very simple definitions, it is not hard to see how Tipping Point is used across multiple disciplines or how it is applied in studying numerous phenomena. Malcolm Gladwell wrote a book titled *The Tipping Point: How Little Things Can Make a Big Difference* (2002), and in this book, he discusses how he uses Tipping Point to describe how events take or have taken placc in history as follows:

> The tipping point is the biography of an idea, and the idea is very simple. It is the best way to understand the emergence of fashion trends, the ebb and flow of crime waves, or, for that matter, the transformation of unknown books into bestsellers, or the rise of teenage smoking, or the phenomena of word of mouth, or any number of the other mysterious changes that mark everyday life is to think of them as epidemics. Ideas and products and messages and behaviors spread just like viruses do (Gladwell, 2002:3).

So Gladwell uses Tipping Point to examine all academic disciplines, making it a very useful concept for all fields of study. Gladwell also argues that no matter how different the disciplines may be, they can still be explained by using Tipping Point. For instance, Gladwell cites the rise of hush puppies and the fall of New York's crime rate as examples:

> The second distinguishing characteristic of these two examples is that in both cases little changes had big effects. All of the possible reasons for why New York's crime rate dropped are changes that happened at the margin; they were incremental changes. The crack trade leveled off. The population got a little older. The police force got a little better. Yet the effect was so dramatic. So too with Hush Puppies. How many kids are we talking about who began wearing the shoes in downtown Manhattan? Twenty? Fifty? One Hundred—at the most? Yet their actions seem to have single handedly started and international fashion trend (Gladwell, 2002:3).

This truism has propelled me to come to the conclusion that indeed Tipping Point can be used in all fields. A profitable question then is the following: Since Tipping Point can be applied in all fields, how exactly is it being used? To examine how the methodology is being used, I look at the field of Ecology because the environment is a big issue these days and is receiving a lot of attention from scholars, policymakers, and activists. In the Ecology field today, numerous studies are being done to determine how much harm is being done to the environment and how we can reverse these effects and live a more sustainable life. Tipping Point is used in these studies to determine what major change will occur from small occurrences. As Jeremy Hance points out,

> Even climate change, which some scientists might consider the ultimate tipping point, does not fit the bill, according to the paper. Impacts from climate change, while global, will not be uniform and hence not a tipping point as such; Local and regional ecosystems vary considerably in their responses to climate change, and their regime shifts are therefore likely to vary considerably across the terrestrial biosphere; from a planetary perspective, this diversity in ecosystem responses creates an essentially gradual pattern of change, without any identifiable tipping points (Hance, 2013).

Scientists are using the methodology of Tipping Point to determine whether or not we will have a great change in our ecosystem. In the article titled "Harnessing Math to Understand Tipping Points" (2013) Mary Lou Zeeman is reported to use mathematics to understand the Earth's climate. According to the article, "Co-director of the Mathematics and Climate Research Network which connects researchers deploying mathematics to better understand Earth's climate, Zeeman wants to empower those she teaches to seek solutions to today's environmental ills" (MAA, 2013:17). Later in the article, Zeeman is noted to have given an example of how small events led to a much bigger change such as how the Earth moved out of its icy state millions of years ago as follows:

> Coming back to a question posed earlier in the lecture, Zeeman explained how Earth emerged from the snowball state it evidently occupied 600 million years ago. While volcanoes spewed greenhouse gases into the atmosphere, the planet's icy blanket prevented these gases from being removed via weathering of rocks or oceanic absorption of carbon. As greenhouse gases increased, the resilience of the stable snowball state decreased until, when it shrunk to nothing, the system tipped. A very hot Earth resulted: all the ice melted, rocks and oceans were exposed, and all those mechanisms for drawing carbon out of the atmosphere [were]...accessible again (MAA, 2013:17).

This example shows how Tipping Point is used to explain how our world is habitable right now. If Tipping Point can be used in something as large and serious as climate change and even how our world survived the ice age, then it can be employed to explain relatively minor phenomena such as things that occur in our daily lives or small changes in business.

Indeed, Tipping Point can be utilized to explore and learn quite a lot about various physical and socioeconomic phenomena, especially the aspect of how things change. How we move from small events to major events and how these events change our lives are pertinent to Tipping Point. It is used to study changes in crime rates and fashion trends like Gladwell shows us, or in climate change such as Zeeman shows us. By using Tipping Point and Sustainability together, we can find a better way to promote Africa's development. Also, by studying numerous fields across the board with these methodologies, we will have a better understanding of each other's disciplines, which will, in turn, give us a better understanding of each other's thoughts. The pace of our world is rapidly increasing with the advancement in technology. It is therefore imperative to have methodologies such as Tipping Point and Sustainability to help us understand our world and how cataclysmic events come about.

To determine whether or not accurate mathematical representations of physical or socioeconomic systems display Tipping Point behavior, bifurcation analysis is employed. As I recount in my book titled *African Mathematics: From Bones to Computers* (2012:121-125), while mathematicians are generally consistent in defining the term "first order differential equation," this is not the case for the term "Bifurcation." Instead, they tend to view Bifurcation as a description of certain phenomena. A common perception is that a system undergoes a bifurcation if and only if the global behavior of the system, which depends on a parameter, changes when the parameter varies. Consequently, a general and a theoretical definition of Bifurcation can be delineated.

Bifurcation can be generally defined as a differential equation system that undergoes a qualitative change in its orbit structure, as one or more parameters of the dynamical system are changed. A Bifurcation is said to occur when a small smooth change made to the parameter values, or the Bifurcation parameters, of a system causes a sudden qualitative or topological change in its short-term dynamical behavior. Bifurcations result in both continuous systems, described as ordinary differential equations (ODEs), delay differential equations (DDEs), and partial differential equations (PDEs); and discrete systems, described by maps.

Theoretically, Bifurcations can be defined as the mathematical study of changes in the qualitative or topological structure of the integral curves of a vector field, or the solutions of a differential equation. Essentially, a Bifurcation is said to occur at a parameter value where a number of solutions changes. In essence, Bifurcation Theory examines structurally unstable dynamical systems. Dynamic stability refers to perturbations in the phase space—i.e., the stability of fixed points and limit cycles, and structural stability refers to perturbations in the function space—i.e., the topological stability of orbit structures.

Bifurcations have been typologized in two ways. One typology, which is less technical, identifies three types of Bifurcations: (1) Subtle Bifurcations, whereby an attractor changes type; (2) Catastrophic Bifurcations, whereby attractors appear out of, or disappear into, the blue: and (3) Explosive Bifurcations, whereby attractors drastically change size. The other typology, which is more technical, identifies two types of Bifurcations: (1) Local Bifurcations that take place through changes in the local stability properties of the equilibria (i.e. fixed points), periodic orbits, or other invariant sets as parameters cross through critical thresholds; and (2) Global Bifurcations that take place when larger invariant sets of a system collide with one another, or with equilibria of a system. These Bifurcations cannot be detected purely by a stability analysis of the equilibria. Since the second typology is predominantly employed in Mathematics, a bit more discussion of their characteristics is in order.

A Local Bifurcation emerges when a parameter change causes the stability of an equilibrium to change. In continuous systems, this corresponds to the real part of an eigenvalue of an equilibrium passing through zero; in discrete systems, this corresponds to a fixed point having a Floquet Multiplier (relating to the class of solutions to linear differential equations) with modulus equal to one. In both cases, the equilibrium is non-hyperbolic at the Bifurcation Point (more on this later). The topological changes in the phase portrait of the system can be confined to arbitrarily small neighborhoods of the bifurcating fixed points by moving the Bifurcation parameter close to the Bifurcation Point—thus, the name "local."

Technically, the continuous dynamical system described by the ODE is considered as follows:

$$\dot{X} = \mathrm{f}(\mathrm{x}, \lambda) \quad \mathrm{f}: \mathbb{R}^{\mathrm{n}} \times \mathbb{R} \rightarrow \mathbb{R}^{\mathrm{n}}$$

A Local Bifurcation emerges at (x_0,λ_0), if the Jacobian Matrix $\mathrm{d}f_{x0,\lambda 0}$ has an eigenvalue with zero real part. If the eigenvalue is equal to zero, the Bifurcation is said to be a steady-state Bifurcation; if the eigenvalue is non-zero, but purely imaginary, then it is said to be a Hopf Bifurcation (more on this later). For discrete dynamical systems, the following system is considered:

$$X_{n+1} = f(X_n, \lambda).$$

Consequently, a Local Bifurcation emerges at (x_0,λ_0), if the Jacobian Matrix $df_{x0,\lambda0}$ has an eigenvalue with modulus equal to one. If the eigenvalue is equal to one, the Bifurcation is either a Saddle-node—often called Fold Bifurcation when it appears in maps, Transcritical or Pitchfork Bifurcation. If the eigenvalue is equal to -1, it is a Period-doubling (or Flip) Bifurcation; otherwise, it is a Hopf Bifurcation. The following are examples of Local Bifurcations:

a) Hopf Bifurcation—a Bifurcation in which a fixed point of a dynamical system loses stability as a pair of complex conjugate eigenvalues of the linearization around the fixed point cross the imaginary axis of the complex plane.

b) (b) Horse Saddle Bifurcation—a saddle point that is a minimax: i.e., a local minimum of maximum depending on the intersecting plane used.

c) Monkey Saddle Bifurcation—an example of an immersion, it is a surface defined by the equation to which it belongs to the class of saddle surface, and its name derives from the observation that a saddle for a monkey requires three depressions: two for the legs and one for the tail.

d) Neimark (Secondary Hopf) Bifurcation—a Bifurcation through which a system loses its stable period one operation.

e) Period-doubling (Flip) Bifurcation—a Bifurcation in which the system switches to a new behavior with twice the period of the original system.

f) Pitchfork Bifurcation—a generic Bifurcation in which a symmetric solution changes its stability; it occurs generically in systems with symmetry.

g) Saddle-node Bifurcation—a Bifurcation in which two fixed points of a dynamical system collide and annihilate each other.

h) Transcritical Bifurcation—a Bifurcation characterized by an equilibrium having an eigenvalue whose real part passes through zero.

A Global Bifurcation emerges when "larger" invariant sets, such as periodic orbits, collide with equilibria. Such a collision leads to changes in the topology of the trajectories in the phase space which cannot be confined to a small neighborhood, as is the case with Local Bifurcations. The changes in topology extend out to an arbitrarily large distance—hence, the name "global." The following are examples of Global Bifurcations:

a) Blue Sky Catastrophe, when a limit cycle collides with a nonhyperbolic cycle.

b) Global Saddle (Fold) Bifurcation, when a system is expressed in polar coordinates.

c) Harmoclinic Bifurcation, when a limit cycle collides with a saddle point.

d) Heteroclinic Bifurcation, when a limit cycle collides with two or more saddle points.

e) Infinite-period Bifurcation, when a stable node and saddle point simultaneously occur on a limit cycle.

It should be noted that Global Bifurcations can also involve more complex sets such as chaotic attractors.

The codimension of a Bifurcation refers to the number of parameters which must be varied for the Bifurcation to emerge. This corresponds to the codimension of the parameter set for which the Bifurcation emerges within the full space of parameters. Saddle-node Bifurcations are the only generic Local Bifurcations which are really codimension-one; the others all have higher codimension. Nonetheless, Transcritical and Pitchfork Bifurcations are also often perceived to be codimension-one, because the normal forms can be written with only one parameter.

An example of a well-studied codimension-two Bifurcation is the Bogdanov-Takens bifurcation, which is characterized by the fact that for the particular parameter value the vector field has a singularity whose linearized field has a double zero eigenvalue, while the other eigenvalues have nonzero real part. The latter condition is imperative.

The Bifurcation Diagram is a very useful tool employed to illustrate Bifurcations. It helps to show the possible long-term values—equilibria or periodic orbits—of a system function of a Bifurcation parameter in a dynamical system. It is normal practice to show stable solutions with a solid line and unstable solutions with a dotted line.

Named after their pioneer, Mitchell Feigenbaum, the Feigenbaum Constants are two mathematical constants used to express ratios in a Bifurcation Diagram. The first Fegenbaum Constant is expressed as follows:

$$\delta = 4.66920160910299067185320328\ldots$$

where sequence A006890 in the On-line Encyclopedia of Integer Sequences (OEIS) is the limiting ratio of each bifurcation interval to the next, or between the diameters of successive circles on the real axis of the Mandelbrot set. While Feigenbaum originally related this number to the Period-doubling Bifurcations in the logistic map, he later showed it to hold for all one-dimensional maps with a single quadratic maximum. As a result of this generality, every chaotic system that corresponds to this description will bifurcate at the same rate. This Feigenbaum's constant is often employed to predict when chaos will arise in such systems even before it does. The second Feigenbaum Constant (sequence A006891 in the OEIS),

$$\alpha = 2.502907875095892822283902873218\ldots$$

is the ratio between the width of a tine and the width of one of its two subtines (with the exception of the tine closest to the fold).

Both of the preceding numbers are applicable to a large class of dynamical systems. They are believed to be transcendental, albeit that claim remains to be proven.

Finally, Catastrophe Theory (which is also a special case of more general Singularity Theory in Geometry) considers the special case in the study of dynamical systems where the long-run stable equilibrium can be identified with the minimum of a smooth, well-defined potential function called Lyapunov Function. Since small changes in certain parameters of a nonlinear system can cause equilibria to appear or disappear, or to change from attracting to repelling and vice versa, this can lead to large and sudden changes of the behavior of a system. Investigated in a larger parameter space, Catastrophe Theory reveals that such Bifurcation points tend to occur as part of well-defined qualitative geometrical structures. Other aspects of Catastrophe Theory include Elementary Catastrophes, the potential functions of one active variable (Fold Catastrophe, Cusp Catastrophe, Swallowtail Catastrophe, and Butterfly Catastrophe), potential functions of two active variables (Hyperbolic Umbilic Catastrophe, Elliptic Umbilic Catastrophe, and Parabolic Umbilic Catastrope) and Arnold's notation, which are beyond the scope of this paper.

As I also discuss in *African Mathematics: From Bones to Computers*, mathematical aspects of Bifurcations are ubiquitous in ancient Egyptian artefacts. Indeed, classic Egyptians were quite involved with symbolism. Thus, their artefacts were designed and aligned cosmologically. The cosmological, astronomical, astrophysical, astrological and mathematical abilities of classic Egyptians were highly organized and quite advanced, and they performed truly remarkable feats of urban planning and architecture (Bangura, 2012:125).

For example, in their work on Auric Time Scale, Sergey Smelyakov, Geoff Stray and Jan Wicherink (2006a & 2006b) assert that in terms of proportions, the scale is reflected in the artefacts of Egypt. According to them, the Auric Time Scale Γ with the radix φ not only describes the crucial epochs in the known history of the Earth and humanity, it also adequately reflects the spectrum of the basic periods of Nature and society, including the geological and Solar activity cycles and artefacts of Egypt. In essence, embedded in Egyptian artefacts are representations of those events that have their impact on worldly events.

In order to get a sense of the connection between Egyptian artefacts and Bifurcations, it makes sense to quote verbatim what Smelyakov and his colleagues mean by Auric Time Scale. According to them, "In the narrow sense, the Auric Time Scale (ATS) is the series G of the Golden section powers (infinite to both ends), which is accompanied by the series G* = 2G of its double values; for its unity, phi^0 = 1 the average Solar cycle length T0 = 11.07 year, or Tropical year is taken" (Smelyakov et al., 2006a:1). They add that "In the broad sense, the ATS is a theory suggesting that for both principal aspects of time (periods and chronology), these series, T and T*, describe the length of the bulk of the basic periods in nature and society, and the major part of the most important historical and natural events in the evolutionary context, respectively" (Smelyakov et al., 2006a:1). I examine two examples of early Egyptian artefacts with aspects of Bifurcation in detail in *African Mathematics: From Bones to Computers*: (1) ancient Egyptian calendars and (2) the pyramid texts of Set (Seth, Seti, Setekh, Setesh, Suty, Sutekh) of Nubet (Bangura, 2012:126-144).

Chapter 13

General Conclusion

The preceding chapters provide ample evidence that African languages exhibit all nine design features that can facilitate the domestication of mathematics. It is therefore only fitting that, as I point out in my works on African Mathematics mentioned in those chapters, Africa was the center of mathematics history for tens of thousands of years. But, as I also add, since European languages have been privileged over African languages in mathematics education in the continent, we must do at least two very important things, if we are to remedy the problems that were stated earlier in the introductory chapter of this book so that Africa can benefit from the tremendous opportunities mathematics offers. First, schools should encourage the use of African languages in order to nurture and promote those languages. Second, more research on the connection between language and the learning and teaching of mathematics from a political point of view is necessary.

What Unites the Preceding Chapters

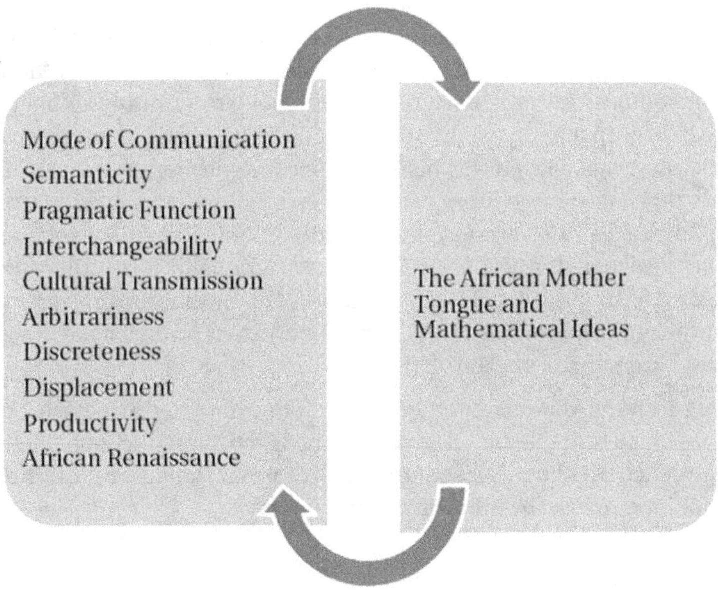

Figure 13.1: African-Centered Linguistics and Mathematics Framework.
Source: Self-generated by the Author

The discoveries in the previous chapters make it possible to delineate the following conceptual framework diagrammatically represented in Figure 13.1. A conceptual framework, according to Chava Frankfort-Nachmias, David Nachmias, and Jack DeWaard, is one in which "descriptive categories are systematically placed in a broad structure of explicit propositions, statement of relationships between two or more empirical properties to be accepted or rejected" (2014:38).

Based on Figure 13.1, the explicit proposition suggested is quite straightforward: i.e. mode of communication, semanticity, pragmatic function, interchangeability, cultural transmission, arbitrariness, discreteness, displacement, and productivity (the nine design features of language), in addition to the need for African mathematicians to play a role in an African Renaissance, are all factors that influence the African mother tongue and mathematical ideas and vice versa. Following Frankfort-Nachmias et al., the conceptual framework therefore constitutes a third level of theory that is amenable to empirical testing.

A salient question that emerges here is the following: What do the design features and the African Renaissance tell us and do not tell us? Beginning with the design features, Anouschka Bergmann, Kathleen Currie Hall, and Sharon Miriam Ross tell us that every language displays all nine design features and that a communication system that lacks these features cannot be a language. They add that only human communication embodies these attributes. They also point to the fact that although what are referred to as "formal" languages such as formal logic employed to write computer programs and mathematical proofs encompass all nine design features, they are critically different from languages like Apache, English, Spanish, etc. Thus, a child cannot acquire a computer language like C++ as his/her native tongue. Bergmann et al. further mention that (a) some people construct as a hobby languages that imitate human languages to be employed in a fictional universe such as the Klingon language designed for the television series *Star Wars*, (b) others develop languages to be utilized for international communication such as the Esperanto language, and still (c) others construct artificial languages for pleasure (Bergmann et al., 2007:23).

Bearing in mind the unsettled question about how to distinguish between languages such as Apache, English, Spanish, etc., on the one hand, and languages such as Klingon, Esperanto, and "formal" languages, on the other, Bergmann et al. make the following distinctions:

> *Natural languages* [are] those languages that have evolved naturally in a speech community. The lexicon and grammar of a natural language have developed through generations of native speakers of that language. A

> *constructed language*, on the other hand, is one that has been specifically invented by a human and that may or may not imitate all the properties of a natural language. Some constructed languages have the potential to become natural languages, if they are learned by native speakers and adopted by a speech community; this is the case with Modern Hebrew, which was reconstructed from ancient Hebrew and then adopted by a particular community (Bergmann et al., 2007:23).

Bergmann and her colleagues add:

> The distinction between constructed languages and formal languages is that formal languages are not the sort of system that a child can acquire naturally...other types of languages do, in fact, exist. Thus, the design features help us distinguish language from other nonlinguistic communication systems, but we need more criteria to ensure that a system is a natural language and not either an artificial language or a man-made code (Bergmann et al., 2007:23).

As it pertains to the African Renaissance, there is a plethora of works on the topic. Here, I focus on the work of Mwalimu Ali Al'Amain Mazrui, the doyen of African Studies, who wrote extensively on the topic. Some of his writings on the subject are discussed in the following paragraphs separately and in the chronological order in which they were published for the sake of cohesion.

To begin with, Mwalimu Mazrui in his article titled "Mazruiana: Between the Renaissance African and the African Renaissance" (1998) suggests that the African Renaissance must be undergirded by Pan-Africanism and the need for strengthening "Aca-Media"—i.e. the nexus between academia and the media. He introduces the concept of *Afrenaissance* and defines it as the African Renaissance involving the re-Africanization of the African continent based on seven imperatives: (1) history, (2) language, (3) talent, (4) dignity, (5) self-reliance, (6) humane self-rule, and (7) creative accommodation to globalization. He then recommends that selections should be made of African Best Books of the Century like the 100 Best Novels published in the English language in the 20th Century (Mazrui, 1998; see also Bemath, 2005:81).

Next, in his "The Language of Francophonie and the Race of the Renaissance: A Commonwealth Perspective" (1999), Mwalimu Mazrui discusses the Francophonie and Commonwealth roots of the African Renaissance, the features and goals of this Renaissance, and the aforementioned seven imperatives of Afrenessaince, and how all of these aspects should be anchored on Afro-optimism and African civilizations and cultures. He iterates his call for the selection of the 100 African Best Books of the Century, and adds that, contrastingly, the African Renaissance would transcend linguistic and racial apartheid of the French and English languages (Mazrui, 1999; see also Bemath, 2005:77).

Thereafter, Mwalimu Mazrui in his article, "The African Renaissance: A Triple Legacy of Skills, Values and Gender" (2001a), identifies two starting points for the African Renaissance. The first starting point is the establishment of a healthy equilibrium between the environment and technology. The second starting point involves three revolutions dealing with (1) gender relations, (2) skills, and (3) values. He suggests that we consider issues such as culture, dependency, development, and modernization and their essence for this African Renaissance. Defining development as modernization without dependency, he mentions the following five factors to be addressed in order to reduce, if not eliminate, dependency: (1) indigenization, (2) domestication, (3) diversification, (4) horizontal integration, and (5) counter-penetration. Central to dealing with dependency, he asserts, is culture, with its following seven attendant functions: (1) as lenses of perception, (2) as means of communication, (3) as a basis of stratification, (4) as a spring of motivation, (5) as a standard of judgment, (6) as a pattern of production and consumption, and (7) as a foundation of identity. He harbors no doubt that if African culture is fully mobilized as an ally in the struggle to combat dependency, a non-Western route to modernity is possible for Africa (Mazrui, 201a; see also Bemath, 2005:53).

Thereupon, in his article titled "Shifting African identities: The Boundaries of Ethnicity and Religion in Africa's Experience" (2001b), Mwalimu Mazrui reiterates his call for an Afrenaissance or the Renaissance in Africa underpinned by the seven imperatives stated earlier. The appeal here is in relation to the connection between ethnicity and religion and the way these two primordial forces impact contemporary African politics, and the push for national and regional integration leading to new identity formation and enlargement of the political scale. Consequently, Mwalimu Mazrui examines the interactions (a) between primordiality and statecraft, (b) between religion and ethnicity, and (c) between domestic forces and international repercussions. He also looks at how within the African context, Christianity and Islam can be a parochial force that splinters a nation while ethnicity can emerge as a transnational and Pan-African exertion. He further explores the following four models as ways politics and religion interact in the African milieu: (1) a theocracy, (2) a religious nation, (3) an ecumenical state, or (4) a secular state, arguing that the African state is more secular compared to the one in the West. He ends by discussing the difference between primary violence ("dispute over the boundaries of a political community") and secondary violence ("disagreement over the goals of a community without wanting to redraw the boundaries"), and the reasons a racial war did not break out in South Africa (Mazrui, 2001b; see also Bemath, 2005:55).

Afterward, Mwalimu Mazrui in his "Linske elite heft Afrika's originaliteit om zeep geholpen" ("The Left Elite Has Killed off Africa's Original Thought") (2001c) directs our attention to the very important roles women can play in the African Renaissance and that due to their positions of power and being less violent than men, we will see less military coups d'état and wars in Africa. This assertion by Mwalimu Mazrui is based on a number of aspects that he broaches. The first aspect is about the issues pertaining to the loss of "intellectual capacity" in post-colonial Africa that he imputes on the "leftist elite" that killed diverse debate and originality in Africa's centers of higher education. The second aspect relates to the exploitation of ethnic differences for political reasons as opposed to class differences proffered by leftist political scientists. The third aspect is in regards to the need for political space to accommodate ethnic differences and the preference of an ethnic political party over an ethnic militia. The fourth aspect is apropos to the need for a less fundamental form of Islam that can mix with the secular state without causing dangerous tensions (Mazrui, 2001c; see also Bemath, 2005:240).

Implications

The ultimate question in this book is therefore the following: What are the implications of the findings in the preceding chapters? In terms of linguistics, at least three implications can be discerned from the discoveries in this book. The first linguistic implication is making very clear the nexus between African languages and culture, thereby leading us to eschew the use of inappropriate terms to characterize African phenomena. Vincent B. Khapoya attends to this issue quite well. For starters, he recounts that during the past millennia and across the African continent, the people established relatively different customs and vocabularies to explicate their lives as families settled and expanded in different ecozones. Variants of African culture and languages were developed as people interacted in their ecozone and migrated to others. Thus, anthropologists and linguists treat culture and language as being virtually the same, as every culture or subculture is comprehensively described in the particular language or dialect that characterizes it. Concomitantly, every language or dialect is a distinct reflection of the name of the culture or subculture. Therefore, for most Africans today, despite the fact that African ethnicity encompasses many variants and has undergone many redefinitions over time, one's home "people" (i.e. those with whom we grow up, know, or are related to via extended networks of kinship groups) continues to be an vital point of reference (Khapoya, 1998:13).

Next, according to Khapoya, a very common and offensive question that Africans in the United States execrate is the following: "Oh, you are from Africa; what tribe are you from?" In addition to Africans, even non-African Africanist scholars find such terms as "tribe," "natives," "primitive," "superstitious," etc. to describe Africans as derogatory. Khapoya cites works that have shown that the intricacies of the every-changing social relationships render such terms nonsensical. He adds that having been popularized by colonial administrators and Western scholars writing about Africans, the term "tribe," in particular, remains a staple in birth certificates, identification cards, and tax records of many independent African states (Khapoya, 1998:13).

Also, Khapoya states that due to their different kinship systems, languages, rituals and traditions, anyone who is familiar with linguistic groups such as Hausa, Igbo, and Yoruba (the three largest linguistic groups in Nigeria) and Kalenjin, Kikuyu, Luhya, and Luo (the four largest linguistic groups in Kenya) can distinguish them from the other linguistic groups in their various countries. He adds that these larger linguistic groups are comprised of many subcultures with each speaking its own variety of the language. He further mentions that these subgroups tend to identify themselves as distinct ethnic groups and would even fight against related and neighboring subcultures (Khapoya, 1998:14).

In addition, Khapoya informs that the term "race" is now regarded by Africanist scholars as being even less accurate than the term "tribe" for distinguishing people. He traces the attempt by Eurocentric scholars to classify Africans and other peoples according to skin color and other physical features such as hair type, head form, lips, nose form, etc., leading them to designate four races of humankind or "Great Chain of Being"—(1) white, (2) yellow, (3) red, and (4) black skin colors in this order—to the 1930s. Other Eurocentric scholars have gone as far as trying to prove that skin color—genetically related to hair and skin color—could possibly be correlated with civilization, head size, intelligence, language, etc. But as Khapoya correctly indicates, after more than many decades of trying to correlate skin color with IQ (intelligence quotient) scores, their hypotheses remain untenable (Khapoya, 1998:14-15).

The second linguistic implication has to do with contributing to Cheikh Anta Diop's postulates for promoting linguistic unity in Africa. He provides three postulates on this issue. His first postulate deals with the "choice of language on a local scale in the framework of a given territory." Citing the case of Senegal, he argues that the relationships among all of the languages in the country must be demonstrated before a choice is made about which one should serve as a choice of language on a local scale for the territory. He suggests that setting out linguistic rules that allow for the systematic passage from Wolof forms to Serer, Peul-Toucouleur and Diola allowed him to show the deep kinship ties that unify the various linguistic groups of Senegal. He states that the same can be done for other African territories (Diop, 1974/1978:9).

In addition, Diop asserts that the essentiality of demonstrating the linguistic nexus hinges upon how we are educated about local particularities that occasionally unify some of the linguistic groups as mini-nations. In a small country such as Senegal, he suggests, Wolof is the logical choice for a national language or a language of government, since all minority groups are at the minimum bilingual, speaking their languages in addition to Wolof. He adds that the local context must be given great consideration when making a choice; in the case of Peul-Toucouleur, for instance, it is a minority language in Senegal but not in Futa Jallon or the Northern Cameroons (Diop, 1974/1978:9).

Diop's second postulate is about the "elevation of the selected language to the level of a modern cultural and government language." He alludes to the fact that it is through the Wolof language that scientists are systematically introducing the necessary concepts to explain the sciences, philosophy, etc. He therefore hopes that the Senegalese government would implement a cultural policy that would develop the language to an optimal level. He believes that the development could be sped up by artificial, albeit effective, means such as establishing literary prizes and a national commission that would develop an academic dictionary and others in the various disciplines, and translating scientific works into the language (Diop, 1974/1978:9-10).

Also, Diop insists that the work must start now if even only on a limited scale to demonstrate that it is possible to elevate the status of any African language to that of any European cultural language. He points to what has been done with the Wolof language as an exemplar and, therefore, calls for similar work to be done for each African territory using the same framework. In order to reduce the possibility of offending regional sensibilities, he enjoins that the same criteria must be applied in selecting a territorial language with the same delicate study beforehand of ethnic, linguistic and other ruminations (Diop, 1974/1978:10).

Furthermore, Diop maintains that even to this day, we Africans still use the languages of the former colonial masters as the prerequisite to hold public office and to conduct the business of government. He also mentions that to this day, the majority of the populations of most of the African states are not literate in the colonial languages. He therefore argues that for voters who are not literate in the European languages to elect representatives who must be literate in those languages is inadequate and unjust. Hence, he proffers that the utilization of the former colonial masters' languages is a tactic to avoid the complaints of the people who, while may be illiterate in those European languages, have a great deal of good sense (Diop, 1974/1978:10).

Diop's third postulate concerns the "choice of language on a continental scale." He begins by stating that when Wolof has been demonstrated to be a viable national language to be used for government and cultural activities, then it could inspire the choice of one language among the major ones to serve similar purposes at the continental level. The selected language will cover all territorial languages the same manner as the Russian language did in the various Socialist Republics during the Soviet Union (Diop, 1974/1978:11) and has continued to this day in most of those former republics. He recommends that the choice of such a language must be made by a competent inter-territorial commission that is deeply patriotic and abnegates concealed jingoism (Diop, 1974/1978:11).

Next, Diop proposes that after the continental language has been selected, it should be taught as a required course in every curriculum of every secondary school in all of the African territories. As textbooks on various subjects have been written in the selected African language, the European languages will be placed in a secondary position as elective courses. While still being able to get secondary and higher education in a territorial language, a citizen of any territory would be obligated to learn and speak the continental language fluently. As Diop also informs us, this framework was officially accepted as a necessity and character of African linguistic unity by Black writers and artists at their Rome convention during Easter of 1959 and by the Federation of African Students in France during their seminar at Rennes in July of 1959 (Diop, 1974/1978:11).

Thereafter, Diop recommends that the European languages should be used during the transitional phase but their utilization should be terminated as soon as it is feasible so that Africa does not become a "super-Switzerland." He avers that there is nothing to be gained by the simultaneous use of the European languages alongside the selected African language (Diop, 1974/1978:11).

Thereupon, Diop urges us, Africans, to be guarded about the subtle project to "Anglo-Americanize Black Africa," especially given the fact that many of the former European colonies in Africa were British. He contends that this joint British and American project is contrary to accepted "intellectual" norms and suggests that the other former French, Portuguese, Spanish and other colonies must make English their lingua franca, thereby propelling the English language as the continental tongue. Thus, for Diop, no matter how one looks at it, linguistic unity based on a foreign language is equivalent to "cultural abortion." Hence, he deftly warns and calls upon us Africans to be vigilant against this Anglo-Saxon machination as follows:

> It would irremediably eventuate in the death of the authentic national culture, the end of our deeper intellectual and spiritual life and reduce us to perpetual copycats, having missed out on our historical mission in this world. Anglo-Saxon cultural, economic, social and even political hegemony would thereby be permanently guaranteed throughout Black Africa. We must remain radically opposed to any attempt at cultural assimilation coming from the outside: none is possible without opening the way to the others (Diop, 1974/1978:12).

Afterward, Diop points out that it is a fallacy for one to state that it makes no difference to a speaker of the Wolof language whether he adopts another African language or a European one. This is because, as he intuitively indicates, an African that is educated in another African language besides his/her own is less culturally alienated than an African who is educated in a European language as a substitute to his/her African language. Similarly, he asserts, a Frenchman who is educated in the Italian language would be less culturally alienated than if s/he were to be educated in the Zulu or the Arabic language. This disparity in cultural interest that prevails between African and European languages, Diop declares, is a reality that we must never ignore (Diop, 1974/1978:12).

Additionally, Diop affirms that the influence of language is so enormous that the European colonialists felt comfortable of withdrawing politically from African countries and still control the continent's cultural, economic and spiritual realms. The Europeans to this day believe that Africans would never do away with the European languages and view such a decision to be disappointing, inacceptable and ungrateful. Therefore, Diop argues that as long as we Africans continue to use European languages in our secondary school curricula, our declarations of independence are not a radical cultural severance from that of the Europeans (Diop, 1974/1978:13).

Also, Diop mentions that due to the significant challenges of dealing with the numerous and different African languages, some Europeans are convinced that Africans are incapable of taking on the enormous task of selecting a continental African language that will replace the European ones for educational and administrative purposes. For these Europeans, Africans lack the creative thinking, human energy, and intellectual cogency to engage in such an endeavor. Collaterally, they believe that we Africans will never achieve the greatness of modern nations in economic infrastructure, much less national culture, based on our own authentic culture. Instead, they envision a continuation of the cultural mix they have established in Africa as the modus operandi. This state of affairs, as Diop avows, is legitimized by "pseudo-dialectical phrases" such as "efficiency, progress, planetary unity," etc. (Diop, 1974/1978:13).

Nonetheless, Diop proclaims that we Africans have no other choice but to take on the great challenge of selecting a continental language, if we are to eliminate from our minds the Eurocentric intellectual formulae and dainty thinking so that we can "enter resolutely upon the only truly dialectical path toward the solution of the problems that history forces upon us" (Diop, 1974/1978:13-14). Diop adds that this calls for us Africans to engage in authentic and active research that would yield effective solutions and seeing them come to fruition without the imposition of "intellectual guardianship" (Diop, 1974/1978:14).

Furthermore, Diop explains that our historical situations compel us Africans of this generation to do all that we can to honestly and comprehensively solve the very important problems that confront the continent. Our failure to do so, he says, will brand us as the generation that failed to insure the unified cultural survival of the African continent. As the generation which would have been intellectually and politically blind to promote Africa's cultural future, we would forever remain the "unworthy generation *par excellence*" (Diop, 1974/1978:14).

Finally, Diop concludes that by selecting one language for the African continent, every foreigner would be forced to learn and use the language in his/her dealings with us. This development, he adds, would not only simplify our communication with the outside world, it would also make international relations less complicated, thereby becoming relatively easier (Diop, 1974/1978:14).

The third linguistic implication concerns Diaspora Africans taking pride in their African dialects. Carter G. Woodson discusses certain aspects of this issue. To start with, he informs us that students in the United States were made to deride the African language when they studied languages in school. Even Black students were made to despise the language as a broken-down African tongue, instead of encouraging them to learn about its linguistic history which is more important to them than those of the European languages they were forced to study. The only time the African language was paid attention was when it came to preparing traders, missionaries, and public functionaries to go to Africa and exploit the people there. The number of people who were trained for the task was so tiny that it hardly attracted attention (Woodson, 1933:19).

Thus, Woodson advocates that after Black students have become proficient in the basics of the English language, the fundamentals of composition, and the main facts of its literature, they should not expend their energies and time on the advanced works on Anglo-Saxons, Chaucer and Shakespeare. Instead, these students should focus on the folklore of the African, the philosophy of his/her proverbs, the development of Blacks in the use of modern languages, and the works of Black writers (Woodson, 1933:150).

As it pertains to mathematics, at least two implications can be espied. The first mathematical implication deals with Ron Eglash's observations on the connection between African mathematics and culture. He commences by pointing out that anthropologists have long since noted that specific design themes typify the patterns produced in different cultures. In the case of Africa, he states that extraordinary examples of fractal structure characterize African settlement patterns and asks the question about whether precolonial Africans actually understood and applied fractal geometry (Eglash, 1999:4).

Next, after defining fractals as "the repetition of similar patterns at ever-diminishing scales," Eglash states that a "self-similar" attribute permeates traditional African settlements: i.e. "circles of circles of circular buildings, rectangular walls enclosing ever-smaller rectangles, and streets in which broad avenues branch down to tiny footpaths with striking geometric repetition" (Eglash, 1999:4). He adds that by employing fractal graphic simulation, this fractal structure is easily discernible when we look at aerial pictures of African cities and villages (Eglash, 1999:4).

In addition, Eglash is quite insistent that while we cannot simply dismiss the notion that precolonial Africans were knowledgeable about fractal geometry, we also must not simply assume that they did possess such knowledge. Therefore, he counsels us to listen to what those who designed and have used these structures say about them. In essence, there might actually be an intentional mathematical element to what might seem like a fortuitous or unintentional pattern (Eglash, 1999:5). Eglash then ends with this cardinal caution:

> If we only use examples where African knowledge systems successfully matched fractal geometry, we would not know its limitations. There are indeed gaps where the family of theories and practices centered around fractal geometry in high-tech mathematics has no counterpart in traditional Africa. Although such gaps are significant, they do not invalidate the comparison, but rather provide the necessary qualifications to accurately characterize the indigenous fractal geometry of Africa (Eglash, 1999:6).

The second mathematical implication pertains to adding to Paulus Gerdes' emphasis on the Africanization of mathematics. In his book titled *Geometry from Africa: Mathematical and Educational Explorations* (1999), Arthur B. Powell in the Foreword informs us that Gerdes in his work stimulates teachers in Africa to Africanize mathematics teaching by addressing questions about the origin of geometrical thinking. Powell adds that it was Gerdes who alluded to the possibility of the African roots of the theorem that indicates "the equality between the sum of squares of the lengths of the two sides of any right angle and the square of the length of its hypotenuse" (Gerdes, 1999:vi). Powell then goes on to make an essential observation that is worth restating here as follows:

> Gerdes demonstrates how one can discover this (i.e. the preceding mathematical notion) and other important mathematical ideas by attending to quantitative, qualitative, and spatial features in a diversity of ancient and modern objects in the multicultural mix of African civilization. Is it not possible, then, to claim African originality of these mathematical ideas? Clearly, cultural groups in other geographical (perhaps even planetary) regions do and can make similar claims. It, therefore, is crucial to strive for a history of mathematics that is unfettered with nationalistic or ethnocentric bias, that acknowledges and valorizes multicultural manifestations of mathematical ideas, and that entertains no primacy claims (Gerdes, 1999:vi).

Next, in his book, *Sipatsi: Basketry and Geometry in the Tonga Culture of Inhambane* (2009), Gerdes maintains that a goal of research on ethno-mathematics is to discover the prospects that would lead to the enhancement of the teaching of mathematics by couching the subject's curricula within the cultural context of students and instructors. Thus, according to him, mathematics curricula should b geared toward valuing the scientific knowledge that is embedded in a culture by employing this understanding to establish the bases to establish quicker and better paths for all students to accurately comprehend humanity's scientific heritage (Gerdes, 1999:19).

Thus, as I have suggested in my works on African Mathematics cited in the preceding chapters, "the essence of an African-centered approach to mathematics education is that it is imperative and urgent for Africans to be concerned about broader development and the linguistic-mathematical nexus as well as approaches to these phenomena that are undergirded by humanity or fellow feeling toward others. When African-centeredness is considered along with the idea of the socialization effects of developmental environments and the possibilities of reinforcement of these notions and contexts, the implications for African development, linguistic and mathematical processes appear vital" (Bangura, 2011a, 2011b, 2014, 2015a, 2019).

As I also mentioned in my works, "although compassion, warmth, understanding, caring, sharing, humanness, etc. are underscored by all the major world orientations, African-centered thought serves as a *distinctly African rationale* for these ways of relating to others. African-centeredness gives a distinctly African meaning to, and a reason or motivation for, a positive attitude towards the other. In light of the calls for an African Renaissance, African-centeredness urges Africans to be true to their promotion of good governance, democracy, peaceful relations and conflict resolution, educational and other developmental aspirations" (Bangura, 2011a, 2011b, 2014, 2015a, 2019).

In addition, as I further advocate in my works, "we ought never to falsify the cultural reality (life, art, literature) which is the goal of African-centeredness. Thus, we would have to oppose all sorts of simplified or supposedly simplified approaches and stress instead the methods which will achieve the best possible access to real life, language and philosophy" (Bangura, 2011a, 2011b, 2014, 2015a, 2019).

Bibliography

Abdoola, Fareeaa, Penelope S. Flack, and Saira B. Karrim. 2017. Facilitating pragmatic skills through role-play in learners with language learning disability. *South African Journal of Communication Disorders* 64, 1:1-12. Retrieved on July 13, 2019 from https://sajcd.org.za/index.php/sajcd/article/view/187/739

Abiodun, Rowland. 2014. *Yoruba Art and Language: Seeking the African in African Art.* Cambridge, UK: Cambridge University Press.

Adler, Ronald B. and Neil Towne. 1993. *Looking Out / Looking In* 7th ed. Fort Worth, TX: Harcourt Brace Jovanovich College Publishers.

Agbinya, Johnson Ilyeh. 2004. *Computer Board Games of Africa: Algorithms, Strategies and Rules.* Bellville, South Africa: Department of Computer Science, University of Western Cape.

Ajayi, Bade. 2016. A new model of *Ifa* binary system. *StudyLib.* Retrieved on July 10, 2019 from https://studylib.net/doc/7443994/a-new-model-of-ifa-binary-system

Akita, Kimi and Mark Dingenmanse. 2019. Ideophones (mimetics, expressives). *Oxford Research Encyclopedia of Linguistics.* Retrieved on July 07, 2019 from https://pure.mpg.de/rest/items/item_3012530/component/file_3060417/content

Akpabot, Samuel. 1975. Random music of the Birom. *African Arts* 8, 2:46-47.

Alamu, F. O., H. O. Aworinde, and W. I. Isharufe. 2013. A comparative study of Ifa divination and computer science. *International Journal of Innovative Technology and Research* 1, 6:524-528. Reterieved on February 01, 2016 from https://www.academia.edu/5315636/A_Comparative_Study_of_Ifa_Divination_and_Computer_Science

Amaglobeli, Givi. 2012. Semantic triangle and linguistic sign. *Scientific Journal of Humanities* 1, 1:37-40. Retrieved on April 5, 2019 from http://citeseerx.ist.psu.edu/viewdoc/download?doi=10.1.1.858.2350&rep=rep1&type=pdf

Andrea, A. J. 2002. The silk road: Afro-Eurasian interchange across the ages. Retrieved on July 14, 2019 from http://www.nerwha.org/files/The_Silk_Road_Essay_HST706.pdf

Animalu, Alexander O. E. and Jonathan Okeke Chimakonam. 2012. 4x4 magic square representation of complementary duality of African three-valued thought logic. *African Journal of Physics* 5:141-168. Retrieved on July 15, 2019 from https://www.researchgate.net/publication/292090546_4x4_Magic_Square_Representation_of_Complementary_Duality_of_African_Three-Valued_Thought_Logic

Arlotto, Anthony. 1972. *Introduction to Historical Linguistics.* Lanham, MD: University Press of America.

Armstrong, Bill and Don Davis. 2014. *Brief Calculus for the Business, Social, and Life Sciences* 3rd ed. Burlington, MA: Jones and Bartlett Learning.

Arwood, Ellyn Lucas. 2011. *Language Function: An Introduction to Pragmatic Assessment and Intervention for Higher Order Thinking and Better Literacy.* London, UK: Jessica Kingsley Publishers.

Asante, Molefi Kete. 1987. *The Afrocentric Idea.* Philadelphia, PA: Temple University Press.

Assaf, George, Eduardo Porta, Ralf Bredel and Cornelius Roschanek. April 2013. Creating a climate-friendly sustainable energy future: The role of mathematics. Mathematics Awareness Month Homepage. Retrieved on February 27, 2013 from http://www.mathaware.org/index.html

Assata Shakur Forums. January 2004. Rules for *Ifa* divination. Retrieved on February 01, 2016 from https://www.assatashakur.org/forum/traditional/34086-rules-ifa-divination

Assmann, Jan. 1994. Ancient Egypt and the materiality of the sign. In H. U. Gumbrecht and K/ L/ Pfeiffer, eds. *Materialities of Communication.* Stanford, CA: Stanford University Press. Retrieved on July 19, 2019 from https://archiv.ub.uni-heidelberg.de/propylaeumdok/2951/1/

Assman_Ancient_Egypt_and_the_Materiality_of_the_Sign_1994.pdf

Association for the Development of Education in Africa (ADEA). 2001. The role of African languages in education and sustainable development. *Newsletter Vol. 8 No. 4.* Retrieved on April 10, 2004 from http://www.adeanet.org/adea/newsletter/Vol8No4/en_n8v4_2.html

Ayoh, Ndi Chick Loveline. May 12, 2019. E-mail: <ayohndi44@gmail.com>.

Ayuninjam, Funwi F. 1998. *A Reference Grammar of Mbili.* Lanham, MD: University Press of America.

Babatunde, Emanuel. January 31, 2016. E-mail: <babemman2000@gmail.com>.

Badejo, B. Rotimi. 1989. Multilingualism in Sub-Saharan Africa. *Africa Media Review* 3, 2:40-53. Retrieved on February 27, 2013 from https://pdfs.semanticscholar.org/e0e4/b62fd1c43b23d507564affbbfca69da77b63.pdf

Bamgbose, Ayo. 2011. African languages today: The challenge of and prospects for empowerment under globalization. E. G. Bokamba et al. *Selected Proceedings of the 40th Annual Conference on African Linguistics.* Somerville, MA: Cascadilla Proceedings Project. Retrieved on July 17, 2019 from http://www.lingref.com/cpp/acal/40/paper2561.pdf

Banda, Felix. 2000. Formal and semantic dimensions of linguistic development: Implications for classroom English second language instruction in multilingual context. *Per Linguam: A Journal of Language Learning* 16, 2:1-14. Retrieved on July 05, 2019 from http://perlinguam.journals.ac.za/pub/article/view/136

Bangura, Abdul Karim. 1990. An analysis and description of the Temne language (preliminary notes). Paper presented at a seminar on Theoretical Linguistics at Georgetown University, Washington, DC, March 1990.

Bangura, Abdul Karim. 1991. *Multilingualism and Diglossia in Sierra Leone.* Lawrenceville, VA: Brunswick Publishing Corporation.

Bangura, Abdul Karim. 2000a. *Chaos Theory and African Fractals.* Washington, DC: The African Institution.

Bangura, Abdul Karim. 2000b. Measurable effects of multilingualism in Africa. *International Journal of the Sociology of Language* 146:111-117.

Bangura, Abdul Karim. 2000c. Book review of George Gheverghese Joseph's *The Crest of the Peacock: Non-* Bangura *European Roots of Mathematics 2nd ed. Nexus Network Journal* 3, 3.

Bangura, Abdul Karim. 2001. Book review of Ron Eglash's *African Fractals: Modern Computing and Indigenous Design. Nexus Network Journal* 2, 4.

Bangura, Abdul Karim. 2005. Ubuntugogy: An African educational paradigm that transcends pedagogy, andragogy, ergonagy and heutagogy. *Journal of Third World Studies* xxii, 2:13-54.

Bangura, Abdul Karim. 2011a. *African Mathematics: From Bones to Computers.* Lanham, MD: University Press of America.

Bangura, Abdul Karim. 2011b. *African-Centered Research Methodologies: From Ancient Times to the Present.* San Diego, CA: Cognella Press.

Bangura, Abdul Karim. 2011c. *The African National Anthem "Nkosi Sikelel' iAfrika": A Pragmatic Linguistic Analysis.* San Diego, CA: Cognella Press.

Bangura, Abdul Karim. 2014. Domesticating mathematics in the African mother tongue. *The Journal of Pan African Studies* 6, 1:12-58.

Bangura, Abdul Karim. 2015a. *Toyin Falola and African Epistemologies.* New York, NY: Palgrave Macmillan.

Bangura, Abdul Karim. 2015b. The colors of the flag of Sierra Leone: An Africancentric analysis. *The Journal of Sierra Leone Studies* 5, 2:11-19.

Bangura, Abdul Karim. 2015c. Bo School symbols: A deep African-centered analysis. *The Patriotic Vanguard.* Retrieved on September 14, 2015 from http://www.thepatrioticvanguard.com/bo-school-symbols-a-deep-african-centered-analysis

Bangura, Abdul Karim. January 31, 2016. E-mail. <theai@earthlink.net>.

Bangura, Abdul Karim, ed. 2017. *Culpability of the Trans-Atlantic Slave Trade: A Multidisciplinary Primer.* Lanham, MD: Hamilton Books/Rowman and Littlefield Publishing Group.

Bangura, Abdul Karim. 2019. *Falolaism: The Epistemologies and Methodologies of Africana Knowledge.* Durham, NC: Carolina Academic Press.

Barakabitze, Alcardo Alex et al. 2019. Transforming African education systems in science, technology, engineering, and mathematics (STEM) using ICTs: Challenges and opportunities. *Education Research International* 2019:1-40. Retrieved on July 17, 2019 from https://www.hindawi.com/journals/edri/2019/6946809/

Bemath, Andul Samed. 2005. *The Mazruiana Collection Revisited: Ali A. Mazrui Debating the African Condition* (an annotated and select thematic bibliography 1962—2003). Pretoria, South Africa: Africa Institute of South Africa; UK, USA, & India: New Dawn Press, Inc.

Benacerraf, Paul and Hillary Putnam, eds. 2004. *Philosophy of Mathematics: Selected Readings* 2nd ed. Cambridge, UK: Cambridge University Press.

Bergmann, Anouschka, Kathleen Curie Hall, and Sharon Miriam Ross. 2007. *Language Files: Materials for an Introduction to Language and Linguistics.* Columbus, OH: The Ohio University Press.

Biava, Christina. 2018. In linguistics, what do they mean by discreteness? And can you give some simple examples? *Quora*. Retrieved on May 30, 2019 from https://www.quora.com/In-linguistics-what-do-they-mean-by-discreteness-And-can-you-give-some-simple-examples

Blume, Werner et al. eds. 2019. *European Traditions in Didactics of Mathematics*. Cham, Switzerland: Springer Open. Retrieved on July 14, 2019 from https://link.springer.com/content/pdf/10.1007%2F978-3-030-05514-1.pdf

Bourderionnet, Oliver. 2008. Displacement in French/Displacement of French: The Reggae and R'n'B of Tiken Jah Fakolay and Corneille. *Research in African Literatures* 39, 4:14-23.

Brice, Henry. 2017. What are some examples of productivity in linguistics? *Quora*. Retrieved on June 18, 2019 from https://www.quora.com/What-are-some-examples-of-productivity-in-linguistics

Brown, Gillian and George Yule. 1983. *Discourse Analysis*. Cambridge, UK: Cambridge University Press.

Budge, E. A. Wallis. 1978. *An Egyptian Hieroglyphic Dictionary* vols. I & II. New York, NY: Dover Publications, Inc.

Butgereit, Laurie. Math on MXit: using MXit as a medium for mathematics education. *Semantic Scholar*. Retrieved on July 10, 2019 from https://pdfs.semanticscholar.org/fca0/234a466359042c59920dcee762566cade895.pdf?_ga=2.212832467.675815575.1562930152-1988515899.1560029087

Byrne, Donn. 1976. *Teaching Oral English*. London, UK: Longman.

Challis, Debbie J. April 27, 2011. Tarot and Ancient Egypt—A connection? *University College London Museums and Collections Blog*. Retrieved on July 19, 2019 from https://blogs.ucl.ac.uk/museums/2011/04/27/tarot-and-ancient-egypt-a-connection/

Chao, Yuen Ren. 1968. *Language and Symbolic Systems*. Cambridge, UK: Cambridge University Press.

Cherry, Brett. February 15, 2011. Egypt's 'tipping point'? *RSS feed*. Retrieved on November 02, 2011 from http://tippingpointsproject.org/2011/02/15/egypts-tipping-point/

Chimakonam, Jonathan M. O. 2013. Quantification in African logic. *Filosofia Theoretica: Journal of African Philosophy, Culture and Religion* 2, 2:409-422. Retrieved on July 12, 2019 from https://www.academia.edu/7819781/QUANTIFICATION_IN_AFRICAN_LOGIC

Chimakonam, Jonathan O. 2019. *Ezumezu: A System of Logic for African Philosophy and Studies*. Cham, Switzerland: Springer.

Chomsky, Noam and Morris Halle. 1968. *The Sound Pattern of English*. New York, NY: Harper and Row.

Clark, Colin W. April 2013. Mathematics and resource management. Mathematics Awareness Month Homepage. Retrieved on February 27, 2013 from http://www.mathaware.org/index.html

Cohen, Steve. February 13, 2012. The growing field of sustainability. *Huffington Post*. Retrieved on November 03, 2013 from http://www.huffingtonpost.com/steven-cohen/the-growing-field-of-sust_b_1272831.html

Contini-Morava, Ellen. 1983. Ranking of participants in Kinyarwanda: The limitations of arbitrariness in language. *Anthropological Linguistics* 24, 4:425-435.

Coster, Helen. January 27, 2010. Ranking the world's most sustainable companies. *Forbes*. Retrieved on November 03, 2013 from http://www.forbes.com/2010/01/26/most-sustainable-companies-leadership-citizenship-100.html

Creissels, Denis. 2000. Typology. In B. Heine and D. Nurse, eds. *African Languages: An Introduction*. Cambridge, UK: Cambridge University Press.

Crystal, David. 1992. *An Encyclopedic Dictionary of Language and Languages*. Oxford, UK: Blackwell.

Dahl, Øyvind. 1999. *Meanings in Madagascar: Cases of Intercultural Communication*. Westport, CT: Bergin and Garvey, an imprint of Greenwood Publishing Group, Inc.

Daily Sun (South Africa). 2018. ANC deletes tweet after backlash! Retrieved on June 23, 2019 from https://www.dailysun.co.za/News/National/anc-deletes-tweet-after-backlash-20180606

DawaNet. 2013. *American Muslim History*. Retrieved on June 01, 2013 from http://www.dawanet.com/history/amermuslimhist.asp

DeGiorgio, Michael, Mattias Jakobsson, and Noah A. Rosenberg. 2009. Explaining worldwide patterns of human genetic variation using coalescent-based serial founder model of migration outward from Africa. *Proceedings of the National Academy of the Sciences of the United States* 106, 38:16057-16062. Retrieved on July 16, 2019 from https://www.pnas.org/content/pnas/106/38/16057.full.pdf

Devlin, Keith. 2000. *The Language of Mathematics: Making the Invisible Visible*. New York, NY: Henry Holt.

Devlin, Keith. 2003. *Mathematics: The Science of Patterns: The Search for Order in Life, Mind and the Universe*. New York, NY: Henry Holt.

Dewey Decimal Classification (DDC). Classification system. Dublin, OH: Online Computer Library Center (OCLC), Inc. Retrieved on March 16, 2019 from https://www.oclc.org/en/dewey/features/summaries.html

Dictionary.reference.org. 2013. Tipping point. Retrieved on November 03, 2013 from http://dictionary.reference.com/browse/tipping+point

Dimmendaal, Gerrit J. 2000. Morphology. In B. Heine and D. Nurse, eds. *African Languages: An Introduction*. Cambridge, UK: Cambridge University Press.

Dingemanse, Mark. 2018. Redrawing the margins of language: Lessons from research on ideophones. Glossa: A Journal of General Linguistics 3, 1, 4:1-30. Retrieved on July 09, 2019 from https://www.glossa-journal.org/articles/10.5334/gjgl.444/

Dinneen, Francis P. 1967. *An Introduction to General Linguistics*. New York, NY: Holt, Rinehart and Winston, Inc.

Dinneen, Francis P. 1995. *General Linguistics*. Washington, DC: Georgetown University Press.

Diop, Cheikh Anta. 1955 & 1967/1974. *Nations nègres et cultures* and *Antériorité des civilizations nègres/The African Origin of Civilization: Myth and Reality*. French edition: Paris, France: Présence Africaine; English edition: Brooklyn, NY: Lawrence Hill Books.

Diop, Cheikh Anta. 1987. *Precolonial Black Africa*. Chicago, IL : Lawrence Hill Books.

Diop, Cheikh Anta. 1974/1978. *Black Africa: The Economic and Cultural Basis for a Federated State*. Paris, France: Présence Africaine (English translation by Harold J. Salemson and published by Lawrence Hill Books, Chicago, IL, 1978).

Diop, Cheikh Anta. 1981/1991. *Civilization or Barbarism: An Authentic Anthropology*. Paris, France: Présence Africaine (English translation by Yaa-Lengi Meema Ngemi and published by Lawrence Hill Books, Chicago, IL, 1991).

Djebbar, Ahmed. 1995. On mathematical activities in North Africa since the 9th Century. *AMUCHA Newsletter 15*. Retrieved on January 25, 2007 from http://www.math.buffalo.edu/mad/AMU/amu+chma_15.html#2

D'Odorico, Laura, Mirco Fasolo, Rosalinda Cassibba, and Alessandro Costantini. 2011. Lexical, morphological, and syntactic characteristics of verbs in the spontaneous production of Italian children. Hindawi *Child Development Research* 2011, 498039:1-17. Retrieved on June 18, 2019 from https://www.hindawi.com/journals/cdr/2011/498039/

Eastwood, Edward B. and Benjamin W. Smith. 2005. Fingerprints of the Khoekhoen: Geometric and handprinted rock art in the Central Limpopo Basin, Southern Africa. *South African Archaeological Society Goodwin Series* 9:63-76. Retrieved on July 15, 2019 from https://www.researchgate.net/publication/261827002_Fingerprints_of_the_Khoekhoen_Geometric_and_Handprinted_Rock_Art_in_the_Central_Limpopo_Basin_Southern_Africa

Egedi, Barbara, 2012. Coptic noun phrases. Doctoral dissertation, Eötvös Loránd Tudományegyetem Bölcsészettudományi Kar, Budapest, Hungary. Retrieved on July 20, 2019 from https://www.academia.edu/6002456/Coptic_noun_phrases

Eglash, Ron. 1999. *African Fractals: Modern Computing and Indigenous Design*. New Brunswick, NJ: Rutgers University Press.

Ehret, Christopher. 2000. Language and history. In B. Heine and D. Nurse, eds. *African Languages: An Introduction*. Cambridge, UK: Cambridge University Press.

Encyclopædia Britanica. 2018. Nominalism. Retrieved on May 21, 2019 from https://www.britannica.com/topic/nominalism

Encyclopædia Britanica. 2019. Mathematics. Retrieved on March 16, 2019 from https://www.britannica.com/science/mathematics

Fajobi, Eunice. 2005. The nature of Yoruba intonation: A new experimental study. In T. Falola and A. Genova, eds., *Yoruba Creativity: Fiction, Language, Life and Songs*. Trenton, NJ: Africa World Press.

Falokun, Awo. April 14, 2018. *Oyeku Meji*. Egbe Iwa-Pele: International Association of Ifa Devotees. Retrieved on July 12, 2019 from http://www.egbeiwapele.com/oyeku-meji-2/

Falola, Toyin. 1999. *Yoruba Gurus: Indigenous Production of Knowledge in Africa*. Trenton, NJ: Africa World Press.

Falola, Toyin. 2005. *A Mouth Sweeter than Salt: An African Memoir*. Ann Arbor, MI: The University of Michigan Press.

Falola, Toyin (ed.). 2006. *Yoruba Identity and Politics*. Rochester, NY: University of Rochester Press.

Falola, Toyin. 2012. *Ibadan: Foundation, Growth and Challenge 1830-1960.* Ibadan, Nigeria: Bookcraft.

Falola, Toyin (ed.). 2013a. *Esu: Yoruba God, Power, and the Imaginative Frontiers*. Durham, NC: Carolina Academic Press.

Falola, Toyin. 2013b. *The African Diaspora: Slavery, Modernity, and Globalization*. Rochester, NY: University of Rochester Press.

Falola, Toyin. 2014. *Counting the Tiger's Teeth: An African Teenager's Story*. Ann Arbor, MI: The University of Michigan Press.

Fasold, Ralph. 1984. *The Sociolinguistics of Society*. New York, NY: Basil Blackwell Inc.

Fay, Nicolas, Simon Garrod, John Lee, and Jon Oberlander. 2003. Understanding interactive graphical communication. In R. Alterman and D Kirsch, eds. *Proceedings of the 25th Annual Conference of the Cognitive Science Society*. Boston, MA: Cognitive Science Society.

Fischer, Henry G. 1978. The evolution of composite hieroglyphs in Ancient Egypt. *Metropolitan Museum Journal* 12:5-19. Retrieved on July 20, 2019 from https://www.metmuseum.org/art/metpublications/evolution_of_composite_hieroglyphs_in_ancient_egypt_the_metropolitan_museum_journal_v_12_1977

Folorunso, Olusegun, Adio T. Akinwale, Rebecca O. Vincent, and Babatunde Olabenjo. 2010. A mobile-based knowledge management system for "Ifa": An African traditional oracle. *African Journal of Mathematics and Computer Science* 317:114-131. Retrieved on February 01, 2016 from https://www.academia.edu/8665370/A_mobile-based_knowledge_management_system_for_Ifa_An_African_traditional_oracle

Ford, Merryl and Adele Botha. 2010. A pragmatic framework for integrating ICT into education in South Africa. P. Cunningham and M. Cunningham, eds. (Innovation, Science and Technology-Africa) *IST-Africa 2010 Conference Proceedings*. Retrieved on July 13, 2019 from https://www.academia.edu/4805703/A_pragmatic_framework_for_integrating_ICT_into_education_in_South_Africa

Frieke-Kappers, Claertje. 2007. The creative use of genre features continuity in patterns of language use in Budu, a Bantu language of Congo (Kinshasha). Doctoral dissertation, Vrije Universiteit, Amsterdam, The Netherlands. Retrieved on July 10, 2019 from https://research.vu.nl/ws/portalfiles/portal/42177465/complete+dissertation.pdf

Gadalla, Moustafa. 2016. *The Egyptian Hieroglyph Metaphysical Language*. Greensboro, NC:L Tehuti Research Foundation. Retrieved on July 18, 2019 from https://egypt-tehuti.org/wp-content/uploads/2018/02/The-Egyptian-Hieroglyph-Metaphysical-Language-sample.pdf

Gaeng, Paul A. 1971. *Introduction to the Principles of Language*. Lanham, MD: University Press of America.

Galbraith, Kate. August 19, 2009. Sustainability field booms on campus. *NY Times.* Retrieved on November 03, 2013 from http://www.nytimes.com/2009/08/20/education/20GREEN.html?_r=1&

Ganiso, Mirriam Nosiphiwo. 2012. Sign language in South Africa: Language planning and policy challenges. Master of Arts Thesis, School of Languages: African Language Studies, Rhodes University, Grahamstown, South Africa.

Gerdes, Paulus. 1994. On mathematics in the history of Sub-Sahara Africa. *Historia Mathematica* 21, 3:345-376. Retrieved on July 15, 2019 from https://core.ac.uk/download/pdf/82352865.pdf

Gerdes, Paulus. 1999. *Geometry from Africa: Mathematical and Educational Explorations.* Washington, DC: The Mathematical Association of America.

Gerdes, Paulus. 2009. *Sipatsi: Basketry and Geometry in the Tonga Culture of Inhambane.* Maputo, Mozambique: Centre for Mozambican Studies and Ethnoscience, Universidade Pedagógica.

Gerdes, Paulus and Ahmed Djebar. 2007. *Mathematics in African History and Cultures: An Annotated Bibliography.* Cape Town, South Africa: African Mathematical Union Commission on the History of Mathematics in Africa (AMUCHMA). Retrieved on July 16, 2019 from http://www.etnomatematica.org/BOOKS_Gerdes/mathematics_in_african_history_and_cultures__an_annotated_bibliography.pdf

Gladwell, Malcolm. 2002. *The Tipping Point: How Little Things Can Make a Big Difference.* New York, NY: Little Brown and Company. January 2002.

Global Community WebNet Ltd. 2004. Measurement of sustainable development. Retrieved on May 03, 2013 from http://globalcommunitywebnet.com/globalcommunity/measurementofsd.htm

Gnaedinger, Franz. 2005. Very early calendars. Retrieved on July 22, 2008. Available at http:// http://www.seshat.ch/home/calendar.htm

Goggins, II, Lathardus. 1996. *African Centered Rites of Passage and Education.* Chicago, IL: African American Images.

Gong, Tao. 2010. Exploring the roles of major forms of cultural transmission in language evolution. A. D. M. Smith, M. Schouwstra, B. de Boer and K. Smith, eds. *The Evolution of Language.* Proceedings of the 8th the World Scientific International Conference, Utrecht, The Netherlands, April 14-17, 2010. https://doi.org/10.1142/7624

Gonsalves, Gregg S., Edward H. Kaplan, and A. David Paltiel. 2015. Reducing sexual violence by increasing the supply of toilets in Khayelitsha, South Africa: A mathematical model. *PLoS One*: 1-12. Retrieved on July 17, 2019 from https://journals.plos.org/plosone/article/file?id=10.1371/journal.pone.0122244&type=printable

Greenberg, Joseph H. 1957. *Essays in Linguistics.* Chicago, IL: The University of Chicago Press.

Grossman, Eitan and Stéphane Polis. 2014. Forms and functions in Ancient Egypt: A short introduction. In E. Grossman, S. Polis, A. Stauder and J. Winand, eds. *On Forms and Functions: Studies in Ancient Egyptian Grammar.* Hamburg, Germany: Widmaier Verlag. Retrieved on July 19, 2019 from https://orbi.uliege.be/bitstream/2268/173893/2/Grossman_Lescuyer_Polis_AllativeFuture_LingAeg-StudMon-15.pdf

Gudykunst, William B. 2004. *Bridging Differences: Effective Intergroup Communication* 4th ed. Thousand Oaks, CA: Sage Publications.

Guglielmino, C. R., C. Viganotti, B. Hewlett, and L. L. Cavalli-Sforza. 1995. Cultural variation in Africa: Role of mechanisms of transmission and adaptation. *Proceedings of the National Academy of Sciences* 92:7585-7589. Retrieved on July 15, 2019 from https://www.ncbi.nlm.nih.gov/pmc/articles/PMC41384/pdf/pnas01494-0458.pdf

Gurib-Fakim, Ameenah and Jacobus Nicolaas Eloff (eds.). 2013. *Chemistry for Sustainable Development in Africa.* Heidelberg, Germany: Berlin Springer.

Gussenhoven, Carlos. 1999. Discreteness and gradience in intonational contrasts. *Language and Speech* 42, 2-3:283-305.

Guthrie, Malcolm. 1948. *The Classification of the Bantu Languages.* London, UK: Oxford University Press.

Hacken, Pius ten. 2008. Prototypes and discreteness in terminology. In E. Bernal and J. DeCesaris, eds. *Proceedings of the XIII Euralex International Congress,* Barcelona, Spain: Institut Universitari de Linguistica Aplicada, Univeritat Pompeu Febra. Retrieved on May 30, 2019 from http://www.euralex.org/elx_proceedings/Euralex2008/022_Euralex_2008_Pius%20ten%20Hacken_Prototypes%20and%20Discreteness%20in%20Terminology.pdf

Hadlock, Charles R. April 2013. Sustaining human society in a rapidly changing world. Mathematics Awareness Month Homepage. Retrieved on February 27, 2013 from http://www.mathaware.org/mam/2013/essays/Hadlock-2013.pdf?utm_source=ThemeEssaysPage&utm_medium=themeEssaysPage&utm_campaign=Theme+Essays

Haikal, Fayza M. H. 200]11. The heritage of Ancient Egypt in the culture of Islamic and modern Egypt. Paper presented at a workshop on "Tradition and Religion: Transmission and Culture versus Similarity of Beliefs" held in Prague, Czechoslovakia, June 07, 2011. Retrieved on July 20, 2019 from https://www.academia.edu/4440038/The_Heritage_Of_Ancient_Egypt_In_The_Culture_Of_Islamic_And_Modern_Egypt

Hakuta, Kenji. 1986. *Mirror of Language: The Debate on Bilingualism.* New York, NY: Basic Books, Inc.

Hall (nee Kruyshaar), Linda. 2005. The –be relative tenses of Zulu. Magister Artium Thesis, Faculty of Humanities, University of Pretoria, South Africa. Retrieved on July 05, 2019 from https://repository.up.ac.za/bitstream/handle/2263/29217/dissertation.pdf;sequence=1

Hammarström, Harald. 2016. Linguistic diversity and language evolution. *Journal of Language Evolution* 1, 1:19-29. Retrieved on July 07, 2019 from https://pure.mpg.de/rest/items/item_2237428_5/component/file_2259449/content

Hammond, Nick. November 24, 2015. Talk like the Egyptians? The rise of the Emojis. *The Digital Filter.* Retrieved on July 19, 2019 from http://www.thedigitalfilter.com/author/nick/page/2/

Hance, Jeremy. March 06, 2013. Warnings of global ecological tipping points. *Environmental Network News.* Retrieved on November 03, 2013 from http://www.enn.com/sustainability/article/45684

Hayward, Richard J. 2000. Afroasiatic. In B. Heine and D. Nurse, eds. *African Languages: An Introduction.* Cambridge, UK: Cambridge University Press.

Hepworth, Samantha Ross. 2008. Absorption or displacement – necessary or gratuitous? The contention between Kiswahili and minority languages in Tanzania. Retrieved on June 07, 2013 from http://www.seameo.org/_ld2008/doucments/Presentation_document/MicrosoftWordAbsorptionOrdisplacement_BKKpaper_SamanthaRossHepworth.pdf

Her, One-Soon and Yan-Ru Chen. 2013. Unification of numerals and plural markers: Empirical facts and implications. *Proceedings of the 27th Pacific Asia Conference on Language, Information, and Computation* 27:37-46. Retrieved on July 16, 2019 from https://www.aclweb.org/anthology/Y13-1002

Hersh, Marion A. 2006. *Mathematical Modelling for Sustainable Development.* Berlin, Germany: Springer.

Hieber, Danny. January 04, 2012. Do languages die? *Mises Institute Mises Daily Articles.* Retrieved on July 10, 2019 from https://mises.org/library/why-do-languages-die

Hinga, Anne Njoki. 2010. Learning and teaching a foreign language in a multilingual setting: The case of German in Kenya. Paper presented at the 5th German African Network of Alumni and Alumnae (GANAA) Workshop at the Cheikh Anta Diop University, Dakar, Senegal, November 26-January 1, 2010. Retrieved on June 30, 2019 from https://ir-library.ku.ac.ke/bitstream/handle/123456789/13515/Learning%20and%20Teaching%20a%20Foreign%20language.pdf?sequence=1&isAllowed=y

Hockett, Charles Francis. 1958. *A Course in Modern Linguistics.* New York, NY: Macmillan.

Hockett, Charles Francis. 1960. The origin of speech. *Scientific American* 203:89-96.

Holloway, Joseph E. and Winifred K. Vass. 1993. *The African Heritage of American English.* Bloomington and Indianapolis, IN: Indiana University Press. http://egyptian-gods.99k.org/isis.html http://www.math.buffalo.edu/mad/Ancient-Africa

Hughes, Mark Alan and Elise Harrington. April 2013. Making greenworks count. Mathematics Awareness Month Homepage. Retrieved on February 27, 2013 from http://www.mathaware.org/index.html

Hunter, Brad. n.d. Tipping points in social networks. Retrieved on November 03, 2013 from. http://www.stanford.edu/class/symbsys205/tipping_point.html

Hurford, James R. 2007. *The Origins of Meaning.* Oxford, UK: Oxford University Press.

Huylebrouck, Dirk, 2006. Mathematics in (central) Africa before colonization. *Anthropologica et Præhistorica* 117:135-162). Retrieved on July 12, 2019 from http://etopia.sintlucas.be/3.14/Ishango_meeting/Mathematics_Africa.pdf

Hyman, Larry M. 2011. The Macro-Sudan Belt and Niger-Congo reconstruction. *Language Dynamics and Change* 1:3-49).

Ibrahim, Mustapha D., Sahand Daneshvar, Mevhibe B. Hocaoğlu, and Olasehinde-Williams G. Oluseye. 2019. An estimation of the efficiency and productivity of healthcare systems in sub-Saharan Africa: Health-centered millennium development goal-based evidence. *Social Indicators Research* 143, 1:371-389. Retrieved on July 17, 2019 from https://econpapers.repec.org/article/sprsoinre/v_3a143_3ay_3a2019_3ai_3a1_3ad_3a10.1007_5fs11205-018-1969-1.htm

Jackendoff, Ray. 2002. *Foundations of Language: Brain, Meaning, Grammar, Evolution.* Oxford, UK: Oxford University Press.

Jespersen, Bjørn and Chris Reintges. 2008. Tractarian sätze, Egyptian hieroglyphs, and the very idea of script as picture. *The Philosophical Forum* 39, 1:1-19. Retrieved on July 20, 2019 from https://onlinelibrary.wiley.com/doi/epdf/10.1111/j.1467-9191.2007.00279.x

Johnson, Marion M. 1976. Toward a definition of the ideophone in Bantu. *Working Papers in Linguistics* 21:240-253.

Johnson, Trevor and Hugh Neill. 2010. *Complete Mathematics.* London, UK: Hodder Education and New York, NY: The McGraw-Hills Companies, Inc.

Joseph, George. Gherverghese. 1991/2000. *The Crest of the Peacock: Non-European Roots of Mathematics.* New York, NY: Penguin Books.

Karenga, Maulana. 1993. *Introduction to Black Studies.* Los Angeles, CA: The University of Sankore Press.

Katz, Victor J. 2008. *History of Mathematics: An Introduction* 3rd ed. Boston, MA: Pearson/Addison-Wesley.

Kerswill, Paul. 2010. Youth languages in Africa and Europe: Linguistic subversion or emerging vernaculars? Lancaster University African Studies Group. Retrieved on July 09, 2019 from https://www.lancaster.ac.uk/fass/doc_library/linguistics/kerswill/Kerswill-African-Studies-19-10-10.pdf

Keto, C. Tsehloane. 1989. *Revision of the African-Centered Perspective of History and Social Science in the Twenty First Century.* Blackwood, NJ: K. A. Publications.

Khapoya, Vincent B. 1998. *The African Experience: And Introduction* 2nd ed. Upper Saddle River, NJ: Prentice Hall.

Kim, Min-Sun. 2002. *Non-Western Perspectives on Human Communication: Implications for Theory and Practice.* Thousand Oaks, CA: Sage Publications.

Kline, Morris. 1985. *Mathematics and the Search for Knowledge.* New York, NY & Oxford, UK: Oxford University Press.

Knox, Jerry, Tim Hess, Andre Daccache, and Tim Wheeler. 2012. Climate change impacts on crop productivity in Africa and South Asia. *Environmental Research Letters* 7:1-8. Retrieved on July 17, 2019 from https://iopscience.iop.org/article/10.1088/1748-9326/7/3/034032/pdf

Kramer, Ruth. 2009. Egyptian. Retrieved on July 20, 2019 from http://faculty.georgetown.edu/rtk8/Egyptian%20Chapter.pdf

Kuehn, Christian, Erik A. Martens and Daniel M. Romero. July 31, 2013. Critical transitions in social network activity. Retrieved on November 02, 2013 from http://arxiv.org/pdf/1307.8250.pdf

Kung, David. April 2013. Mathematics, social justice & sustainability. Mathematics Awareness Month Homepage. Retrieved on February 27, 2013 from http://www.mathaware.org/index.html

Labov, William. 1984. *Sociolinguistic Patterns* 10th ed. Philadelphia, PA: University of Pennsylvania Press.

Lafleur, Gwenyth J. via Cynthia L. Hallen. 1998-1999. Exploring the Niger-Congo languages. Retrieved on June 01, 2013 from http://linguistics.byu.edu/classes/ling450ch/reports/niger-congo.html

Leech, Geoffrey. 1983. *Principles of Pragmatics.* New York, NY: Longman Inc.

Lekov, Teodor. 2004. The formula of the "Giving of the Heart" in Ancient Egyptian texts. *The Journal of Egyptological* Studies 1:33-60. Retrieved on July 19, 2019 from https://core.ac.uk/download/pdf/12037163.pdf

Levinson, Stephen C. 1983. *Pragmatics.* Cambridge, UK: Cambridge University Press.

Liszkowski, Ulf, Mark Schafer, Malinda Carpenter, and Michael Tomasello. 2009. Pre-linguistic infants, but not chimpanzees, communicate about absent entities. *Psychological Science* 20, 5:654-660).

Loprieno, Antonio. 1995. *Ancient Egyptian: A Linguistic Introduction* Cambridge, UK: Cambridge University Press.

Loux, Michael J. 1998. Nominalism. *Routledge Encyclopedia of Philosophy.* Retrieved on May 21, 2019 from https://www.rep.routledge.com/articles/thematic/nominalism/v-1

Lumpkin, Beatrice. 1986. Africa in the mainstream of mathematics history. I. V. Sertima, ed. *Blacks in Science: Ancient and Modern.* New Brunswick, NJ: Transaction Books.

Lüdeling, Anke and Stefan Evert. 2003. Linguistic experience and productivity: Corpus evidence for fine-grained distinctions. *Semantic Scholar.* Retrieved on June 18, 2019 from https://www.semanticscholar.org/paper/Linguistic-experience-and-productivity-%3A-corpus-for-L%C3%BCdeling-Evert/987996d3aebaab8d95ace4613d657f8b51a455e4

Lyons, John. 1971. *Introduction to Theoretical Linguistics* (3rd reprint). Cambridge, UK: Cambridge University Press.

Lyons, John. 1977. *Semantics* vols. 1 & 2. Cambridge, UK: Cambridge University Press.

MacFadyen, Sandra, Nick Zabatis, Astrid J. A. Van Teeffelen, and Cang Hui. 2018. Long-term rainfall regression surfaces for the Kruger National Park, South Africa: A spatio-temporal review of patterns from 1981 to 2015. *International Journal of Climatology* 38:2506-2519. Retrieved on July 17, 2019 from https://www.researchgate.net/publication/333056336_Long-term_rainfall_regression_surfaces_for_the_Kruger_National_Park_South_Africa_a_spatio-temporal_review_of_patterns_from_1981_to_2015

MacLeish, Andrew. 1971. *A Glossary of Grammar and Linguistics.* New York, NY: Grosset & Dunlap, Inc.

Madugu, Isaac S. 1985. Complex verbs in Nupe and Yoruba. *Studies in African Linguistics.* 16, 3:295-321.

Makonye, Judah Paul. 2011. Learner mathematical errors in introductory differential calculus tasks: A study of misconceptions in the senior school certificate examinations. Doctoral thesis, University of Johannesburg, South Africa. Retrieved on July 16, 2019 from https://ujcontent.uj.ac.za/vital/access/services/Download/uj:3390/CONTENT1

Mankiewicz, Richard. 2001. *The Story of Mathematics*. Princeton, NJ: Princeton University Press.

Marewo, Gerald T. and D. J. Henwood. 2006. A mathematical model for supplying air-cooling for a building using a packed bed. *Building Science Engineering* 27, 1:11-26.

Margulis, Jennifer and Peter Nowakoski. Spring 1996 (last updated July 2017). Language. *Postcolonial Studies @ Emory*. Retrieved on July 10, 2019 from https://scholarblogs.emory.edu/postcolonialstudies/2014/06/21/language/

Mark, Joshua J. January 19, 2013. Ancient Egyptian burial: Definition. *Ancient History Encyclopedia*. Retrieved on July 20, 2019 from https://www.ancient.eu/Egyptian_Burial/

Marzagora, Sara. 2015. Literatures in African languages. *Journal of African Cultural Studies* 27, 1:1-6. Retrieved on July 09, 2019 from https://www.tandfonline.com/doi/pdf/10.1080/13696815.2014.978842

Mathematical Association of America. March 28, 2013. Harnessing math to understand tipping points. Retrieved on November 03, 2013from http://archive.is/zGOtm

Maxwell, Kori Lloyd Hugh and Iman C. Chahine. 2013. Cultural immersion and mathematics teacher education: Explorations in Morocco and South Africa. *Journal of Humanistic Mathematics* 3, 2:62-75. Retrieved on July 15, 2019 from https://pdfs.semanticscholar.org/1973/a1776e6767ff3dcd1dde797ab8b859906474.pdf

Mazrui, Ali Al'Amin. 1998. Mazruina: Between the Renaissance African and the African Renaissance. Speech at the book launch of *The Mazruina Collection*, Zimbabwe International Book Fair, Harare, Zimbabwe, August 04, 1998.

Mazrui, Ali Al'Amin. 1999. The Language of Francophonie and the race of the Renaissance: A Commonwealth perspective. Occasional Paper, Institute of Global Cultural Studies, State University of New York, Binghamton, New York.

Mazrui, Ali Al'Amin. 2001a. The African Renaissance: A triple legacy of skills, values and gender. IN S. C. Saxena, ed. *Africa beyond 2000: Essays on Africa's Political and Economic Development in the Twenty-First Century*. Delhi, India: Kalinga Publications.

Mazrui, Ali Al'Amin. 2001b. Shifting African identities: The boundaries of ethnicity and religion in Africa's experience. In S. Bekker, M. Dodds, and M. M. Khosa, eds. *Shifting Afrian Identities*. Pretoria, South Africa: Human Sciences Research Council.

Mazrui, Ali Al'Amin. 2001c. Linske elite heft Afrika's originaliteit om zeep geholpen ("The Left Elite Has Killed off Africa's Original Thought"). *De Volkskrant* 23:4-6.

Meehan, Andrew. 1912. Subreption. *The Catholic Encyclopedia*. New York, NY: Robert Appleton Company. Retrieved on April 6, 2019 from http://ccm.chaldeans.uk/library/advent/cathen/14322a.htm

Meierkord, Christiane. 2005. Black South African Englishes—towards a variationist account. *Erfurt Electronic Journal in English 1/2005*. Retrieved on July 05, 2019 from http://webdoc.sub.gwdg.de/edoc/ia/eese/artic25/meierk/1_2005.html

Meli, Francis. 1988. *South Africa Belongs to Us: A History of the ANC.* Harare, Zimbabwe: Zimbabwe Publishing House (republished in 1989, London, UK: James Currey).

Merzbach, Uta C. and Carl B. Boyer. 2011. *A History of Mathematics* 3rd ed. San Francisco, CA: Josey Bass.

Minger, David. 2018. In linguistics, what do they mean by discreteness? And can you give some simple examples? *Quora.* Retrieved on May 30, 2019 from https://www.quora.com/In-linguistics-what-do-they-mean-by-discreteness-And-can-you-give-some-simple-examples

Mochesela, Palesa Rebecca. 2007. The role of the problem-based approach in the performance of grade 9 learners in solving word problems. Master's dissertation, University of South Africa. Retrieved on July 14, 2019 from https://pdfs.semanticscholar.org/b241/c7967730b2b3d4628f41c24a9066fade7c2e.pdf

Molema, Karabo. 2016. Are African languages in South Africa mutually intelligible to each other? *Quora.* Retrieved on July 05, 2019 from https://www.quora.com/Are-African-languages-in-South-Africa-mutually-intelligible-to-each-other

Montague, Richard. 1974. *Formal Philosophy* (Richmond H. Thomason, ed.). New Haven, CT: Yale University Press.

Moore, Will H. and David A. Siegel. 2013. *A Mathematical Course for Political and Social Research.* Princeton, NJ: Princeton University Press.

Mous, Maarten. 2003. Loss of linguistic diversity in Africa. Retrieved on June 07, 2013 from http://www.ddl.ish-lyon.cnrs.fr/Fulltext/philippson/Mous_LossLinguisticsDiversityAfrica.pdf

Mufwene, Salikoko S. 2013. What African linguistics can contribute to evolutionary linguistics. In Olanike Ola Orie and Karen W. Sanders, eds. Selected Proceedings of the 43rd Annual Conference on African Linguistics: Linguistic Interfaces in African Languages. Somerville, MA: Cascadilla Proceedings Project. Retrieved on July 10, 2019 from http://mufwene.uchicago.edu/publications/WHAT_AFRICAN_LINGUISTICS_CAN_CONTRIBUTE_TO_EVOLUTIONARY_LINGUISTICS_-_ACAL_43.pdf

Mulder, Nicola J. et al. 2016. The development of computational biology in South Africa: Success achieved and lessons learnt. *PLoS Computational Biology* 12, 2. Retrieved on July 14, 2019 from https://journals.plos.org/ploscompbiol/article?id=10.1371/journal.pcbi.1004395

Muysken, Pieter. 2011. Bridges over troubled waters: Theoretical linguistics and multilingualism research. *Stellenbosch Papers in Linguistics* 40:20-38. Retrieved on July 07, 2019 from https://www.ajol.info/index.php/spl/article/viewFile/114442/104136

Muysken, Pieter. 2017. Learn more about Afrikaans: Linguistic categories in language contact. In H. Cohen and C. Lefebvre, eds. *Handbook of Categorization in Cognitive Science* (2nd ed.). London, UK: Elsevier ScienceDirect. Retrieved on July 05, 2019 from https://www.sciencedirect.com/topics/psychology/afrikaans

Nabudere, Dani Wadada. 2003. Towards the establishment of a Pan-African university: A strategic concept paper. *African Journal of Political Science* 8, 1:1-30.

Nachmias, Chava-Frankfort, David Nachmias and Jack DeWaard. 2014. *Research Methods in the Social Sciences* 7th ed. New York, NY: Worth Publishers/Palgrave Macmillan.

Ndlovu-Gatsheni, Sabelo J. 2012. Rethinking *Chimurenga* and *Gukurahundi* in Zimbabwe: A critique of partisan national history. *African Studies Review* 55, 3:1-26.

Nicolaï, Robert. 2007. A Songhay-Mande convergence area? Facts, questions, frames. In B. Heine and D. Nurse, eds. *Africa as a Linguistic Area/Areal Typology and African Languages.* Cambridge, UK: Cambridge University Press. Retrieved on July 05, 2019 from http://sites.unice.fr/ChaireIUF-Nicolai/TextesRN/Songhay-mande_convergence.pdf

Nobles, Wade W. 1990. African philosophy: Foundations of Black psychology. In R. E. Jones, ed. *Black Psychology.* Berkeley, CA: Cobb and Henry.

Nordquist, Richard. 2018. Displacement in language. *ThoughtCo.* Retrieved on June 08, 2019 from https://www.thoughtco.com/displacement-language-term-1690399

Nordquist, Richard. 2019a. Productivity: Glossary of grammatical and rhetorical terms—definition and examples. *ThoughtCo.* Retrieved on June 18, 2019 from https://www.thoughtco.com/productivity-language-1691541

Nordquist, Richard. 2019b. The cultural transmission of language. *ThonghtCo.* Retrieved on April 26, 2019 from https://www.thoughtco.com/what-is-cultural-transmission-1689814

North Carolina State University (NCSU). 2013. What is sustainability? *Sustainability at NC State.* Retrieved on November 03, 2013 from http://sustainability.ncsu.edu/about/what-is-sustainability

Nunberg, Geoffrey. 1988. *The Field of Linguistics.* Washington, DC: Linguistic Society of America.

Obiero, Ogone John. 2010. A case of a mother tongue and another mother tongue in school: Efforts at revitalization of Olusuba language of Kenya. *Journal of Third World Studies* 27, 2:267-291.

O'Connor, John J. and Edmund F. Robertson. 2019. A mathematical chronology. *MacTutor History of Mathematics Archive,* University of St. Andrews, UK. Retrieved on March 16, 2019 from http://www-history.mcs.st-and.ac.uk/

O'Grady, William, Michael Dobrovolsky, and Mark Aronoff. 1989. *Contemporary Linguistics: An Introduction.* New York, NY: St. Martin's Press.

Ogundiran, Akinwumi. January 31, 2016. E-mail. <ogundiran@uncc.edu>.

Okediji, Moyo. March 10, 2013. *Ifa* mathematics: *Oyeku Meji.* Retrieved on February 01, 2016 from https://www.facebook.com/1601674946732004/posts/oyekun-mejioyeku-odu-meji-means-sadness-and-unhappiness-and-warns-about-death-si/2170902166475943/

Olanipekun, Shola Sunday, Ebenezer Omoniyi Onabanjo, and Clara Modupe Olayemi. 2016. Linguistics innovation of the twenty-first century African writing. *Journal of Studies in Social Sciences and Humanities* 2, 1:28-33. Retrieved on July 05, 2019 from http://www.jssshonline.com/wp-content/uploads/2016/02/JSSSH_Vol.2_No.1_2016_28-33_4.pdf

Olivier, Jacobus Alwyn Kruger. 2009. Multilingualism in South Africa.*SAlanguages.com*. Retrieved on November 26, 2010 from http://www.salanguages.com/multilingualism.htm

Olofsson, Joel. 2018. Motion in Swedish—on productivity from a construction grammar perspective. Doctoral thesis. Retrieved on June 18, 2019 from https://www.academia.edu/36771687/Motion_in_Swedish_on_productivity_from_a_construction_grammar_perspective_doctoral_thesis_summary_

Omachonu, Gideon S. and David A. Abraham. 2012. Compounding in Igala. *Unizik Journal of Arts and Humanities* 13, 2:184-206.

Ordway, Elsa M., Gregory P. Asner, and Eric F. Lambin. 2017. Deforestation risk due to commodity crop expansion in sub-Saharan Africa. *Environmental Research Letters* 12:1-14. Retrieved on July 17, 2019 from https://iopscience.iop.org/article/10.1088/1748-9326/aa6509/pdf

Pakin, Scott. January 19, 2017. The comprehensive LaTeX symbol list. Retrieved on July 10, 2019 from http://tug.ctan.org/info/symbols/comprehensive/symbols-a4.pdf

Papakonstantinou, Joanna M. and Richard A. Tapia. 2013. Origin and evolution of the secant method in one dimension. *The American Mathematical Monthly* 120, 6:500-517).

Patil, Rujuta. 2018. A thorough understanding of cultural transmission with examples: The seven principles of Keanzaa festival exemplify cultural transmission of African traditions. *Historyplex* Retrieved on April 26, 2019 from https://historyplex.com/understanding-cultural-transmission-with-examples

Perek, Florent. 2019, in press. Using distributional semantics to study syntactic productivity in diachrony: A case study. *Linguistics*. Retrieved on June 18, 2019 from http://www.fperek.net/pdfs/PerekToAppear-using-distributional.pdf

Pereltsvaig, Asya. 2011. Six things we have and they don't—part 1. *Languages of the World*. Retrieved on June 08, 2019 from https://www.languagesoftheworld.info/language-and-mind/six-things-we-have-and-they-dont-part-1.html

Perniss, Pamela, Robin L. Thompson, and Gabriella Vigliocco. 2010. Iconicity as a general property of language: Evidence from spoken and signed languages. *Frontiers in Psychology* 1, 227:1-16. Retrieved on July 07, 2019 from https://pdfs.semanticscholar.org/a4fe/b77ef8a4b529c190f860caf2b0f3d155853f.pdf?_ga=2.98934366.1978000856.1562590119-1988515899.1560029087

Phillips, Jason. May 2010. A mathematical model of sustainable development using ideas of coupled environment-human systems. *The Pelican Web's Journal of Sustainable Development* 6, 5.

Plag, Ingo. 2004. Productivity. In B. Aarts and A. McMahon, eds. *Handbook of English Linguistics*. Oxford, UK: Blackwell. Retrieved on June 18, 2019 from http://www.anglistik.hhu.de/fileadmin/redaktion/Fakultaeten/Philosophische_Fakultaet/Anglistik_und_Amerikanistik/Ang3_Linguistics/Dateien/Detailseiten/Plag/Plag2006_Productivity.pdf

Port, Robert F. 1996. Letter to the editor: The discreteness of phonetic elements and formal linguistics: Response to A. Manaster Ramer. *Journal of Phonetics* 24:491-511.

Prah, Kwesi Kwaa. 2010. Multilingualism in urban Africa: Bane or blessing. *Journal of Multicultural Discourses* 3, 2:169-182.

Pullum, Geoffrey K. and András Kornai. 2003. Mathematical linguistics. *The Oxford International Encyclopedia* 2nd ed., 17-20. Oxford, UK: Oxford University Press. Retrieved on June 08, 2013 from http://www.kornai.com/MatLing/matling3.pdf

Quine, Willard Van Orman. 1961. *From a Logical Point of View*. New York, NY: Harper Torchbooks, Harper & Row, Publishers.

Rambane, Daniel Thanyani and Mashudu C. Mashige. 2007. The role of mathematics and scientific thought in Africa: A Renaissance perspective. *International Journal of African Renaissance Studies* 2, 2:183-199. Retrieved on July 15, 2019 from https://www.researchgate.net/publication/233265928_The_role_of_mathematics_and_scientific_thought_in_Africa_A_Renaissance_perspective

Ramsey, Scott. November 03, 2015. Ancient Egyptian religion. *Awareness of Nothing*. Retrieved on July 20, 2019 from http://www.awarenessofnothing.com/ancient-egyptian-religion.html

Rassafi, Amir Abbas and Manouchehr Vaziri. Fall 2003. African sustainable transport by numbers. *Journal of Sustainable Development in Africa*Retrieved on May 03, 2013 from http://www.jsd-africa.com/Jsda/Fall2003/articlepdf/ARC-African%20Sustainable%20Transport%20by%20number%20%28article%29.pdf

Rensselaer Polytechnic Institute. July 26, 2011. Minority rules: Scientists discover tipping point for the spread of ideas. *Science Daily*. Retrieved on November 02, 2013 from http://www.sciencedaily.com/releases/2011/07/110725190044.htm

Roberts, Fred S. April 2013. Energy as a contributor to human well-being: Electric power grids. Mathematics Awareness Month Homepage. Retrieved on February 27, 2013 from http://www.mathaware.org/index.html

Robinson, David. February 06, 2019. Words as feelings: A special class of vivid, textural words defies linguistic theory: Could 'ideophones' unlock the secrets of humans' first utterances? *Aeon Essays*. Retrieved on July 07, 2019 from https://aeon.co/essays/in-the-beginning-was-the-word-and-the-word-was-embodied

Rohrer, Katherine E. 2008. Language and dialects. *Gale Library of Daily Life: Slavery in America. Encyclopedia.com*. Retrieved on June 30, 2019 from https://www.encyclopedia.com

Rossi, Corinna. 2007. *Architecture and Mathematics in Ancient Egypt*. New York, NY: Cambridge University Press.

Rutledge, Brett. 2011. Clever Hans Phenomena and the origins of nonverbal communication. *The Articulate CEO*. Retrieved on April 26, 2019 from https://thearticulateceo.typepad.com/my-blog/2011/12/clever-hans-phenomenon-the-origins-of-non-verbal-communication-.html

SamForay, Alfred Munda. May 03, 2019. E-mail: <alsamforay@gmail.com>.

Sampson, Geoffrey. 1980. *Schools of Linguistics.* Stanford, CA: Stanford University Press.

Sampson, Geoffrey. 1994. Chinese script and the diversity of writing systems. *Linguistics* 32, 1:117-132.

Sankoff, Gillian. 2001. Linguistic outcomes of language contact. In P. Trudgill, J. Chambers, and N. Schilling-Estes, eds. *Handbook of Sociolinguistics.* Oxford, UK: basil Blackwell. Retrieved on July 10, 2019 from https://www.ling.upenn.edu/~gillian/Interlang.doc.pdf

Savini, Marina. 1991. On the (non-)discreteness of morphological categories with special reference to affix categories in Afrikaans. Doctoral thesis, University of South Africa. Retrieved on July 09, 2019 from http://uir.unisa.ac.za/handle/10500/24835

Schouwstra, Marieke. 2007. Displaced reference in the evolution of language and cognition. *CORE Research Papers.* Milton Keynes, UK: The Open University. Retrieved on June 08, 2019 from https://core.ac.uk/download/pdf/28975298.pdf

Schuh, Russel G. 1980. Paradigmatic displacement. *UCLA Linguistics Papers.* Retrieved on June 07, 2013 from http://www.linguistics.ucla.edu/people/schuh/Papers/A32_1980_paradigmatic_displ.pdf

Schuh, Russel G. 1989. The reality of Hausa "low tone raising": A response to Newman & Jaggar. *Studies in African Linguistics* 20, 3:257-262.

Schuh, Russel G. 2002a. Palatalization in West Chadic. *UCLA Linguistics Papers.* Retrieved on June 07, 2013 from http://www.linguistics.ucla.edu/people/schuh/Papers/A78_2002_WC_palatalization.pdf

Schuh, Russel G. 2002b. The locus of pluractional reduplication in West Chadic. *UCLA Linguistics Papers.* http://www.linguistics.ucla.edu/people/schuh/Papers/ms_2002_WC_redup.pdf

Serageldin, Ismail and June Taboroff, eds. 1992. *Culture and Development in Africa.* Washington, DC: The World Publications.

Setati, Mamokgethi. 1998. Code-switching in a senior primary class of second-language mathematics learners. *For the Learning of Mathematics* 18, 1:34-40.

Setati, Mamokgethi. 2002. Researching mathematics education and language in multilingual South Africa. *The Mathematics Educator* 12, 2:6-20.

Setati, Mamokgethi. 2003. 'Re'-presenting qualitative data from multilingual mathematics classrooms. *ZDM* (Zentralblatt für Didaktik der Mathematik)*: The International Journal on Mathematics Education* 35, 6:294-300.

Setati, Mamokgethi. 2005a. Teaching mathematics in a primary multilingual classroom. *Journal for Research in Mathematics Education* 36, 5:447-466.

Setati, Mamokgethi. 2005b. Researching teaching and learning from "with" *or* "on" teachers to "with" *and* "on" teachers. *Perspectives in Education* 23, 1:91-101.

Setati, Mamokgethi. 2008. Access to mathematics versus access to the language of power: The struggle in multilingual mathematics classrooms. *South African Journal of Education* 28:103-116.

Shapiro, Michael. 2013. Nominalism and realism in linguistics from a neostructuralist perspective. *Language Lore.* Retrieved on May 21, 2019 from http://languagelore.net/2013/02/15/nominalism-and-realism-in-linguistics-from-a-neostructuralist-perspective/

Shupak, Nili. 1992. A new source for the study of the judiciary and law of Ancient Egypt: "The Tale of the Eloquent Peasant." *Journal of Near Eastern Studies* 51, 1:1-18.

Sloan, Derek. September 27, 2005. The alphabet. *Development of the Alphabet.* Retrieved on July 20, 2019 from http://www.sfu.ca/~ramccall/DevelopmentoftheAlphabet.pdf

Smelyakov, S. V., G. Stray and J. Wicherink. 2006a. The Auric Time Scale. *Astrotheos.* Retrieved on August 9, 2008. Available at http://www.ASTROTHEOS.narod.ru

Smelyakov, S. V., G. Stray and J. Wicherink. 2006b. The last multi-turns of the spiral of time before it rolls up to appear in new reality. *Astrotheos.* Retrieved on August 9, 2008. Available at http://www.ASTROTHEOS.narod.ru

Smith, David Eugene. 2012. *A Source Book in Mathematics.* Mineola, NY/North Chelmsford, MA: Dover Publications/Courier Corporation.

Smith, Kenny, Michael L. Kalish, Thomas L. Griffiths, and Stephan Lewandowsky. 2008. Introduction, cultural transmission and the evolution of human behavior. *Philosophical Transactions of the Royal Society B* 363:3469-3476. Doi: 10.1098/rstb.2008.0147. Retrieved on April 27, 2019 from https://royalsocietypublishing.org/doi/full/10.1098/rstb.2008.0147

Smith, Tony. 2015. *FA,* the god with 16 eyes. *African Math: Ifa Drums.* Retrieved on February 01, 2016 from http://www.tony5m17h.net/VodouFA.html

Speaks, Jeff. 2005. Quine on the analytic/synthetic distinction. Retrieved on April 23, 2019 from https://www3.nd.edu/~jspeaks/courses/mcgill/415/quine-2dogmas.pdf

Stahl, Saul and Paul E. Johnson. 2007. *Understanding Modern Mathematics* Sudbury, MA Jones and Bartlett Publishers.

Stankiewicz, Edward. 1957. On discreteness and continuity in structural dialectology. *Word* 13, 1:44-59.

Stauder, Andréas. 2018. On system-internal and differential iconicity in Egyptian hieroglyphic writing. *Signata: Annals of Semiotics* 9:365-390. Retrieved on July 20, 2019 from https://journals.openedition.org/signata/1927?lang=en

Stewart, Ian. 1995. *Concepts of Modern Mathematics.* New York, NY: Dover Publications, Inc.

System Analysis & Decisions. 2009. Sustainable development: Gauging matrix mathematical model. Retrieved on May 03, 2013 from http://systemdecisions.com/research/sdgm-mathematical-model

Taljard, E. G, Faaß, U. Heid, and D. J. Prinsloo. 2008. On the development of a tagset for Northern Sotho with special reference to the issue of standardisation. *Literator* 29, 1:111-137. Retrieved on July 05, 2019 from https://www.researchgate.net/publication/224851991_On_the_development_of_a_tagset_for_Northern_Sotho_with_special_reference_to_the_issue_of_standardization

Tamura, Kaori and Takashi Hashimoto. 2012. Displacement in communication. In T. C. Scott-Phillips and M. Tamariz, eds. *The Evolution of Language*. Proceedings of Evolang IX, the 9th International Conference on the Evolution of Language, Tokyo, Japan, March 13-16, 2012. Retrieved on June 08, 2019 from https://www.researchgate.net/publication/268602295_DISPLACEMENT_IN_COMMUNICATION

Tanács, János. 1999. From analyticity to cognitive synonymy on Quine's path. *Periodica Polytechnica Series Social and Management Sciences* 7, 1:77-89. Retrieved on April 23, 2019 from https://pp.bme.hu/so/article/download/1755/1073/

Tannenbaum, Peter. 2014. *Excursions in Modern Mathematics*. Boston, MA: Pearson Education, Inc.

Teclu, Dagnachew. 2006. PanAfrica: AU calls for preservation of African languages. *The Daily Monitor* (Addis Ababa). Retrieved on August 12, 2006 from https://allafrica.com/stories/200608090479.html

The Free Dictionary. Available at http://www.thefreedictionary.com

The Internet Encyclopedia of Science. Lebombo bone. Retrieved on July 25, 2007 from http://www.daviddarling.info/encyclopedia/L/Lebombo_bone.html

The Sheward Partnership, LLC. April 2013. Calculating return on investment in sustainable design. Mathematics Awareness Month Homepage. Retrieved on February 27, 2013 from http://www.mathaware.org/index.html

Trudell, Barbara. 2009. Multilingualism in Sub-Sahara Africa: Describing a phenomenon, leveraging a resource. Paper presented at the International Conference on Multilingualism, Bamako, Mali, January 19-21, 2009. Retrieved on February 27, 2013 from http://citeseerx.ist.psu.edu/viewdoc/download;jsessionid=4E87A644F0C6D2351271A10B2A4030A8?doi=10.1.1.501.5983&rep=rep1&type=pdf

Trudgill, Peter. 1987. *Sociolinguistics: An Introduction to Language and Society*. Middlesex, UK and New York, NY: Penguin Books Ltd.

Tyldesley, Joyce. 1996. *Hatchepsut: The Female Pharaoh*. New York, NY: Viking/Penguin Group.

Ugwa, Kalu A. and Agwu A. April 2012. Mathematical modeling as a tool for sustainable development in Nigeria. *International Journal of Academic Research in Progressive Education and Development* 1, 2:251-258.

Ukaga, Okechukwu and Osita George Afoaku. 2005. *Sustainable Development in Africa*. Trenton, NJ: Africa World Press.

United Nations News Centre. November 21, 2011. Sustainable development key to Africa's socio-economic challenges. Retrieved on October 24, 2013 from http://www.un.org/apps/news/story.asp/html/story.asp?NewsID=40463&Cr=sustainable+development&Cr1=#.UoCXYSfAZfQ

United States Environmental Protection Agency (USEPA). 2013. What is sustainability. Retrieved on November 03, 2013 from http://www.epa.gov/sustainability/basicinfo.htm

Venant, Fabienne. 2010. Meaning representation: From continuity to discreteness. *Proceedings of the Seventh International Conference on Language Resources and Evaluation* (LREC), May 2010, Malte, Malta, hal-

00526811. Retrieved on May 30, 2019 from https://hal.archives-ouvertes.fr/hal-00526811/document

Verwey, Johanna Cornelia (Hanlie). 2010. Investigating the interaction of mathematics teachers with learners' mathematical errors. Master's dissertation, University or Pretoria, South Africa. Retrieved on July 16, 2019 from https://pdfs.semanticscholar.org/23f9/33f0367fcddde20cf6eff55453a296418dbc.pdf

Voice of America. November 01, 2012. 'Future Earth' discusses sustainable development challenges for Africa. Retrieved on October 24, 2013 from http://www.voanews.com/content/experts-discuss-sustainable-development-for-africa/1537550.html

Wacewicz, Slawomir and Przemyslaw Zywiczyński. 2015. Language evolution: Why Hockett's design features are a non-starter. *Biosemiotics* 8:29-46. DOI 10.1007/s12304-014-9203-2. Retrieved on April 23, 2019 from https://link.springer.com/content/pdf/10.1007%2Fs12304-014-9203-2.pdf

Wade, Nicholas. 2011. Phonetic clues hint language is Africa-born. *The New York Times.* Retrieved on April 27, 2011 from https://www.nytimes.com/2011/04/15/science/15language.html

Walters, Ronald W. 1993. *Pan-Africanism in the African Diaspora: An Analysis of Modern Afrocentric Political Movements.* Detroit, MI: Wayne State University Press.

Wariboko, Nimi. January 31, 2016. E-mail. <nimiwari@bu.edu>.

Watters, John R. 2000. Syntax. In B. Heine and D. Nurse, eds. *African Languages: An Introduction.* Cambridge, UK: Cambridge University Press.

Williams, Robert S. and Jade Comfort. 2007. Language documentation and description among refugee populations. In P. K. Austin, O. Bond and D. Nathan, eds. *Proceedings of Conference on Language Documentation and Linguistic Theory.* London, UK: School of Oriental and African Studies, University of London, UK, December 7-8:261-268.

Wilson, Penelope. 2003. *Sacred Signs: Hieroglyphs in Ancient Egypt.* Oxford, UK: Oxford University Press.

Winterbach, Ingrid. 2010. *Die benedenryk.* Kaapstad/Pretoria, South Africa: Human and Rousseau.

Winters, Clyde-Ahmad. 1986. The ancient Manding script. In I. V. Sertima, ed. *Blacks in Science: Ancient and Modern.* New Brunswick, NJ: Transaction Books.

Wolff, H. Ekkehard. 2000. Language and society. In B. Heine and D. Nurse, eds. *African Languages: An Introduction.* Cambridge, UK: Cambridge University Press.

Wolff, H. Ekkehard. 2018. How the continent's languages can unlock the potential of young Africans. *The Conversation.* Retrieved on July 05, 2019 from http://theconversation.com/how-the-continents-languages-can-unlock-the-potential-of-young-africans-90322

Woodson, Carter G. 1933. *Mis-education of the Negro.* Washington, DC: The Associated Publishers, Inc.

Wright, Lawrence. 2004. Language and value: Towards accepting a richer linguistic ecology for South Africa. *Language Problems and Language*

Planning 28, 2:175-197. Retrieved on July 10, 2019 from http://citeseerx.ist.psu.edu/viewdoc/download?doi=10.1.1.629.9310&rep=rep1&type=pdf

Yang, Charles. 2005. On productivity. *Linguistic Variation Yearbook 5.* Amsterdam, The Netherlands: John Benjamins Publishing Company. Retrieved on June 18, 2019 from https://www.ling.upenn.edu/~ycharles/papers/1.pdf

Zahrt, Kim R. W. 2000. Thoughts on Ancient Egyptian mathematics. *Indiana University South Bend Undergraduate Research Journal* 3:90-93. Retrieved on July 18, 2019 from https://scholarworks.iu.edu/journals/index.php/iusburj/article/view/19842/25919

Zaslavsky, Claudia. 1999. *Africa Counts: Number and Pattern in African Cultures* 3rd ed. Chicago, IL: Lawrence Hill Books.

Zawada, Britta and Mtholeni N. Ngcobo. 2008. A cognitive and corpus-linguistic re-analysis of the acquisition of the Zulu noun class system. *Language Matters: Studies in the Languages of Africa* 39, 2:316-331.

Zeldes, Amir. 2009. Applying morphological productivity measures in syntactic constructions: German comparatives and the *je…desto* constructions. Retrieved on June 18, 2019 from https://corpling.uis.georgetown.edu/amir/pdf/DocDay2009_poster.pdf

Index

A

B

C

D

E

F

G

H

I

J

K

L

M

www.ingramcontent.com/pod-product-compliance
Lightning Source LLC
LaVergne TN
LVHW010723210926
848485LV00007B/1430/J